T0140759

New Solutions for an Old Challenge: Chances and Limitations of Optical, Non-Invasive Acquisition and Digital Processing Techniques for the Age Estimation of Latent Fingerprints

DISSERTATION

zur Erlangung des akademischen Grades

Doktoringenieur (Dr.-Ing.)

angenommen durch die Fakultät für Informatik
der Otto-von-Guericke-Universität Magdeburg

von Dipl.-Inform. Ronny Merkel

geb. am 07.05.1983 in Leipzig

Gutachterinnen/Gutachter

Prof. Dr. Jana Dittmann, O.-v.-G.-University Magdeburg, Germany
Prof. Dr. Nikola Pavešić, University of Ljubljana, Slovenia
Prof. Dr. Slobodan Ribarić, University of Zagreb, Croatia

Magdeburg, den 07.07.2014

Bibliografische Information der Deutschen Nationalbibliothek

Die Deutsche Nationalbibliothek verzeichnet diese Publikation in der Deutschen Nationalbibliografie; detaillierte bibliografische Daten sind im Internet über http://dnb.d-nb.de abrufbar.

ISBN 978-3-8325-3823-1

Logos Verlag Berlin GmbH
Comeniushof, Gubener Str. 47,
10243 Berlin
Tel.: +49 (0)30 42 85 10 90
Fax: +49 (0)30 42 85 10 92
INTERNET: http://www.logos-verlag.de

ABSTRACT

Estimating the time a latent fingerprint has remained at a crime scene is in an important challenge to forensic experts, which could not be solved for 80 years. Potential suspects often admit to have been at the scene of a crime, but claim to have done so at a time before or after the crime has happened, significantly reducing the evidentiary value of latent prints. Furthermore, the age of a fingerprint might provide important information about the sequence of events and even the sequence of overlapped prints. Research conducted on this issue in the last decades has failed to provide reliable results, partially due to the invasiveness of used fingerprint lifting techniques (e.g. alteration of the trace during development) as well as the lack of using digital processing methods. However, the improvement of non-invasive capturing devices in the last decades has enabled significant potential for transferring fingerprint age estimation to the digital domain.

In the scope of this thesis, a Chromatic White Light (CWL) sensor and a Confocal Laser Scanning Microscope (CLSM) are for the first time applied to the challenge of estimating a latent fingerprints age. Using such devices, high-resolution time series of fingerprints in the form of intensity and topography images are captured. They allow to digitally study the degradation behavior of a fingerprint over time, while at the same time not altering the original trace. In particular, this thesis studies the fingerprint degradation behavior for short-tem (24 hours) and long-term (CWL: 2.3 years, CLSM: 30 weeks) aging periods. Basic settings as well as limitations of the devices are identified and summarized.

Based on such captured high-resolution fingerprint images, a digital processing pipeline is designed, including (apart from data acquisition) image preprocessing, feature extraction and age estimation. Such novel way of processing time series enables a much more accurate, objective and automated observation and evaluation, due to its digital properties. Algorithms for alignment, noise reduction, temporal normalization and segmentation are designed or adapted from the literature. Furthermore, 23 different binarization methods are evaluated and the best performing techniques are selected in respected to each capturing device and data type.

Statistical image properties are proposed as novel aging features, which have so far not been considered for age estimation. Evaluating seven exemplary statistical (pixel-based) measures, their general ability to capture characteristic changes during fingerprint aging is shown, where best features exhibit a strong correlation to a logarithmic function in nine of ten cases. Additional comparisons to prior investigated morphological (shape-related) aging features (ridge thickness, pores) as well as the definition and segmentation of additional ones (dust, big and small particles, puddles) show that statistical features seem to be more reliable in capturing characteristic changes during fingerprint degradation.

Applying the well known concepts of regression and machine-learning to the age estimation challenge, automated schemes are designed to estimate a fingerprints age. Consequently, objective performance measures are provided for the first time in this field. Huge ranges of estimated ages and a median deviation of three to five hours to the real fingerprint age are observed for a formula-based approach, while a classification kappa of 0.52 - 0.71 (equivalent to an accuracy of 76% - 86% for equally distributed test sets) is reached for a machine-learning based two class age estimation (based on indoor locations). At this point of time, five (CWL) and three (CLSM) consecutive scans of a print are required, equaling total capturing times of five (CWL) and three (CLSM) hours for the age estimation. Although these results are so far limited to short-term aging periods and might partially be subject to systematic influences of the capturing device (especially for the CLSM sensor), they provide a valuable measure of comparison for future work in the field.

Studying the influences of sweat composition, substrate, temperature, humidity, UV radiation, wind, contact time, contact pressure, smearing, dust, gender of the donor, inter-/intra donor variation, inter-/intra print variation as well as influences from the capturing devices, a first qualitative influence model is created for the short-term aging, providing a starting point for future quantitative research.

Even though the work of this thesis cannot provide a final and highly accurate age estimation scheme, it demonstrates the potential of applying non-invasive capturing devices in combination with a digital processing pipeline towards this challenge. Furthermore, additional benefits from the here investigated age estimation approach are identified, such as the potential to enhance data privacy in daily police work by a preselection of prints based on their age (computed from very small areas, insufficient for identification). Also, fingerprint age estimation might be used in preventive application scenarios, which has so far been prohibited by data protection laws.

ZUSAMMENFASSUNG

Die Altersbestimmung latenter Fingerspuren an Tatorten ist eine bedeutende Herausforderung für forensische Untersuchungen, welche seit etwa 80 Jahren nicht gelöst werden konnte. Verdächtige räumen oft die Anwesenheit am Tatort ein, beschränken diese jedoch auf Zeitpunkte deutlich vor oder nach der Tat, was den Beweiswert einer Fingerspur stark reduziert. Weiterhin beinhaltet das Alter von Fingerspuren wertvolle Informationen über den Tathergang (Abfolge von Ereignissen) sowie die Reihenfolge überlagerter Fingerspuren. Bisherige Forschungen der letzten Jahrzehnte haben auf diesem Gebiet kaum zuverlässigen Ergebnisse hervorgebracht, hauptsächlich bedingt durch die Nutzung von spurenverändernden Methoden zur Erfassung (wie z.B. die Vorbehandlung zur Sichtbarmachung) sowie dem mangelnden Einbezug digitaler Verarbeitung. Die erhebliche Weiterentwicklung zerstörungsfreier Erfassungsmethoden in den letzten Jahrzehnten eröffnet mittlerweile jedoch die Möglichkeit, die Altersbestimmung latenter Fingerspuren in die digitale Domäne zu überführen.

Im Rahmen der vorliegenden Arbeit wird erstmalig ein chromatischer Weißlichtsensor (CWL) sowie ein konfokales Laserscanningmikroskop (CLSM) auf die Herausforderung der Altersbestimmung latenter Fingerspuren angewendet. Diese Geräte ermöglichen es, hochaufgelöste Zeitreihen in Form von Intensitäts- und Topographiedaten zu erfassen und erlauben somit die digitale Untersuchung der Degeneration von Fingerspuren über die Zeit, ohne dabei die Spur zu verändern. Die vorliegende Arbeit studiert die Kurzzeitalterung (24 Stunden) sowie die Langzeitalterung (CWL: 2,3 Jahre, CLSM: 30 Wochen) von Fingerspuren. Grundlegende Einstellungen sowie Limitierungen der genutzten Erfassungsgeräte werden identifiziert und zusammengefasst.

Basierend auf den erhobenen, hochaufgelösten Fingerspurbildern wird im Rahmen dieser Arbeit eine digitale Verarbeitungskette konzipiert, welche neben der Datenerfassung die Schritte der Vorverarbeitung, Merkmalsextraktion und Altersbestimmung enthält. Diese neuartige Vorgehensweise bei der Verarbeitung von Fingerabdruck-Zeitreihen erlaubt durch ihre digitalen Eigenschaften eine deutlich genauere, objektive und automatisierte Beobachtung sowie Auswertung. Algorithmen zur Ausrichtung, Rauschunterdrückung, temporaler Normalisierung und Segmentierung werden entworfen oder adaptiert. Die Untersuchungen beinhalten die Auswahl geeigneter Verfahren zur Binarisierung bezüglich beider Erfassungsgeräte und Datentypen aus einer Menge von 23 unterschiedlichen Verfahren.

Statistische (pixelbasierte) Bildeigenschaften werden erstmalig zur Nutzung im Rahmen der Altersbestimmung von Fingerspuren vorgeschlagen und diesbezüglich untersucht. Anhand sieben beispielhafter Merkmale wird ihre generelle Eignung zur Erfassung charakteristischer Veränderungen während der Degeneration von Fingerspuren evaluiert. Dabei ergibt sich in neun von zehn Fällen eine starke Korrelation der besten Merkmale zu einer logarithmischen Funktion. Weiterhin werden Vergleiche zu bisherigen morphologischen (die Form betreffenden) Alterungsmerkmalen gezeigt (Rillenbreite, Poren) sowie weitere morphologische Merkmale vorgeschlagen (Staub, große und kleine Partikel, Pfützen). Die Auswertungen legen nahe, dass statistische Metriken zur Erfassung charakteristischer, altersbedingter Veränderungen zuverlässiger als morphologischen Merkmale sind.

Unter Anwendung der bekannten Konzepte von Regression und maschinenbasiertem Lernen werden automatisierte Verfahren zur Altersbestimmung von Fingerspuren vorgeschlagen, aus welchen erstmalig objektive Performanzmaße für die Adressierung dieser Herausforderung hervorgehen. Im

Rahmen der formelbasierten Altersbestimmung ergeben sich große Schwankungen des bestimmten Spurenalters, wobei Abweichungen des berechneten Alters zum realen Alter im Median von drei bis fünf Stunden erreicht werden. Im Rahmen der maschinenlernbasierten Klassifikation von Fingerspuren basierend auf Zweiklassenproblemen ergeben sich Kappa-Werte von 0,52 - 0,71 (76% - 86% Klassifikationsgenauigkeit für gleichverteilte Testmengen), wobei die Evaluation auf Innenstandorte beschränkt ist. Zum gegenwärtigen Zeitpunkt werden drei (CWL) sowie fünf (CLSM) konsekutive Messungen für die Altersbestimmung benötigt, was einem Zeitaufwand von drei (CWL) sowie fünf (CLSM) Stunden bei der Altersbestimmung entspricht. Obgleich die erzielten Ergebnisse der Altersabschätzung im Rahmen der vorliegenden Arbeit auf die Kurzzeitalterung beschränkt sind und teilweise dem Einfluss systematischer Veränderungen der genutzten Erfassungsgeräte (insbesondere bei Nutzung des CLSM-Gerätes) unterliegen, stellen sie einen wertvollen Ausgangspunkt für zukünftige Arbeiten auf diesem Gebiet dar.

Im Rahmen erster Untersuchungen des Einflusses der Schweißzusammensetzung, Oberfläche, Temperatur, Luftfeuchte, UV-Strahlung, Wind, Kontaktzeit bei Aufbringung, Kontaktdruck bei Aufbringung, Verschmieren, Staub, Geschlecht des Spenders, Inter-/Intra-Spender-Variation, Inter-/Intra-Finger-Variation sowie des Einflusses der genutzten Erfassungsgeräte auf die Alterung der Spuren wird ein erstes, qualitatives Einfluss-Model erstellt, welches als Ausgangsbasis für zukünftige, quantitative Studien genutzt werden kann.

Obwohl im Rahmen der vorliegenden Arbeit kein abschließendes und umfängliches Verfahren zur präzisen Altersbestimmung bereitgestellt werden kann, wird das Potential zerstörungsfreier Erfassungsgeräte in Kombination mit einer digitalen Verarbeitungskette zur Adressierung dieser Herausforderung deutlich. Des Weiteren ergeben sich zusätzliche Vorteile des hier vorgeschlagenen Vorgehens, wie beispielsweise die Verbesserung des Datenschutzes in der täglichen Polizeiarbeit im Rahmen einer Vorauswahl von Fingerspuren, basierend auf ihrem Alter. Dabei werden ausschließlich kleine Fingerausschnitte genutzt, welche für eine Identifikation nicht ausreichend sind. Weiterhin ergibt sich auch die Möglichkeit, die Altersbestimmung latenter Fingerspuren in präventiven Szenarien einzusetzen, welches bisher aus datenschutzrechtlichen Gründen gesetzlich untersagt war.

ACKNOWLEDGEMENTS

This work was carried out at the Workgroup on Multimedia and Security (Otto-von-Guericke-University Magdeburg) during the years 2010 - 2014. I wish above all to express my deepest gratitude to my supervisor, Prof. Jana Dittmann, for our joint work as well as her support and valuable feedback throughout the years of research and writing of this thesis. Furthermore, I am deeply grateful to Prof. Pavešić (University of Ljubljana) and Prof. Ribarić (University of Zagreb) for having the kindness of participating in the review process.

The here conducted work has been funded in part by the German Federal Ministry of Education and Research (BMBF) through the Research Programme under Contract No. FKZ: 13N10818, "Digi-Dak Digitale Fingerspuren". I would like to show my gratitude to the BMBF for funding this work. I especially thank the initiators of this project Jana Dittmann, Claus Vielhauer, Michael Ulrich, Christian Krätzer and Tobias Hoppe for enabling this thesis as well as our join work. Also, I am indebted to all project partners for their support and contribution to the Digi-Dak project and hereby to this thesis. Particular thanks are due to Anja Bräutigam (State Office of Criminal Investigation Saxony-Anhalt), Stefan Gruhn (University of Applied Sciences Brandenburg) and Michael Quinten (FRT GmbH) for their assistance in realizing the numerous scans and the joint discussion. Furthermore, I would like to thank the partners Stefan Kiltz, Ina Großmann, Michael Ulrich, Marcus Leich, Tobias Kiertscher, Stefan Gruhn, Thomas Fries, Martin Schäler, Sebastian Pocs, Benjamin Stach and Silke Jandt for their input and discussion during the project.

I am deeply grateful to all my colleagues from the Workgroup on Multimedia and Security and from the University of Applied Sciences Brandenburg for our joint work, plentiful discussions and valuable exchange. It has been a pleasure as well as a great support to work with Claus Vielhauer, Christian Krätzer, Mario Hildebrandt, Robert Fischer, Stefan Kiltz, Andrey Makrushin, Maik Schott and Kun Qian.

I am honored to thank Rainer Herrmann, Ernie Hamm and Ed German. They have supported this thesis by providing not readily accessible literature on the state of the art in fingerprint age estimation. Furthermore, I would like to thank my students Andy Breuhan and Andriy Krapyvskyy for their participation in the form of bachelor and master theses. Also, my gratitude is given to Karen Otte, Robert Clausing, Ayneta Adege and Martin Linnemann, who have contributed to this thesis in the scope of different courses.

TABLE OF CONTENT

List of Figures

List of Tables

LIST OF ACRONYMS

ABGV	Average background gray value
AC	Auto correlation
Ac	Age class
AE	Age estimation
AEF	Age estimation feature
AFM	Atomic Force Microscopy
Ap	Age estimation performance
Ar	Fingerprint area
As	Aging speed
AS	Artificial sweat
B	Background mask
B'	Background mask after one dilation
b_err	Binarization error
BG	Background
BIN	Binarize
c	Center point
CCD	Charge-coupled device
CD	Cumulative dust mask
CFM	Chemical Force Microscopy
Cl	Classifier
CLSM	Confocal Laser Scanning Microscopy
CN	Cumulative null value mask
Conf	Configuration
CWL	Chromatic White Light (sensor)
d	Diameter
DA	Data acquisition
DW	Distilled water
EFM	Electric Field Microscopy
EPS	Electric Potential Sensor
ESDA	Electrostatic Detection Apparatus
est	Estimated (experiments)
exp	Exponential
F	Feature
Fb	Best performing feature
FBS	Flat bed scanner
FE	Feature extraction
Fk	Feature with number k
FP	Fingerprint
Fr	Feature representation

FTIR	Fourier-Transform Infrared (spectrometer)
G	Gradient image
GT	Ground truth
h	Relative frequency
Hu	Relative humidity
I	Intensity image
I_{CLSM}	Binarized CLSM intensity image
I_{CWL}	Binarized CWL intensity image
IMG	Image
Inf	Influences
Int	Intensity
INV	Invert
Ip	Intersecting point
J48	Name of a machine-learning classifier
Java	Name of a programming language
kap	Kappa value
lin	Linear
LMT	Name of a machine-learning classifier
log	Logarithmic
max	Maximum
MH	Mexican hat
min	Minimum
MPLSM	Multi Photon Laser Scanning Microscopy
Npk	No prior knowledge
O	Optimized
OCT	Optical Coherence Tomography
Op1	Optimization of parameters (phase 1)
Op2	Optimization of parameters (phase 2)
PC	Preprocessing challenge
PES	Performance evaluation set
Pk	Prior knowledge
PM	Particle mask
PP	Preprocessing
Pr	Property of an experimental aging curve (lin, log, exp)
Pri	Printed
Ps	Preselection
PT	Particle type
Qt	Name of a software development framework
Qwt	Name of a software library used with Qt (Qt Widgets)
R	Raw
r	Pearson correlation coefficient

ra	Classification performance range
Re	Regression type
real	Real (ground truth)
Reg	Regression
RS	Real (fingerprint) sweat
S	Selection
Sas	Statistically significant amount of samples
SERS	Surface Enhanced Raman Spectroscopy
SFM	Scanning Force Microscopy
Sim	Similar
SKPM	Scanning Kelvin Probe Microscopy
spl	Split value
Ss	Statistical significance
ST	Segmentation target
T	Topography image
t	Point of time
t_diff	Time difference
T_{CLSM}	Binarized CLSM topography image
T_{CWL}	Binarized CWL topography image
Te	Temperature
TG	Test goal
thresh	Threshold
Ti	Temporal images
t_{max}	Total time period
TN	Temporally normalized
Topo	Topography
Tp	Time period
TRS	Model training set
Ts	Time series
TSA - TSF	Test set A - Test set F
Tt	Time threshold
TW	Tap water
UV	Ultra violet (light)
val_cir	Valid circle
Var	Variance
VIS	Visible (light)
VP	Valid pixels mask
WEKA	Waikato Environment for Knowledge Analysis
WSH	Windows Script Host
Δt	Time offset

1 Introduction

Age is an important issue in biometrics as well as forensics. In the area of biometrics, age is often referred to as either the biological age of a human in regard to a certain trait, or the amount of time which a biometric template has been enrolled in a database [1]. In forensics, the biological age of a person is also of interest, e.g. when deriving the biological age of an offender from traces left at a crime scene [2]. However, most important to forensic experts is the age of various traces found at crime scenes, meaning the time they have been present at the scene [3]. Such traces include tool marks, bullets, fibers, shoe or tire impressions or latent palm and fingerprints [4]. Their age can provide valuable information about the time and sequence of events. For example, the age of a trace might be used to link a particular piece of evidence to a time span associated with a crime or to determine a sequence in which different pieces of evidence have been handled. In the scope of this thesis, the age of latent fingerprints is investigated in a forensic trace context, based on the description of the Digi-Dak project [5]. Such age can be defined as the time which has passed from the initial application of a latent fingerprint by touching an object until its acquisition by forensic experts.

Age estimation of latent fingerprints has been a topic of interest for about 80 years [6]. Many cases are known from practical police work, where a successful age estimation of latent prints would have provided valuable information to the forensic investigation. For example, a restaurant burglary was reported in 2011 [7], after which a latent print was found in a cracked open slot machine. The identified suspect stated to have worked for the slot machine company earlier and had collected coins from that particular slot machine six months ago. Because the age of the fingerprint could not be determined, the charges had to be dropped. Further examples are given in [7], [8], [9], [10], [11], [12], [13], [14] and [15]. Taking into account the variety of such reports, a significant amount of criminal inquires can benefit from age estimation of fingerprints.

The overall challenge of latent print age estimation is unresolved so far. Many forensic experts can be cited on this issue. Moenssens is quoted in [16] to have said in 1972: "I would simply say that I cannot tell with any degree of precision because there is no known way to determine positively, or to even closely approximate by opinion testimony the length of time a latent has been on an object." His statement was confirmed in 1992 by Detective James F. Schwabenland: "Determining the 'age' of the latent finger or palm print residue on a particular piece of physical evidence is generally not possible. [11]" In the year 2012, still no effective and reliable method for determining a fingerprints age has been established: "Although fingerprint identification has been repeatedly proven as one of the most robust and definite forensic techniques, a measure of the rate at which latent fingerprints degrade over time has not been established effectively. [7]"

During many different efforts to provide age estimates for practical criminal cases, important issues of this challenge have emerged. A first issue can be seen in the fact that in classical and current police work, latent fingerprints are generally developed prior to being lifted, which creates a competitive situation between different trace examination methods, such as identification of a person, age estimation of a print and DNA analysis.

A second issue can be seen in the rather fuzzy and inaccurate methods currently used for observing the aging behavior of a fingerprint over time, e.g. aging fingerprints under reproduced crime scene conditions. Such approach is described by Detective Schwabenland in 1992 [11], who reproduced the crime scene conditions of a second degree murder case in his backyard, using aluminum beer cans and his own fingerprints. He derived conclusions about the age of the real prints from these studies, which he subsequently included in his testimony in court as an expert witness. He was later criticized by McRoberts and Kuhn for this inaccurate approach. They state that "the scientific certainty and acceptability must be gained within the field of Friction Ridge Skin Identification prior to presenting testimony in a court of law. [17]" Observation methods of much higher accuracy are required to achieve such "scientific certainty and acceptability." Digital observation methods and age estimation strategies exhibit the potential to provide such increased accuracy.

A third issue can be seen in the very complex set of influences on the aging process of latent prints and the uncertainty of their initial state directly after being deposited. In 1993, Midkiff comes to the conclusion that "from the studies and cases examined, it is apparent that wide variations exist in the ability of a latent print to survive, even under rather harsh conditions. As a result, development of a latent print at a crime scene is no guarantee of its having been recently placed. In addition, the studies suggest that no reliable indication of a print's freshness can be obtained from its rate of development or appearance after it is developed. Because of their importance in criminal cases, reliable approaches to the estimation of the age of a latent print are needed. Until they are tested and shown to be reliable, however, speculation or court testimony concerning the time when a latent print was placed is fraught with danger and may be hazardous to the reputation of the examiner [18]." Several cases have been reported confirming the very different durability of latent prints under different environmental conditions. In one example, fingerprints of a police officer were found on the window glass of a burglarized house [10]. Claiming to have helped the house owner earlier to gain entry to his house by pushing against all windows, the officer testified that his prints dated back to an event two years before. The story was later confirmed, showing that the prints had survived for two years on the window glass and still appeared to be fresh.

In the scope of this thesis, a novel approach for the age estimation of latent fingerprints is proposed, aiming at addressing the three issues identified above. The approach is based on the usage of optical, contactless and non-invasive capturing devices in combination with digital processing techniques. Because the challenge of estimating a latent prints age is a very complex task that would have to involve, amongst others, chemical analyses, a comprehensive age estimation scheme cannot be investigated in the limited scope of this thesis. The aim here is rather to analyze the potential of non-invasive capturing devices in combination with digital processing techniques, providing a basis for the future exploration of this promising area. The approaches and concepts proposed in this thesis are based on digital acquisition and processing methods and can be considered as a part of the emerging fields of computational forensics [19] and digitized forensics [20].

1.1 Relevance of the Age Estimation Challenge

When exploring the relevance of latent fingerprint age estimation, linking a latent print to the time of a crime is not the only aspect of this multi-facetted application area. Possible promising applications are the following:

Linkage of fingerprints to the time of a crime: As has been illustrated earlier, it often is of utmost importance for forensic experts to link a latent fingerprint to a certain time period, to provide evidence proving or disproving the testimony of a suspect [7], [8], [9], [10], [11], [12], [13], [14], [15]. This can greatly enhance the evidentiary value of latent fingerprints at court hearings.

Data privacy protection: If age estimation can be performed successfully and if the time of a crime is known (which is often the case in forensic investigations), fingerprints not belonging to this time period might be excluded from identification [21]. Therefore, subjects which have been present at the crime scene prior to or after the crime are not considered as suspects, which is important considering the psychological effects and loss of reputation when being falsely accused. Such timely distinction between subjects which have been at a crime scene would greatly enhance privacy protection. In particular, if age estimation can be performed using only small areas of the fingerprint (which exhibits insufficient dactyloscopic elements to be used for identification), no person-related data of subjects outside the time period of the crime are collected at all [21], [22]. This could be seen as a major improvement to data privacy protection in daily police work, where at this point of time, all fingerprints found at a crime scene are captured. Such method would enforce the data minimization principle of collecting as little person-related data as possible to solve a crime (§ 3a of the German Data Protection Act [23], article 19 of the draft of a European Union law on a Police Data Protection Directive [24]). Furthermore, it would also enforce the privacy by design principle, by which privacy protection mechanisms are to be included into the design of a method or technique (§ 3a of the German Data Protection Act [23], article 19 of the draft of a European Union law on a Police Data Protection Directive [24] and article 23 of the draft of a European Union law on a General Data Protection Regulation [25]).

Sequence of events: If several fingerprints are found at a crime scene, the sequence of events might be determined by sorting touched objects according to the age of their fingerprints. However, such sequencing might be difficult, if objects are touched shortly after each other, depending on the temporal resolution of the used age estimation scheme. Also, a combination with the age of other forensic traces can be useful in determining such sequence.

Separation and sequencing of overlapped fingerprints: Overlapped fingerprints provide another challenge to forensic experts. If the same area of a surface is touched several times, two or more fingerprints might overlap. Currently, such fingerprints are often considered not usable for investigation. It is of interest to forensic experts to separate these fingerprints and to determine the sequence in which they were applied. While the age of already separated prints might be determined individually, followed by an age-based sequencing [26], there might also be the possibility of separating prints by determining the age of selected ridge parts and assigning parts with a similar age to certain fingerprints. However, the practical feasibility of such approach remains subject to future research.

Preventive application scenarios: So far, fingerprints are barely captured in preventive application scenarios, because the capture of person-related data without any suspicion is forbidden by data protection laws in the European Union (established case law since 2006 in Germany [27], article 7 of the draft of a European Union law on a Police Data Protection Directive [24]). However, if age estimation can be performed on only small areas of a latent print (providing insufficient dactyloscopic elements for identification), it can be assumed not capturing person-related data, as identification is not possible [21], [22]. In such case, age estimation might be introduced to preventive application scenarios, where a suspicious time period is defined in advance and only fingerprints belonging to this period (i.e. where a specific suspicion is given) are then captured and identified. This method might allow for latent fingerprints to be used in preventive application scenarios without violating data protection laws.

Because of such great relevance, the age estimation of latent fingerprints has been researched for many decades and has so far not lost its popularity: "The question of 'age determination' in latent prints is one that will be talked about for many years... (Greenlees, 1994) [10]."

1.2 The Research Gaps

To identify the research gaps addressed in this thesis, obstacles to successful age estimation of latent fingerprints need to be explored. Previous research on the topic fails to provide accurate and reliable age estimation results due to the following major reasons:

(1) **Invasive capturing methods:** If fingerprints are developed prior to their capture or if the lifting process itself is invasive, a fingerprint has been altered after being lifted once. Lifting it again at a later point of time can therefore not give unbiased information on whether certain observed changes are a result of fingerprint aging or the lifting process. A prints aging behavior can therefore not be accurately studied over time. Researchers often circumvent this restriction by applying a substantial amount of prints onto a surface and lifting a fixed subset of these prints at each investigated point of time. However, because the complex nature of different influences leads to every single fingerprint exhibiting a unique aging behavior, this method cannot be considered accurate.

(2) **Lack of using digital preprocessing methods:** In comparison to a potential digital processing, classical age estimation techniques suffer severe limitations in respect to distortion reduction, fingerprint enhancement and segmentation as well as a significantly worse accuracy and degree of automation. Furthermore, statistical features and machine-learning based classification can only be applied in the digital domain. Therefore, the design of digital preprocessing, feature extraction and age estimation strategies seems to have a significant potential. At this point of time, digital methods are barely applied to the field of fingerprint age estimation.

(3) **Focus on chemical substances and morphologic changes rather than statistical features:** Studies have been conducted in the area of chemistry concerning changes of certain substances or substance groups during fingerprint aging. Additionally, first studies target morphological (shape-related) features like ridge thickness and pores. However, statistical (pixel-based) features of captured fingerprints like the degradation of fingerprint image contrast over time have, to the best of the authors knowledge, not been investigated so far.

(4) **Lack of digital age estimation strategies:** No approach is known so far utilizing digital age estimation strategies for mapping fingerprint data into age information, such as rule- or machine-

learning based approaches. Digital age estimation strategies might provide a more accurate and deterministic mapping than a human observer and can incorporate more features at the same time into the final decision. Furthermore, statistical approaches might allow for probability-based statements about the age of a print, which are not practically used at this point of time.

(5) **Insufficient knowledge about influences on the aging process and the initial state of features:** When examining an aging feature, its initial state during print application is usually unknown. This applies to the qualitative and quantitative composition of certain chemical substances in the same way as to any other feature, such as the morphology of ridges and pores or statistical features such as image contrast. Furthermore, a wide set of different influences seems to impact the aging of latent fingerprints, such as temperature, humidity, surface properties and sweat composition, to name only a few. The variety of influences results in a significant variation of the speed of fingerprint degradation and therefore the aging behavior. A detailed understanding of the underlying processes and influences is necessary to find aging features or feature combinations with a similar initial value or to successfully extract age information regardless of the initial state.

While the variety of different influences on a fingerprints degradation process as well as the uncertainty of its initial state concerning certain features remains an issue, recent advancements in non-invasive capturing and imaging devices as well as digital processing methods offer great potential towards solving the age estimation challenge, which have barely been explored so far. Non-invasive capturing and imaging methods have been recently introduced to the domain of lifting latent fingerprints [28], [29], [30], [31], [32]. The contactless acquisition of fingerprints enables researchers for the first time to observe a single fingerprint over arbitrary time periods by capturing it in regular intervals, creating time series.

Digital image-based preprocessing methods in combination with (automatically calculated) statistical and morphological features have not yet been explored towards their feasibility for estimating the age of a fingerprint. Potential features might target statistical properties based on the overall image histogram as well as morphological traits, which can be segmented from a fingerprint image using automated and semi-automated methods. Because of the high resolution and precision of the currently available optical devices, also small morphological structures like skin particles, droplets, dust or pores can be investigated. Furthermore, automated digital age estimation strategies can be applied, such as formula- or machine-learning based strategies, with the potential to achieve objective age estimation performance measures for the first time. Using non-invasive acquisition and digital processing therefore seems to be able to achieve significant advancements in the speed, accuracy and objectivity of a potential age estimation approach.

1.3 Contribution of this Thesis

The challenge of estimating the age of a latent fingerprint is very complex and so far unresolved. However, the topic is of great relevance to forensic investigators, as has been discussed in sections 1.1 and 1.2 above. The aim of this thesis is to advance the knowledge on fingerprint age estimation in the digital domain to build a basis for achieving a reliable age estimation in the medium-term future. This is done by exploring the potential and limitations of two exemplary non-invasive capturing devices in combination with digital processing methods towards this challenge, as proposed in [5]. The contribution of this thesis can be summarized in five main points, addressing the research gaps introduced earlier:

(1) **Non-invasive capturing methods:** In the scope of this thesis, two optical, non-invasive capturing devices are applied to the challenge of latent print age estimation. In particular, a Chromatic White Light (CWL) sensor and a Confocal Laser Scanning Microscope (CLSM) are used. The contribution of applying these devices can be seen in their ability to consecutively capture latent prints without invasive operations, therefore allowing for the first time to create time series of a single print and to observe its degradation behavior over time. Furthermore, they achieve high capturing resolutions (CWL: 2 µm, 12700 dpi; CLSM: 1.3 µm, 20007 dpi) and provide digitized, image-based intensity and topography data of a print. As a result, the digital processing of high-resolution fingerprint time series is now possible. Furthermore, a contribution of this thesis lies in the determination of minimum required sensor settings (CWL dot distance: 20 µm, CWL measured area size: 4 x 4 mm; CLSM dot distance: 1.3 µm, CLSM measured area size: 1.3 x 1 mm). Also, major limitations of the devices are identified, such as: inaccurate spatial alignment mechanisms of the measurement tables of the sensor hardware setups, changes in overall image brightness over time, sensitivity to environmental influences of temperature and vibrations, noise (especially on the topography data of both devices), scan artifacts at droplets leading to null values (CWL), distortions of the measurement table propulsion (CWL), geometrical (barrel) distortions of the objective lens (CLSM) as well as influences when starting the device (CLSM).

(2) **Digital preprocessing methods:** A general digital processing pipeline for latent print age estimation, including data acquisition (digitization), preprocessing, feature extraction and age estimation, does not seem to exist at this point of time. It is therefore the contribution of this thesis to adapt the general processing pipeline from the area of biometrics for the age estimation of latent prints. A major effort can be seen in the selection, design and arrangement of adequate image preprocessing techniques into the steps and sub-steps of such pipeline. In particular, several methods for the reduction of sensor-related distortions are designed, such as methods for spatial alignment, temporal normalization, dust exclusion and masking of sensor artifacts. Furthermore, 17 different binarization techniques (mainly from the literature) for captured intensity images are compared for selecting the optimal algorithm in respect to each capturing device. For acquired topography images, three binarization techniques are designed for each capturing device and similarly evaluated. For the best performing techniques, the following binarization errors (b_err) are achieved in comparison to a manual segmentation: CWL intensity data: b_err = 0.11, CWL topography data: b_err = 0.29, CLSM intensity data: b_err = 0.07, CLSM topography data: b_err = 0.15. To segment certain structures of interest, different techniques are designed or adapted from the literature, to extract complete fingerprints, ridges, pores or certain particles.

(3) **Statistical aging features:** Statistical image-based features have not been used so far for the purpose of fingerprint age estimation. In this thesis, seven statistical measures are for the first time applied to fingerprint intensity and topography time series of both capturing devices and are compared between short-term (24 hours) and long-term (CWL: 2.3 years, CLSM: 30 weeks) aging periods. Out of 770 time series, 88% (CWL) and 92% (CLSM) of experimental aging curves are observed to exhibit a strong correlation to a logarithmic function (Pearson correlation coefficient $r \geq 0.8$) for the best performing feature in the scope of short-term aging. Long-term aging curves exhibit in 70% (CWL) and 95% (CLSM) of cases a strong logarithmic curve, however are of less amount (20 series per device) and potentially subject to systematic influences of the capturing device (especially for the CLSM data). Furthermore, several morphological long-term aging features are investigated, targeting changes in ridge thickness, pores, dust, big and small particles as well as fingerprint puddles. Comparing them with a state of the art study on ridge thickness and pores, it is shown that degradation seems to be a process occurring throughout the complete

image rather than at specific locations, such as ridge edges or pores. Consequently, statistical features are shown to deliver more reliable results than morphological ones at this point of time.

(4) **Digital age estimation techniques:** So far, no digital age estimation scheme for latent fingerprints is known from the literature. A contribution of this thesis can therefore be seen in the first automated application of digital age estimation strategies, however limited to the short-term aging of 24 hours. In particular, four different formula-based age estimation approaches (using regression) as well as a machine-learning based classification approach (into well defined age classes) are investigated and objective performance measures are provided for the first time. For the formula-based approaches, the lowest median deviation of the estimated fingerprint age from the original age is observed to be 5 hours (CWL) and 3.4 hours (CLSM). However, age estimates are not normally distributed and vary widely, rendering such approach comparatively unreliable. Machine-learning based classification achieves a maximum kappa performance of kap = 0.52 (equivalent to an accuracy of 76% for equally distributed test sets) for the CWL sensor and kap = 0.71 (86%) for the CLSM device when separating prints into those younger as or older than 6 hours (CWL) or 7 hours (CLSM), using a combination of all introduced statistical features. Different parameters are optimized and the statistical significance of the sample set size is shown. After such optimizations, five (CWL) and three (CLSM) consecutive scans of a print are required for age estimation, equaling total capturing times of five (CWL) and three (CLSM) hours.

(5) **Influences on the Aging process and initial state of features:** Different influences on the aging process of latent prints have been studied in state of the art work. However, their impact might be significantly different depending on the used capturing resolution, targeted properties (morphological, statistical or substance-specific) as well as aging period. Furthermore, it seems unclear at this point of time to which extent such influences impact the short-term aging (maximum of 24 hours). Therefore, known influences on the latent print aging process, such as sweat composition, different substrates, temperature, humidity, UV exposure and wind are re-evaluated for the newly applied two capturing devices. Others are less well explored, such as the influence of the fingerprint application process (contact time, contact pressure, smearing), dust, gender of the donor, inter-/intra-donor and intra-/inter print variation as well as influences from the capturing devices themselves. A contribution of this thesis lies in the realization of first qualitative studies on all these influences and their aggregation into a first short-term aging influence model, which aims at providing a starting point for future investigations. The experiments performed here show that contact time, contact pressure, smearing as well as the gender of a donor seem to be rather negligible, while dust, temperature, humidity, UV exposure and wind are considered to be of importance (wind and UV exposure might be largely excluded for indoor scenarios). The sweat composition seems to have an important impact on the aging behavior, however cannot be properly investigated with the devices used here. Significant influences of the capturing device as well as the substrate are shown to exist, however might be controlled in future work by adequate climatic capturing conditions and specific considerations of different surface types.

Summarizing the contribution of this thesis, the aim is to provide first steps of basic research on the challenge of latent print age estimation, enabling further advancements. When considering the little progress made on the issue within the last 80 years of research, the potential of non-invasive capturing devices and digital processing techniques seems to be high. It is here demonstrated to motivate further research in the area, hopefully leading to significant advancements in the next decades. However, the aim of this thesis is not to comprehensively solve the research challenge of latent print age estimation.

1.4 Topics Outside the Scope of this Thesis

Because the age estimation of latent fingerprints is a very complex challenge, it can by no means be the aim of this study to investigate the topic in all detail or to solve all its issues. The aim here is rather to investigate comparatively small, well defined areas of the topic, to show the general potential of non-invasive capturing devices in combination with digital processing methods, statistical features and automated age estimation strategies as well as their limitations. Consequently, some areas are only subject to brief qualitative evaluations, while others remain completely untouched. The following issues are outside the scope of this thesis and remain topics for future research:

Investigation of additional non-invasive capturing devices: Additional, non-invasive capturing devices, especially those capturing substance-specific information (such as FTIR or Raman spectroscopes) seem to be promising. They might be investigated and combined with the here examined devices in future work. However, substantial research and financial efforts are required.

Adaptation and optimization of sensor properties and parameters: Numerous optimizations and adaptations of the capturing hard- and software are required in future work. Different specific requirements are identified in the scope of this thesis. However, their realization seems to remain a task for future research.

A comprehensive, inter-disciplinary study of potential compounds: The identification of different compounds or substances responsible for the observed aging behavior is a valuable research topic, especially in combination with experts from the area of chemistry and biology. Such findings might lead to a significantly improved aging model in the future.

Combination of aging information from different capturing devices: The combination of different capturing devices in the form of sensor fusion approaches seems to be promising. Especially when using devices from different capturing domains (e.g. those capturing morphological and those capturing substance-specific properties of a print), fusion can be very effective. Future research in this area might be able to achieve valuable improvements in the accuracy of potential age estimation schemes.

Quantitative investigations of influences: The first qualitative studies of influences on the aging process conducted in this thesis require confirmation and extension by a substantial amount of quantitative studies in future work. Efforts in this field might lead to a significantly improved influence model. Interdependencies between the specific influences as well as relations to certain capturing devices need to be examined as well.

Development of a fully functioning, accurate age estimation scheme: To develop an age estimation scheme that achieves acceptance in court, e.g. by fulfilling Daubert-compliance [33], can by no means be achieved in the scope of this thesis. Extensive research of the different influences on the degradation behavior of prints is required, apart from the variety of necessary optimizations in respect to the capturing devices and digital processing methods. To develop an age estimation scheme with high accuracy seems to be realistic only in the mid- or long-term future. Considering the significant impact of the age estimation challenge and the huge complexity of the issue, this thesis can only be seen as a first step of basic ground research, opening the door for future studies, combining different disciplines and approaches.

1.5 Structure of the Thesis

This thesis is divided into seven chapters, containing an introduction (chapter 1), state of the art review (chapter 2), presentation of the used methodology, assumptions and research objectives (chapter 3), general design of the used processing pipeline (chapter 4), experimental setup (chapter 5), presentation of results (chapter 6) as well as summary, discussion and identification of future work (chapter 7). The main structure of this thesis is visualized in Figure 1, depicting the basic relations between different sections.

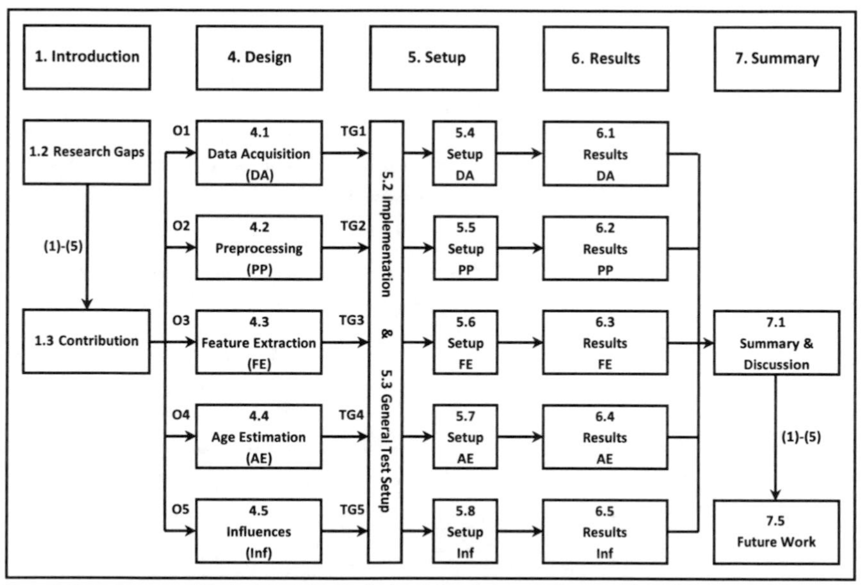

Figure 1: Main structure of this thesis including the basic relations between different sections of chapters 1, 4, 5, 6 and 7.

In chapter 1 of this thesis, an introduction has been given to the field of fingerprint age estimation and its relevance for different applications. Five major research gaps (1) - (5) have been introduced in section 1.2 and were mapped to corresponding contributions in section 1.3. They form the basic structure of this thesis and all studies are conducted in respect to them. Therefore, they are depicted in Figure 1 as the starting elements of the structure graph. Topics outside the scope of this thesis have been introduced in section 1.4.

Chapters 2 and 3 are not depicted in Figure 1, as they are not part of a specific structure, but rather provide the foundation for all following chapters of this thesis. Chapter 2 summarizes relevant state of the art work in the scope of fingerprint age estimation (section 2.1), contactless capturing devices (section 2.2), digital processing methods (section 2.3) as well as potential influences on the fingerprint aging (section 2.4). Chapter 3 presents the main methodology of this thesis (section 3.1) and introduces the basic assumptions and research objectives (section 3.2). In particular, the proposed processing pipeline based on the general biometric pipeline is introduced in section 3.1.6 (Figure 3), containing the four steps of data acquisition (DA), preprocessing (PP), feature extraction (FE) and age estimation (AE), forming the basic structure of this thesis. Furthermore, the five identified research gaps (1) - (5) are mapped to five corresponding research objectives O1 - O5 in section 3.2.2.

Chapter 4 proposes the design of the digital processing pipeline for the realization of the research objectives O1 - O4. As shown in Figure 1, each research objective corresponds to one step of the processing pipeline. Therefore, a separate section (4.1 - 4.4) is assigned to each objective, where addressing the objective is considered as being similar to designing the corresponding step of the processing pipeline. The objective O5 consists of different influences on the aging of fingerprint. As such, it is relevant to all steps of the processing pipeline and is investigated separately in section 4.5.

After the design of procedures for the capture and digital processing of latent prints, the experimental test setup is given in chapter 5 (Figure 1). The five research objectives are mapped to specific experimental test goals TG1 - TG5 in section 5.1, followed by an introduction into the software implementation of the designed methods in section 5.2. The general test setup is specified in section 5.3, while experiment-specific setups are introduced separately for each test goal (sections 5.4 to 5.8). Chapter 6 presents and discusses the experimental results in respect to the test goals TG1 - TG5 (sections 6.1 to 6.5).

Chapter 7 summarizes and discusses the findings of this thesis and outlines potential subjects of future research. All five research gaps (1) - (5) are re-visited in section 7.1 and are evaluated in respect to the corresponding experimental findings. Prospects and limitations of the approach are discussed in section 7.2 and their implications to practical age estimation scenarios (section 7.3) as well as to the general forensic domain (section 7.4) are elaborated. Future work is identified in section 7.5.

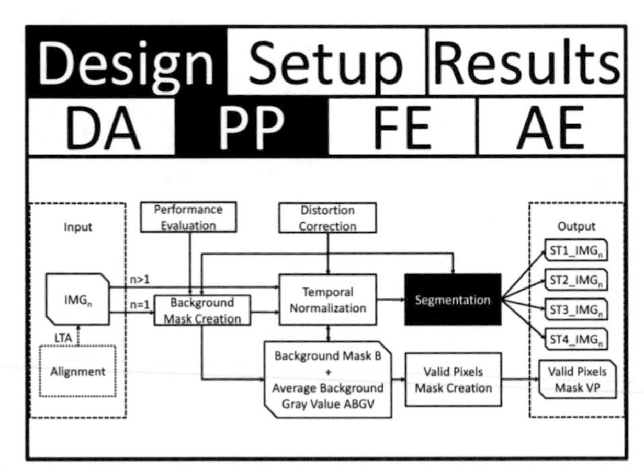

Figure 2: Exemplary enlarged pictogram for navigation through chapters 4 to 6. The depicted section belongs to chapter 4 (Design) and is located in the preprocessing step (PP) of the processing pipeline. The introduced sub-step is focused on image segmentation.

Because chapters 4 to 6 are comparatively complex in their structure, they are marked at the beginning of each section by a pictogram designed to facilitate navigation. The pictograms are well structured in three levels, allowing to mark the three top-levels of each chapter. An exemplary pictogram is given in Figure 2 (enlarged).

The exemplary pictogram of Figure 2 visualizes the general principle of navigation through chapters 4 to 6. The main chapters (top level) are referred to according to their main functionality of designing digital processing methods (Design, chapter 4), describing the experimental test setup (Setup, chapter 5) as well as presenting and discussing the experimental results (Results, chapter 6). The current chapter is marked in white letters on black background (Design in case of Figure 2). The second level is oriented at the four steps of the processing pipeline. Because all acquisition and digital processing steps are included in such pipeline, each presented topic can be assigned to one or more of the four processing steps of data acquisition (DA), preprocessing (PP), feature extraction (FE) and age estimation (AE). The assigned processing step of Figure 2 is the preprocessing (PP) step. The investigation of influences applies to all four steps at the same time.

For the third level, the pictogram visualization is flexible and depicts the detailed structure of each step where necessary. In the case of Figure 2, image segmentation is investigated as a sub-step of the overall preprocessing step. While all sub-steps of this step are depicted for providing context information, only the sub-step depicted in white letters on black ground is discussed in the corresponding section. The third level is only required for chapter 4, where the detailed structure of each processing step is introduced.

2 State of the Art

Because the age estimation of latent fingerprints is an inter-disciplinary topic, a wide variety of state of the art work can be considered relevant in the scope of this thesis. In this chapter, a brief overview of these publications is given. A structured presentation of state of the art approaches towards fingerprint age estimation is given in section 2.1, followed by a summary of contactless optical capturing devices in section 2.2. Basic image processing methods are briefly introduced in section 2.3, followed by a summary of potential influences on the fingerprint aging process in section 2.4.

2.1 Latent Fingerprint Age Estimation

In the scope of this thesis, state of the art work on fingerprint age estimation is divided into morphological, substance-specific and statistical approaches. Methods mainly aiming at the shape and size of the fingerprint (e.g. ridge lines, pores and particles) are referred to as morphological approaches and their features as morphological features. Approaches explicitly targeting certain chemical substances or substance groups (e.g. the qualitative and quantitative composition of a print) are referred to as substance-specific methods and their features as substance-specific features. Approaches calculating statistical, pixel-based measures of captured fingerprint images (e.g. mean, variance, contrast, roughness and coherence) are referred to as statistical methods and their features as statistical features. This distinction is made for a better categorization of different age estimation techniques. However, it is to note that this separation is not always trivial and that intersections may occur. For example, capturing devices such as FTIR spectroscopes (section 2.1.4), can create morphological representations of certain substance groups. A method or feature is therefore only called morphological, if no specific distinction between the targeted substances is made, acknowledging the fact that different substances might react differently to certain capturing techniques. Furthermore, calculating image-based statistical measures is always referred to as a statistical method, regardless if morphological or substance-specific images are used, with the only exception of the qualitative and quantitative composition of the residue, which is not considered a statistical method, but rather a substance-specific approach.

Furthermore, morphological, substance-specific and statistical approaches for age estimation are divided by their type of acquisition into contact-based and contactless approaches. Contact-based capturing approaches are altering the sample either by a physical or chemical development prior to the lifting process or by physical contact or other changes of the print by the imaging device during capture. Therefore, the main characteristic of contact-based approaches can be seen in the alteration of the print. They can thus also be referred to as invasive lifting techniques. In contrast, contactless approaches neither use development of a fingerprint sample prior to the lifting process nor is the lifting process itself altering the print. They are therefore also referred to as non-invasive lifting techniques. However, it is to note that electromagnet waves of certain wavebands, such as UV radiation, might carry sufficient energy to change the physical state of molecules, electrons or atoms of the residue and therefore increase the speed of the degradation process. In the scope of this thesis, only light in the visible range is used. If UV light is to be used in future studies, additional experiments have to be conducted on whether a specific UV radiation source induces degradation to a fingerprint trace and to which magnitude. Here, great care needs to be taken during experiments to limit the radiation intensity of the light source as well as the spatial and timely exposure of the fingerprint residue to a minimum.

2.1.1 Contact-Based Morphological Approaches

Most state of the art experiments investigating the aging behavior of fingerprints are contact-based, because contactless capturing techniques have only recently become available in sufficient precision. Most approaches are based on the visual inspection of developed fingerprints and the subjective opinion of an expert on how long a print has remained on a certain surface under specific conditions. Such approaches often focus on the quality of the ridges or minutiae and are therefore considered to be morphology based. A numerous amount of case-related experiments has been conducted in the field of forensic crime scene investigation, mainly focusing on the controlled aging of fingerprints under reconstructed crime scene conditions to prove or disprove a suspects version of being at the crime scene at a certain time [7], [9], [10], [11], [12], [13], [14], [15], [34]. Even though such experiments provide an interesting insight into the different possibilities of fingerprint aging behavior, they are always case specific and can barely provide generalizable findings for the creation of a common age estimation scheme. Also, their validity as evidence is questionable [17]. However, it is interesting to note that these experiments often include additional information into the age estimation, such as the last time of cleaning, the amount and identity of people having access to a crime scene or the last time a window was replaced. Such indirect age indicators can often provide valuable information to the age estimation [35].

Early morphology based age estimation approaches include the usage of an Electrostatic Detection Apparatus (ESDA) in 1980 [36] and the healing stage of wounds on a finger [14], [37]. Such approaches have a very narrow application area and are limited in their prospects. ESDA is destructive, possibly preventing identification after age estimation and can only visualize prints younger than 24 hours [35]. Using the healing state of wounds can only be applied if wounds are indeed present. Both experiments do neither provide specific rules for age estimation, nor objective quality measures or error rates. Their practical application therefore remains doubtful.

Holyst and Baniuk investigated the age estimation challenge from a practical police point of view. Holyst describes in 1983 three factors, which are of importance to a successful age estimation of prints [38]. These factors include the development of practical aging features as well as the consideration of creational circumstances and storage conditions. Smooth and non-porous surfaces, such as glass, metal, veneer and plastic were investigated over time periods between one day and several years. By visual inspection, Holyst observes the fading, drying as well as loss of adhesive properties and stickiness of the prints over time. Also, a decrease in ridge thickness is reported. Discussing different influence factors on the speed of the degradation process, he presents average times of permanency for prints of different sweat compositions (eccrine vs. sebaceous sweat), applied to different surfaces and stored under different conditions. He emphasizes the fact that only when the various different influences are considered, similar aging behavior of a fingerprint can be assumed. A specific procedure on how to estimate a fingerprints age is not given.

Based on this work, Baniuk describes in 1990 a practical scheme for fingerprint age estimation [39]. The method was developed and used at the Department of Dactyloscopy of the Civic Militia headquarters in Warsaw since 1970 and is based on the manual examination of about 20.000 fingerprint samples during 20 years. The technique is limited to smooth and non-absorptive surfaces and can be applied to indoor as well as outdoor crime scenes. The approach is based on the reconstruction of the environmental conditions in which a fingerprint was found. At such reconstructed crime scene, fingerprints of the suspect are applied and consecutively lifted at different aging stages. They are then manually compared by an expert to the print found at the crime scene, using a video comparator. Such comparison includes the evaluation of certain "age traits," representing characteristic hints to a certain aging stage as well as magnification and a possible overlay of the prints. The method is therefore based on a subjective assessment. No information is given on the used techniques for fingerprint development and the properties or parameters of the capturing device. It is therefore assumed here that development has been applied if necessary and the process is therefore considered as contact-based. A database provides additional information to the examiner, such as the general qualitative composition of sweat, environmental conditions of the crime scene, general properties of certain surfaces as well as normative fingerprint life-spans during certain environmental conditions. Results of the study are presented in terms of expert opinions being accepted at court hearings over a time period of 10 years (a total of over 100 expert opinions were given). On the base of such expert hearings, suspect statements of the time being present at the crime scene was rejected in 88% of cases and was confirmed in 10% of cases, while no clear opinion could be formed in 5% of cases according to Baniuk.

Although this study seems to be relevant for the practical police work in Poland, it is based on a subjective assessment of forensic experts and can therefore not provide an objective measure of performance. It remains doubtful concerning the complex set of influences on the aging process of a latent print as well as the huge variety of initial sweat compositions that such method is accurate. Furthermore, the presentation given in [39] is not systematic in terms of investigated aging time spans, the used fingerprint enhancement, technical details of the capturing device, the personal properties (age, gender) of the investigated suspects as well as the environmental conditions and specific surfaces examined. Therefore, it remains unclear to which extent the study is accurate, to which scenarios it might be applied, which aging time spans can be distinguished and if the process is non-invasive (which is important for further evaluation techniques, such as DNA analysis). Baniuk acknowledges the incompleteness of the method by concluding her report stating that "the problem has not been solved anywhere in the world." The issue is further discusses by Marcinowski and Baniuk in [40], [41] and [42].

Alcaraz-Fossoul et al. describe in [7] an approach to visually observe changes in developed (dusted with titanium dioxide based powder) and photographed latent fingerprints under varying environmental conditions over six months. The approach is based on one single test subject (Caucasian male) and varies in the application scenario (indoor vs. outdoor), substrate (glass vs. plastic), sweat type (eccrine vs. sebaceous), light conditions (light, medium, dark), ambient temperature (between 12°C and 36°C) as well as ambient humidity (between 20% and 100%). Results are obtained by two different investigators, manually judging the fingerprints quality according to powder adherence, ridge and valley thickness, ridge continuity and contour, number of minutiae and diffusion of the residue into six quality classes. They conclude that fingerprints left on glass are preserved much better than on plastic, that sebaceous sweat ages slower than eccrine sweat, that not necessarily a significant difference does occur between the aging speed of fingerprints stored in the dark and those exposed to sunlight through a double-glazed window and that environmental factors can significantly influence the aging process if occurring in specific combinations, while barely influencing it in other combinations.

Because a fingerprint is altered during development, the approach is invasive and the print might not be available for any further study, potentially required in a forensic context (such as DNA analysis). Furthermore, due to the destructive nature of the lifting process, a single fingerprint cannot be observed over time, as would be possible with non-invasive techniques. The authors circumvent this restriction by applying numerous fingerprint samples at the beginning of the experiments, of which a dedicated number is developed and lifted at each investigated point of time. However, this cannot be considered similar to observing a specific fingerprint over time, because each of the different fingerprints might have a different sweat composition, depending on the part of the body (e.g. forehead) which was touched before fingerprint application. Also, a single test subject cannot be considered representative to derive valid conclusions towards fingerprint aging. Most important, a subjective evaluation of the quality of degraded fingerprints has similar issues than the earlier introduced approach of Baniuk. The correctness of the scheme cannot be objectively verified and remains doubtful. Error rates cannot be measured. Results cannot be compared with other approaches. Especially concerning the ridge and valley thickness, no specific results are presented, preventing a comparison to other studies. No aging curves are presented and no information is provided about potential aging classes.

Several additional studies have been conducted towards the degradation of a fingerprints visibility over time when using different development techniques, such as [43], [44], [45], [46], [47], [48], [49], [50] and [51]. They are mainly focused on the visual quality of latent fingerprints after development (e.g. the visibility of ridges and minutiae) and come to the general conclusion that the visibility of a print decreases over time, depending on the used developer. These studies do not provide any objective results about statistical or morphological changes occurring within the fingerprint over time and are therefore of minor relevance in the scope of this thesis.

Because age estimation of latent fingerprints has been a challenge for many decades, different researchers have published summaries concerning state of the art work, case-related experiments and respective court rulings. Such summaries include a review of Edelmann in 1982 [52], Howorka in 1989 [3], Midkiff in 1992 [53] and 1993 [18], Kohlepp in 1994 [36], Schottenheim in 1996 [54], Stamm in 1997 [37], Herrmann in 2000 [55], Oppermann in 2000 [56], Aehnlich in 2001 [35], Wertheim in 2003 [57] Barcikowski et al. in 2004 [58] and Champod et al. in 2004 [34]. The reviews include also substance-specific approaches, which are discussed in the next section.

Summarizing the properties of contact-based morphological fingerprint aging investigations, very different approaches have been examined. However, results are limited, not being able to provide any objective information about possible age classes, accuracy and error rates. The invasive nature of such approaches, preventing the consecutive capture of single fingerprints over time, as well as the lack of using digital processing and evaluation methods seem to be major disadvantages of these schemes. In the scope of this thesis, contact-based invasive age estimation approaches are therefore discarded.

2.1.2 Contact-Based Substance-Specific Approaches

By far the most fingerprint aging research has been conducted in the field of chemistry, where the qualitative and quantitative chemical composition of print residue as well as the degradation of numerous contained substances or substance groups has been studied over time. It is not possible in the scope of this thesis to provide a comprehensive overview over each of these studies. Therefore, only a few recent exemplary articles are introduced here. For additional studies, the reader may refer to [59], which provides a comprehensive summary of state of the art work in this field.

First approaches include the investigation of Chloride ion diffusion from silver nitrate developer through paper [60]. Such early approach is very narrow in its scope and cannot provide reliable age estimation information. It suffers from severe drawbacks, such as being limited to paper surfaces and exhibiting significant differences depending on paper type, thickness and sweat composition [60].

Recent substance-specific studies often use techniques such as chromatography or mass-spectrometry or a combination of both [61], [62], [63], [64], [65], [66], [67], [68], [69]. Such methods are invasive, requiring the residue (or certain parts of it [68]) to be dissolved and therefore prevent any further analysis of the print. Substances such as fatty acids, amino acids, lactic acid, squalene, cholesterol, wax and cholesterol esters, triglyceride or urea are targeted. Also, the ratio of different substances or substance groups to each other is evaluated. The studies show significant differences in the initial quantity of certain substances between different individuals and even for single individuals depositing fingerprints at different times. Such differences in the initial composition of substances and its behavior over time seem to be caused, amongst others, by the type of sweat (e.g. eccrine, sebaceous), the used surface (e.g. porous, non-porous) or different environmental conditions (e.g. temperature, light; section 2.4).

Duff and Menzel observed in [70], [71] and [72] that latent fingerprints excited by an argon laser as well as UV light seem to differ in their luminescent behavior between fresh and old prints. They conclude that such change is a result of the chemical degradation of luminescent components within the residue (parts of the findings were later withdrawn in [73]). The approach was revisited by Aehnlich [35] and Barcikowski et al. [58]. They investigate the luminescence behavior of the fingerprint constituents lumiflavin, riboflavin, l-tryptophan, cholesterol, oleic acid, squalene, stearic acid, palmitic acid and artificial sweat as well as natural sweat from one test subject. Because substances are prepared in a cuvette, the approach is of invasive nature. Being excited with UV laser light at 355 nm, the fluorescence behavior is studied over time. The radiation is also regarded as a means to accelerate the substance aging, where 10 minutes of laser radiation are considered equivalent to 1670 hours of sunlight exposure. Fluorescent reactions were observed for lumiflavin, riboflavin, squalene, stearic acid, tryptophan, artificial sweat and human sweat, with the fluorescence behavior decreasing over time. A non-invasive technique using Multi Photon Laser Scanning Microscopy (MPLSM) for observing changes in fingerprint luminescence over time is also proposed, which is introduced in section 2.2.

The method of using luminescence for fingerprint age estimation seems to have severe limitations. Apart from not being able to produce an age estimation procedure or any objective age estimation results at this point of time, observed luminescence can be subject to severe distortions, e.g. by fluorescing background materials, different fingerprint compositions leading to a wide variety of fluorescence behavior or contamination of the print with external fluorescing substances (such as mineral oil) [35], [55], [71]. Furthermore, it remains doubtful that UV radiation is an adequate means of simulating fingerprint aging, because numerous other influences seem to impact the aging of a latent print (section 2.4).

Contact-based substance-specific investigations have provided many details about the different chemical substances present in latent fingerprints, which can be over 300 [74]. Although the different studies have successfully investigated the degradation process of certain substances, substance groups, or ratios between substances, they have also shown the great variability of such degradation, mainly influenced by the chemical composition of the prints as well as environmental influences. Therefore, no research beyond the presentation of degradation tendencies is known from the field, such as specific age estimation approaches, possible age classes or objective measures of performance and error. The lack of capturing consecutive images of a specific fingerprint as well as the lack of digital processing and evaluation methods seems to be a major disadvantage of these schemes, similar to the contact-based morphological approaches. Contact-based substance-specific approaches are therefore not examined in the scope of this thesis.

2.1.3 Contactless Morphological Approaches

A few first studies have investigated the changes occurring within contactless captured latent fingerprints over time [32], [75]. Such examinations have significant advantages over earlier contact-based methods, because they allow single fingerprints to be captured and studied continuously. However, their potential for the age estimation has barely been explored at this point of time.

Popa et al. conducted in [75] a study at the Campina Police School (Romania), where fingerprints of over 800 students were observed on sterile glass in indoor and outdoor scenarios every 15 days over a time period of 180 days. Investigated features include the thickness of ridges and valleys, changes of pores, the amount of minutiae and epithelial cells as well as a quantification of DNA. Fingerprints were visualized using different light sources and captured using microscopes and cameras. The method seems to be contact-based only for the microscopy evaluation of epithelial cells (samples are enhanced with Nuclear Fast Red) and the DNA quantification (7500 fast Real-Time PCR System is used). In the scope of this thesis, mainly changes in ridges and pores are of interest, which seem to be captured without prior enhancement. The method is therefore regarded as being contactless and non-invasive in respect to the features of interest.

Concerning the thickness of ridges and valleys, Popa et al. report a decrease in ridge thickness and an increase in valley thickness. However, only ranges are given without the mean value or variance (ridge thickness indoor in mm: [0.3,0.34] after 1 day, [0.24,0.28] after 180 days; ridge thickness outdoor in mm: [0.28,0.32] after 1 day, [0.22,0.26] after 180 days; valley thickness indoor in mm: [0.24,0.28] after 1 day, [0.28,0.30] after 180 days; valley thickness outdoor in mm: [0.26,0.28] after 1 day, [0.26,0.28] after 180 days). Apart from only giving ranges, Popa et al. do not specify how the thickness of ridges or valleys was measured, e.g. at how many points along a ridge the thickness was obtained, which technique was used for measurement and if the same positions were used for thickness determination in each temporal image of a fingerprint sample. Despite these limitations, the provided intervals can be used for comparison within this thesis. However, for performing such comparison, the average change of ridge and valley thickness needs to be approximated from the given ranges. The mean value of the given minimum and maximum is therefore used as an approximate measure for the average ridge thickness, leading to an approximated ridge thickness decrease of

60 µm (from 0.32 mm to 0.26 mm) for indoor and 60 µm (from 0.3 mm to 0.24 mm) for outdoor scenarios, as well as an average valley thickness increase of 30 µm (from 0.26 mm to 0.29 mm) for indoor and 0 µm (from 0.27 mm to 0.27 mm) for outdoor scenarios. It is acknowledged that such approximation might be subject to errors, because the statistical distributions of the thickness values are unknown and their real mean values can be far different from the ones computed here from the ranges. This error might also be the reason for the absolute value of the approximated ridge thickness decrease not being equal to the absolute valley width increase, which are expected to be of equal amount under the assumption that the overall size of the fingerprint remains constant. However, because the distributions are not given by Popa et al., this approximation seems to provide the only possibility for obtaining a means of comparison from their study.

From the microscopic evaluation of pores Popa et al. conclude that pores after one day have a density greater than 10 pores per centimeter, can be in the center or close to the periphery of a ridge and can be closed (the pore is completely enclosed by ridge material) or opened (part of the pore is merged with the ridge valley). According to them, pores start changing their appearance after five days, where closed pores open up, adjacent pores unite and other pores disappear. The process is not quantized, specific measures of the shape, state or size of pores are not given. A comparatively steady decrease of minutiae was observed from 41 to 12 over the course of 150 days. However, it is not specified at which point the continuously degrading minutiae are considered as not being visible any more. A specific age estimation scheme with possible age classes, accuracy measures or error rates is not given. Despite the shortcomings of this paper, it suggests specific morphological aging features and gives measures of their changes over time using non-invasive capturing methods. Especially the reported changes of ridge thickness and pore size seem to be important to be further investigated in the scope of this thesis. They are therefore digitally computed and re-evaluated for comparison (section 6.3.3.1).

Watson et al. present in [32] first results of their fingerprint aging studies on electric charges induced to insulating surfaces by contact electrification. Using an electric field microscopy system, the electric surface charge induced to 50 µm thick plastic (PTFE) sheets by finger contact (triboelectric charging) is measured. The system consists of an electric potential sensor, which is mounted onto a scanning apparatus, moved over the sample at a constant height. Measurement seems to not alter the traces charge and can therefore be considered as contactless and non-invasive. Furthermore, because the ridge structure of prints is targeted, the approach is considered morphological. In the study, eccrine fingerprints of a single test subject were applied to the plastic surface with a contact time of about one second. An area of 36 mm x 36 mm was imaged with a dot distance of 120 µm. A single fingerprint ridge was examined every 30 minutes over a total period of 110 hours. The measured peak surface charge density shows a near exponential decay. Furthermore, the profile of the ridge is imaged, showing a similar decay over time.

An important unknown factor for most age estimation approaches can be seen in the uncertainty of the initial state of the investigated characteristic. Watson et al. claim that the initial amount of charge seems to be dependent mainly on the surface material and the fingerprint application and therefore can possibly be estimated. Whether the initial state of the charge can indeed be obtained has yet to be shown and remains a crucial aspects for the feasibility of the scheme. However, the application influence of contact pressure and -time seems to be significant. Furthermore, the method is limited to insulating surfaces. The decay of the surface charge is dependent on the surface characteristics and environmental influences, according to the authors. A time period of days or weeks is given for common plastics, before the charge has dropped below measureable levels. The method is therefore only suitable for comparatively short aging periods. Despite the mentioned drawbacks, the method seems to be promising, if the reported findings can be confirmed by quantitative tests.

Summarizing the two introduced morphological approaches, the use of non-invasively captured, consecutive fingerprint images for the evaluation of aging properties seems to be promising, yet barely explored. It is therefore investigated in this thesis based on two selected non-invasive capturing devices. Potential contactless devices are introduced and discussed in section 2.2.

2.1.4 Contactless Substance-Specific Approaches

Recent improvements in the performance of contactless capturing devices include the area of structure determination in chemistry. Here, the Fourier-Transform Infrared (FTIR) spectrometer is one of the most important capturing devices, non-invasively measuring the amount and wavelength of middle infrared light absorbed by a substance. The absorbed wavelength provides information about the presence of a certain substance or substance group, while the amount of absorbed light reveals information about its quantity. The device can therefore acquire image representations of the spatial distribution of selected chemical substances and is able to provide information about their qualitative and quantitative composition. In comparison to morphology based approaches, specific substance groups are targeted, which are commonly found in fingerprints.

Contactless substance-specific approaches based on FTIR technology are so far mainly used for the detection and visualization of fingerprints, e.g. as demonstrated in [76] and [77]. Also, recently handled materials can be identified [78]. Only few approaches have utilized the technique for observing changes of the fingerprint residue during aging. First studies show a comparatively constant behavior of salt and a decrease of esters [31] as well as changes in lipids, proteins and their ratio [30]. No age estimation method, possible age classes or performance measures have been provided so far. The highly variable qualitative and quantitative composition of different fingerprints furthermore poses a major challenge to this approach.

Aehnlich [35] and Barcikowski et al. [58] proposed a non-invasive technique for the age estimation of latent prints by targeting luminescent substances within the residue using Multi Photon Laser Scanning Microscopy (MPLSM). The focused area of the print is excited using two or three photons from a Titan-Sapphire laser, which emits electromagnetic pulses with a wavelength of 780 - 900 nm. Because the excitation is limited to the focused area and the emitted pulses are in the near infrared waveband, the technique can be considered non-destructive to the print. Using point by point rastering, fingerprint images can be captured. Three fingerprints were deposited on glass surfaces [35], showing a shift in the emission spectra during an aging period of one week, especially when exposed to sunlight. No quantitative studies are performed and no age estimation procedure, possible age classes or performance measures are given. Furthermore, the approach faces similar restrictions than the earlier introduced invasive luminescence technique (section 2.1.2) in terms of varying fluorescence properties based on the qualitative and quantitative sweat composition, variability in aging speed due to environmental conditions (especially sunlight) as well as potential errors induced by the contamination of fingerprints with fluorescing external substances.

Other potential contactless, substance-specific capturing techniques exist, such as UV/VIS spectroscopy, Raman spectroscopy and infrared or UV imaging using cameras. A discussion of such capturing devices is given in section 2.2. Apart from the non-invasiveness of UV radiation on a fingerprint sample, which remains questionable [58], such capturing devices exhibit potential for the age estimation. At this point of time, no research is known on utilizing these techniques for the contactless observation of fingerprint aging behavior.

Although the contactless nature of certain substance-specific capturing devices offers significant potential, such approaches face similar restrictions than their contact-based counterparts in terms of variability of the qualitative and quantitative residue composition and the uncertainty of its initial state. Furthermore, the lack of using digital processing methods poses a significant limitation, similar to the contactless morphological schemes. As a result, current approaches do not present any results beyond showing a certain aging trend of a selected substance or substance group over time. No age estimation scheme is proposed. In the scope of this thesis, no FTIR spectroscope is available, preventing substance-specific experimental investigations. However, the digital processing pipeline proposed here might be applied to non-invasive substance-specific capturing devices in future work, extending the findings of this thesis to the substance-specific domain.

2.1.5 Image-based Statistical Approaches

At this point of time, no approaches seem to exist using statistical image-based features for the age estimation of fingerprints, neither in combination with contact-based, nor with contactless capturing devices. Statistical features seem to provide a significant potential to the area of fingerprint age estimation, because they can measure the overall changes occurring within a print, which a human expert would observe during visual inspection, e.g. the loss of contrast over time. Therefore, they might even provide more reliable results than their morphological counterparts, which often target only the edges of structures, e.g. when evaluating the ridge thickness or size of pores.

Reasons for this type of feature not being investigated in earlier research work can be seen in the lack of applying digital processing techniques. Because statistical features are mainly based on digital image representations, including an automated processing of their pixel values, they can only be investigated once a digital processing pipeline has been established. The introduction of statistical features, which are well known from other areas, to the domain of latent fingerprint age estimation, is a major contribution of this thesis. In this scope, statistical features are exclusively based on contactless and non-invasive capturing devices, because only they allow for a consecutive capture and examination of a single fingerprint over time.

2.2 Contactless Capturing Devices

In recent decades, a wide variety of accurate and high-resolution capturing devices has emerged. Such devices offer great potential to be transferred to the domain of fingerprint age estimation and to significantly enhance the prospects of solving this challenge. Many of these capturing devices are contactless, offering the benefit of creating consecutive fingerprint images and to precisely observe changes occurring over time. In this section, a few of such capturing devices are introduced and discussed in terms of their invasiveness (which is not always trivial to decide) and other aspects concerning their feasibility. The investigation of different capturing devices is important for many reasons. Finding the best performing device is an important issue along with creating the possibility of combining devices from different capturing domains (e.g. measuring morphological and substance-specific traits) in a future fusion approach. Furthermore, several other requirements are important in practical police applications, such as mobility, acquisition time and precision. This summary can by no means be complete and might be further extended by additional state of the art devices or emerging new sensors. A summary of potential capturing devices based on the state of the art discussion of this thesis has been published in [79].

Chromatic White Light (CWL) sensors are optical, contactless and high-resolution devices, which are often used in surface measurement. An example for such device is the FRT MicroProf200 CWL600 [80], which is studied by Leich et al. [28] and Hildebrandt et al. [29] for the non-invasive capture of latent fingerprints, including challenging surfaces. CWL sensors are based on the chromatic aberration of white light at lenses and can deliver intensity as well as topography representations of a surface. For capturing intensity data, light is emitted onto a surface, matching its focal length to the distance between sensor and surface. Due to the often different reflection properties of fingerprint residue and substrate, the reflected light will differ in its intensity and create a contrast between fingerprint and surface. In the topography mode, the white light is refracted at a lens into wavebands of different colors. Different wavebands have different focal points, where the waveband whose focal point matches the distance between sensor and measured object reflects with the highest intensity and its corresponding distance is used to compute the height of the object. Because only visible light is emitted, CWL sensors can be considered as being non-invasive. In combination with a measurement table of adequate quality, a high precision can be achieved, with a minimum lateral dot distance of 2 µm. However, such precision comes at the expense of mobility (the device described here is bulky and heavy, but mobile devices using similar technology seem to be available [80]). Capturing times are dependent on resolution and captured image size and range between a few minutes and a few hours for a latent print. The MicroProf200 CWL600 sensor is available in the scope of this thesis and is used for the experimental investigations (sections 3.1.4 and 5.3).

Optical Coherence Tomography (OCT) is a technique based on interferometry and usually used with near-infrared light, therefore being considered non-invasive. It can provide 3D-volume representations of latent fingerprints. Dubey et al. utilize swept-source OCT in [81] for capturing fingerprints from glass surfaces, which can also be detected from beneath a dust layer. Such technique only works for optically scattering materials, but might be feasible for the age estimation of latent prints. However, capturing time and mobility need to be considered.

Gloss measurement is another non-invasive technique for capturing latent prints, as exemplary demonstrated by Kuivalainen et al. in [82]. Using a diffractive glossmeter, they show that fingerprints can successfully be separated from the background using gloss. A helium-neon laser is utilized with a waist diameter of 10 µm. The applicability of gloss to the age estimation of latent fingerprints is unknown at this point and seems worth investigating.

Confocal microscopy (as described in [83]) emerged from conventional light microscopy by the invention of point illumination and exclusion of non-focused light by a pinhole. Different focus levels can be triggered for the creation of topographic representations. The technique is usually based on white light or red laser light and is therefore considered to be non-invasive. For example, the Keyence VK-X110 Confocal Laser Scanning Microscope (CLSM) [84] can produce intensity as well as topography representations of a print in reflectance mode. The capturing resolution is dependent on the magnification of the used object lens, which achieves magnifications of up to 100x (dot distance: 65 nm). Furthermore, the device can also produce color representations of a print using a common CCD detector. The high resolution of the CLSM device seems promising for the age estimation of fingerprints. However, moderate magnifications (e.g. using a 10x objective lens) seem to be sufficient for the purpose of age estimation. Its drawback can be seen in its comparatively long capturing time, which can be in the double-digit hour range for a complete print. Concerning mobility, the device might be transported to a crime scene, however requires a trace to be fixed to the measurement table, preventing a mobile application. The Keyence VK-X110 Confocal Laser Scanning Microscope (CLSM) is available in the scope of this thesis and is used for the conducted experiments (sections 3.1.4 and 5.3).

Photoluminescence measurement, which has been applied in a destructive way to fingerprint analysis before [35], [58], [85], is proposed by Aehnlich [35] and Barcikowski et al. [58] to be used in a non-invasive technique utilizing multi photon excitation at 780 - 900 nm by a Multi Photon Laser Scanning Microscope (MPLSM). Such approach is of interest in future studies, in case a Multi Photon Laser Scanning Microscope can be obtained. However, it faces similar restrictions than the CLSM device in terms of capturing time and mobility.

Atomic Force Microscopy (AFM), also called Scanning Force Microscopy (SFM) can provide topographic print visualization and has been utilized for capturing fingerprints from metallic surfaces [86] and glass [87]. Even after wiping a print from a surface by cleaning, fingerprints might still be reconstructable by small cavities from corrosion processes caused by the residue [88]. Atomic Force Microscopy can be used in different modes. In its non-contact mode, a vacuum is required to capture topographic information. Such changes in atmospheric pressure can lead to alterations of the deposit. The technique is therefore regarded as being invasive and infeasible for consecutive fingerprint scans. Its resolution is much higher than that of the previously discussed devices, which might be disproportionately high. Other modes include **Scanning Kelvin Probe Microscopy** (SKPM), which combines a Kelvin probe with an ATM. It allows for the investigation of conductive materials using a potentiometric measurement and was used by Williams et al. in [86] and [89] to visualize prints on iron and brass surfaces. An **Electric Potential Sensor** (EPS) was used by Watson et al. [87] for capturing prints from a polytetrafluoroethylene surface. The technique utilizes the electrostatic properties of prints and is considered contactless and non-invasive. **Chemical Force Microscopy** (CFM) represents another potential capturing technique [90], which is not known to have been applied to fingerprint visualization before, but might be of interest. Unfortunately, the technique seems to be invasive. Although Atomic Force Microscopy seems to be an interesting technique, only the Electronic Potential Sensor seems to be non-invasive, excluding other devices from further studies.

Electric Field Microscopy (EFM) has subsequently been proposed by Watson et al. [32] to measure the electric charge induced in insulating surfaces during fingerprint application (section 2.1.3). Utilizing an Electric Potential Sensor (EPS), a fingerprint image was reconstructed from measured electric charges with a lateral dot distance of 100 µm. Such technique is considered non-invasive, because the surface charges of the fingerprint are not altered during measurement. The technique seems promising for the age estimation of fingerprints. However, it also faces severe limitations. The scheme is limited to insulating surfaces and to aging periods of a few days or weeks, after which the surface has been mainly discharged. Furthermore, it needs to be investigated to what extent the lateral dot distance of 100 µm can be improved, e.g. using overlapped dots. However, the investigation of such device might be of interest for the age estimation of fingerprints.

Structured light illumination is another potential method for capturing latent fingerprints, by projection of a striped light pattern onto the print. The structure of the print causes deformations of the stripes, which can be used by a camera system to reconstruct a topographic image. Exemplary scanners are offered in [91]. It remains unclear if the comparatively small contrast between fingerprint and surface as well as the tiny changes in the residue during aging can be captured with the amount of precision offered by structured light illumination. However, such devices are significantly faster and cheaper than the CWL or CLSM device and might be used for mobile detection of prints at crime scenes (e.g. by scanning complete walls). Therefore, structured light illumination is an interesting technique to be investigated in future work.

Visible light scanners and cameras are used in the area of biometrics to capture live traits. In most cases, off-the-shelf low cost scanners are applied, with resolutions slightly above 500 dpi for fingerprint images (e.g. as used by Pavesic and Ribaric et al. in [92]) and even lower resolutions for other biometric traits [93]. Modern cameras can provide comparatively high resolutions and are non-invasive, fast, cheap and mobile, possibly being feasible for the acquisition of latent fingerprints. However, it is difficult in most cases to capture the small difference in contrast between a latent fingerprint and a surface with such a device. Lin et al. have proposed in [94] an approach to visualize latent fingerprints with a standard camera using specular reflection and polarization under well defined illumination angles. However, it remains doubtful if the captured prints are of sufficient quality for the age estimation of latent prints, which relies on the accurate measurement of tiny changes in the residue.

Flat bed scanners (FBS) are a capturing method utilizing the visible range of light for scanning objects in a cheap, fast and mobile way. They also seem to be promising. For the technique to be non-invasive, the fingerprint has to be fixed upside down on the capturing area without touching the surface and the lid should be kept open to prevent the possible influence of increased temperature from the comparatively powerful light source on the aging process. In own studies [95], the Epson Perfection 1660 Photo flat bed scanner [96] was investigated towards its feasibility for the age estimation. The device can capture intensity images with a maximum lateral dot distance of about 8 µm. A complete fingerprint can be captured within a few seconds. The study shows that such device has the general potential to capture characteristic changes during fingerprint aging, however, the limited precision of the device seems to severely limit its prospects.

Ultraviolet spectroscopy is based on the reflection and absorption properties of materials towards UV radiation in the range between 50 nm and 380 nm. Chemical substances or substance groups absorb or reflect different amounts of radiation, leading to a contrast between them at the detector, e.g. between fingerprint residue and surface. UV spectroscopy is often coupled with visible light spectroscopy (380 nm - 780 nm), resulting in an UV/VIS spectrometer. An exemplary device is the FRT FTR 330 UV/VIS spectroscope [80], which was investigated in a first study by Hildebrandt et al. [97] for the acquisition of latent prints from different substrates. Such spectroscope offers the potential of targeting fingerprint specific substance groups and their degradation over time. However, it does not seem to allow a precise differentiation between substances, as can be provided by FTIR spectroscopy discussed later. The FRT FTR 330 spectroscope operates in the range of 163 nm and 826 nm and provides 2048 different spectra for points of 100 µm in size. Evaluating the invasiveness of such UV/VIS spectroscope is not a trivial task. Only wavelengths above 200 nm should be considered, because lower wavebands are absorbed by ambient air and therefore require vacuum spectroscopy, which is assumed invasive. Furthermore, prints should only be captured in reflection mode, because for other modes (such as transmission), the residue is usually altered (e.g. when being moved to a transparent surface for transmission measurement). However, even with such restrictions, the technique might not be considered non-invasive, because UV radiation is high in energy and can initiate photo-degradation, polymerization of organic material or degradation of DNA [98]. Therefore, it needs to be evaluated first to which degree the short pulses of UV radiation used for spectroscopic measurement accelerate the fingerprint degradation. Only if the influence of such radiation on the observed aging behavior is shown to be negligible, the technique can be considered feasible for fingerprint age estimation. Other limitations of UV spectroscopes can be seen in their immobility and high integration time.

Ultraviolet cameras are considered to be the less accurate, but fast, cheap and mobile alternative to UV/VIS spectroscopes. They are usually equipped with common CCD area detectors, which can provide real-time images of crime scenes without requiring the photographed object to be planarized or transported to a laboratory. An UV radiation source might be used to illuminate the sample during the process of photographing. Applying such technique, different kinds of forensic traces can be visualized, as demonstrated by Richards et al. in [99]. Because CCD detectors are sensitive to a wide range of radiation, the cameras need to be equipped with specific bandpass filters, such as the Baader-Venus filter [99]. Similar to the UV/VIS spectroscope introduced earlier, the negligibility of UV radiation in respect to the aging behavior of fingerprints needs to be demonstrated before considering the technique non-invasive. Furthermore, the advantages of a low price, high capturing speed and great mobility involves a significant loss in precision, where the single spectral frequencies cannot be differentiated any more.

Infrared spectroscopy is mainly performed in the form of Fourier-Transform Infrared spectroscopy (FTIR) [30], [31], [76], [77], [78], [100], which is based on the reflection and absorption properties of certain chemical substance groups to mid-infrared illumination. In contrast to UV/VIS spectroscopy, certain substances absorb specific frequencies of infrared radiation, allowing for a precise determination of their qualitative and quantitative composition. Because infrared radiation leads only to the vibration of molecules but not to their disintegration, the technique is regarded as non-invasive. However, this applies only to the reflectance mode, because transmission measurement requires an invasive preparation of the fingerprint (lifting onto a transparent slide) and the ATR mode requires contact with the measurement crystal. Evaluating the degradation of selected chemical components of latent fingerprints in a non-invasive way seems to have significant potential for the age estimation, as is shown in first examinations of Antoine et al. [30] and Williams et al. [31], who have examined the degradation of certain fingerprint constituents in first studies (section 2.1.4). FTIR spectroscopy is therefore strongly recommended to be investigated towards the age estimation of latent fingerprints. Unfortunately, an FTIR device is not available in the scope of this thesis. Limitations of the technique can be seen in its slow capturing speed and immobility.

Infrared cameras provide a cheap, fast and mobile alternative to FTIR spectroscopy. However, their loss in precision is dramatic. Single spectral frequencies might not be differentiated any more, therefore specific substances and their qualitative and quantitative compositions seem to not be evaluable. However, because infrared cameras often work in the near-infrared band, they have the potential to capture certain changes in temperature, e.g. the time a fingerprint needs to assume the ambient temperature. In [100] infrared cameras were used to visualize different types of forensic evidence. In preliminary own studies, the infrared camera FLIR SC305 [101] was investigated towards its potential for the age estimation of fingerprints. So far, no reliable results could be produced, due to the lack of precision of the device. The technique can be considered as being non-invasive, in case the fingerprint is not artificially illuminated with a high energy light source. In case of manual illumination (heating of the fingerprint), the impact of the illumination on the latent print needs to be considered.

Raman spectroscopy is based on the inelastic scattering of near-infrared, ultraviolet or visible monochromatic light at the molecules of the residue. Photons of monochromatic light excite electrons of the residue to higher energy levels, leading to the emission of photons at similar or different frequencies when the electrons fall back. Using adequate detectors, the presence and amount of certain chemical substances can be derived from the received frequencies. Connaster et al. have proposed to use Surface Enhanced Raman Spectroscopy (SERS) in [102] to capture latent fingerprints. However, silver and gold nanocomposite substrates were used, which are Raman active, but do barely occur at real crime scenes. The feasibility of Raman spectroscopy for the effective lifting of non-developed latent prints without using Raman active surfaces has yet to be shown. If the excitation wavelengths are limited to non-destructive ranges (e.g. visible or infrared light), the technique can be considered non-invasive. Raman spectroscopes face similar limitations than other spectroscopic techniques, which are comparatively long capturing times and immobility. However, studies of this capturing technique seem promising for future work.

2.3 Digital Processing Methods

In the scope of this thesis, a digital processing pipeline for the automated preprocessing, feature extraction and age estimation of non-invasively captured (digitized) latent prints is designed. While most parts of this pipeline have to be newly designed, certain state of the art digital processing methods can be re-used and adapted. They act as basic building blocks for the more complex methods designed here. The preprocessing of digitized fingerprint images is partially based on general image processing techniques as well as selected biometric algorithms from the literature. These methods are introduced in section 2.3.1. Adequate metrics for feature extraction are summarized in section 2.3.2. Digital decision-making strategies relevant in the scope of this thesis are presented in section 2.3.3. In the scope of designing each step of the processing pipeline (chapter 4), a more detailed insight is given on how certain state of the art methods are combined for the creation of adequate processing algorithms.

2.3.1 Image Preprocessing Methods

Preprocessing of captured fingerprint images is an important prerequisite for extracting characteristic aging features. Features of sufficient quality depend to a significant extent on an accurate normalization, noise reduction, fingerprint enhancement, segmentation as well as binarization. In the scope of this thesis, basic image processing methods are used as building blocks for fingerprint preprocessing methods, which are briefly introduced in this section. Well known general methods include standard convolution filters, such as blur or Difference or Gaussian (DoG) filters, as well as morphological operations, such as dilation and erosion. An introduction into these techniques is given in [103]. Several of such methods have been implemented and are publicly available in the OpenCV framework [104].

Best-fit plane subtraction is a well known method for the planarization of angular surfaces. Especially in the case of topography images, which measure the height of a fingerprint, small angles in the underlying surface might significantly distort the results. Therefore, the plane best fitting a captured image is calculated (using least squares approximation [105]) and is subtracted from the original topography image, as proposed by Leich et al. for Chromatic White Light (CWL) sensors [28].

Binarization is an important issue in fingerprint preprocessing, to adequately separate fingerprint from background (surface) pixels. Different methods have been proposed in the literature on how to find optimal thresholds for different application fields. Several of them have been included into the distribution package Fiji [106], [107], of the open source image processing tool ImageJ [108], [109], [110]. These approaches are introduced in [111] and include the methods Default [111], Huang [112], Intermodes [113], IsoData [114], Li [115], MaxEntropy [116], Mean [117], MinError(I) [118], Minimum [113], Moments [119], Otsu [120], Percentile [121], RenyiEntropy [116], Shanbhag [122], Triangle [123] and Yen [124]. For more detailed information about the specific methods, the reader may refer to the corresponding literature. In the scope of this thesis, adequate fingerprint binarization methods are selected or designed, depending on the used capturing device and data type.

In their natural form, fingerprint ridges consist of small droplets or puddles of fluids and particles. However, in some cases it is of interest to segment them in the form of continuous, filled lines (fingerprint enhancement). For such purpose, the algorithm of Hong et al. [125] was proposed in the area of biometrics to be used with exemplary fingerprints. The algorithm calculates the ridge orientation and frequency of non-overlapping image blocks, which are then applied in combination with a Gabor filter to locally enhance the fingerprint ridges. The algorithm is only available in the form of a Matlab implementation by Kovesi et al. [126] and is designed especially for fingerprint images with a resolution of 500 dpi. In the scope of this thesis, the algorithm is re-implemented in C++/OpenCV and adapted to high-resolution images of latent fingerprints. Although the algorithm seems to perform well on exemplary fingerprint images, it faces challenges when being applied to latent prints, which are significantly more distorted than their exemplary counterparts. For the segmentation of pores, a Mexican Hat filter was proposed by Jain et al. [127] for exemplary fingerprints used in biometrics. The idea is transferred here to the area of forensics and applied to high-resolution latent fingerprints.

2.3.2 Image Features

For the evaluation of statistical image-based features, many concepts from the areas of statistics and image processing are available, some of which are exemplary used here as aging features. Basic statistical metrics include the mean and standard deviation of an image as well as its local (block-based) variance, describing a general pixel distribution. Others are based on certain surface characteristics, such as image gradients [103], roughness [128] or coherence [129].

Biometric features are often based on the techniques of principal component analysis, most discriminant features and regularized-direct linear discriminant analysis, e.g. as compared by Pavesic and Ribaric et al. in [92] for fingerprints and digitprints (area between first and third phalanges of a finger). These techniques are not investigated in the scope of this thesis, but are interesting research targets for future work.

For the evaluation of morphological aging features, the size and shape of certain fingerprint structures need to be computed. In the scope of this thesis, the size of a structure is in most cases analyzed by computing the relative amount of pixels belonging to it. However, for the computation of the mean ridge thickness of a latent print, the concept of skeletonization is used, as described by Toennies in [130].

2.3.3 Decision-Making Strategies

In the scope of this thesis, two main strategies are applied for deciding on the age of latent fingerprints. Both strategies are based on well known concepts and seem to be most promising for first evaluations in the context of fingerprint aging. In a first attempt, age estimation performance is evaluated using a formula-based approach. Such technique is based on the approximation of mathematical functions from the computed experimental aging curves. Regression is a feasible method for providing such approximation. Exemplary literature describing this mathematical technique can be found it [131].

In a second attempt, fingerprints are classified into age classes using machine-learning techniques (pattern classification). Exemplary literature for further reading about this topic can be found in [132]. For the experimental model training and classification performance evaluation, the common toolbox WEKA is used [133], providing a collection of different classifiers and evaluation routines, which are well-suited for the here conducted first studies on the aging of latent fingerprints.

2.4 Influences on Fingerprint Aging

Influences on the aging of latent fingerprints are illustrated using the following two examples. Identification officer Balloch was very surprised in his 1977 article [13] that without prior knowledge, he could not differentiate between different fingerprints he had applied onto a glass window of his police department, where prints varied in age between one and three days and between one and three months. In the scope of a bathroom murder, it was furthermore observed that prints classified as not being dactyloscopically evaluable, suddenly became visible after a time period of nine weeks [36], potentially having undergone a refreshment by high ambient humidity in the bathroom.

Numerous influences on the aging behavior of latent prints have been identified in the literature, which are considered as being one of the main challenges to successful age estimation. The two examples illustrate the impact these influences might have, which can range from being negligibly small to a complete reversal of the aging process. Furthermore, influences do not apply to all situations in the same way. Some influences might only be relevant for short aging periods, while others only manifest themselves over long time period. Some influences only become relevant when certain other conditions are present. Also, certain influences do not only impact the fingerprint, but also the capturing devices. Influences proposed by state of the art work are summarized here into five main categories, comprising the sweat influence, environmental influence, surface influence, application influence and scan influence. From the significant amount of investigations, only exemplary studies are listed here.

The sweat influence describes the impact of the qualitative and quantitative sweat composition of an investigated fingerprint region on its degradation behavior. Such influence can vary between different people, different fingers of a single person, the same finger or a person at different points in time or even between different locations within a print. The qualitative and quantitative composition of sweat has been widely studied in the literature and seems to be dependent on age, gender, race, diet, health, blood group, pharmaceuticals, drugs, metabolism, type of secretion (eccrine vs. sebaceous), body temperature, ambient temperature, general occupation, level and type of current activity, duration of sweating, psychological state (stress, nervous, afraid, anxious, exited), frequency of earlier contact with other objects as well as external contamination (drugs, food, cosmetics, blood, dirt, oil, paint, mucous, ink, dust, wax, seminal fluid, explosives, firearm residues) [7], [11], [15], [16], [17], [18], [32], [35], [36], [37], [38], [39], [40], [41], [42], [45], [46], [48], [49], [50], [53], [55], [57], [58], [59], [62], [75], [134], [135], [136], [137], [138].

The environmental influence is composed of different ambient conditions, which might impact a prints degradation process, such as temperature, humidity, precipitation, radiation, mechanical stress, bacterial activity, atmospheric contamination (dust, sand, emissions), air circulation (wind, air currents), air pressure, condensation, friction or submersion [7], [9], [11], [12], [15], [16], [18], [32], [35], [36], [37], [38], [39], [40], [41], [42], [45], [46], [48], [49], [50], [53], [55], [57], [58], [59], [62], [67], [75], [134], [135], [136], [138], [139], [140], [141], [142].

Surface influences on the aging of latent fingerprints describe all characteristics of a certain substrate impacting the speed of degradation. Such influences include surface structure (smooth, rough), texture, absorptiveness (porous, non-porous, semi-porous), chemical reaction potential (corrosion), inherent moisture, thickness, curvature, surface temperature, electrostatic properties (surface charge, insulation properties), surface tension, adhesive forces, surface coating and surface deformability [7], [15], [16], [18], [32], [35], [36], [37], [38], [39], [45], [46], [48], [49], [50], [53], [55], [57], [59], [62], [134], [135], [136], [138], [140], [143].

The application influence describes all factors related to the moment when a fingerprint is applied to a surface. Such factors might impact the amount, form and distribution of the deposited material. Application influences include contact pressure, contact time, smearing, ridge skin damage (skin disease, scars, injuries, abrasions) and the size of fingertip area in contact with the surface [14], [15], [16], [32], [48], [59], [134], [136], [137], [138].

Scan influences are considered those arising from the specific parameterization as well as technical characteristics of a used capturing device. Such influences include the used light source, capturing resolution, size of measured area, location of measured area, type of captured data, capturing technique or proprietary internal preprocessing. They might differ greatly between different capturing devices and often cannot be compared. At this point of time, no studies seem to exist investigating the impact of capturing properties on the age estimation of latent fingerprints. However, it is generally known that most optical devices are sensitive to temperature fluctuations and vibrations.

The different influences summarized in this section are not considered to be exhaustive, possibly requiring additional influences to be added. However, there is also a high probability of several influences to be negligible in specific application scenarios. The identification and consideration of relevant influences therefore seems to be an important task for any age estimation scheme.

3 A NEW APPROACH TOWARDS LATENT FINGERPRINT AGE ESTIMATION

The challenge of estimating a latent fingerprints age has been an issue to forensic investigators for about 80 years [6], which has not been adequately solved so far. With the significant advances, contactless and non-invasive capturing devices have undergone in the last decades, a new potential arises for addressing this challenge. In this thesis, a novel approach is proposed, combining two exemplary selected non-invasive, high-resolution capturing devices with the design of a digital processing pipeline for the automated data acquisition, preprocessing, feature extraction and age estimation. The new approach can provide a significantly increased amount of precision to the processing of latent prints for age estimation and is for the first time able to provide a computer-based, objective age estimation performance evaluation. In this chapter, the main methodology and concepts of this thesis are introduced in the first part (section 3.1). The second part (section 3.2) describes main assumptions, forming the basis of the later conducted evaluations. It furthermore includes the definition of five research objectives, addressing the earlier identified research gaps.

3.1 Methodology and Concepts

The general idea of investigating optical non-invasive capturing devices towards their potential for the age estimation of latent fingerprints is derived from the Digi-Dak project proposal [5]. Contactless sensors enable researchers for the first time to investigate fingerprint aging behavior using consecutive scans of a single fingerprint. Due to the non-invasive nature of such devices, a latent print can be captured repeatedly over arbitrary time periods. Because prints are available in digitized form, a complete digital processing pipeline can be designed and applied, which does not exist so far in the domain of forensic fingerprint age estimation.

In this section, the main concept and basic methodology of this thesis is presented. While the concepts of absolute and relative age are described in section 3.1.1, a differentiation between short- and long-term aging is performed in section 3.1.2. The concept of time series is summarized in section 3.1.3, followed by the selection of two exemplary non-invasive capturing devices in section 3.1.4. The indoor application scenario used in the scope of this thesis is introduced in section 3.1.5 and the proposed digital processing pipeline is presented in section 3.1.6. Section 3.1.7 summarizes the quality measures chosen for investigating selected aspects of the designed processing pipeline in the scope of the experimental evaluations.

3.1.1 Absolute vs. Relative Age

The terms absolute and relative age have been used earlier, e.g. in [34] and [55]. The absolute age describes the total time which has passed from fingerprint deposition onto a surface until it is lifted by a forensic expert or captured by a certain device. The relative age describes the time offset between the applications of two or more fingerprints. Such relative age is of interest when determining the sequence in which people have been present at a crime scene or in which events have occurred. It might also be of interest in the scope of sequencing overlapped fingerprints [26], or even separating them.

In the scope of this thesis, the main focus lies on the estimation of the absolute age of a fingerprint, which is the more comprehensive information, however more challenging to obtain. The relative age can be seen as a subset of the absolute age, because from the knowledge of absolute ages of certain prints, their relative time offsets can be derived. Methods for estimating the absolute ages of latent prints can therefore also be used to estimate their relative ages. Nevertheless, the relative age should not be neglected, because in case the absolute age cannot be estimated, it might still provide valuable information about the sequence of latent prints and therefore the sequence in which objects were touched.

3.1.2 Short- and Long-Term Aging

A specific separation into fingerprint short- and long-term aging does not seem to exist at this point of time. Previous work has focused on arbitrary time intervals, which often contain several weeks, months or years (section 2.1). In the scope of this thesis, optical non-invasive capturing devices are investigated, which require capturing times between a few seconds and several hours for a single fingerprint, depending on the used resolution and measured area size. Such devices therefore allow to investigate fingerprint aging behavior within a few hours after deposition. To also address such very short aging periods, fingerprint aging is specified here into short- and long-term periods.

Short-term aging in the scope of this thesis is referred to as comprising an aging period of up to 24 hours. During such period, significant changes seem to happen to a fingerprint, which can be described as 'drying' of the print [42]. Large amounts of water and other volatile substances seem to evaporate from the print (which is comprised to approximately 99% - 99.5% of water, according to [38]). Therefore, this period contains the most discriminative characteristics, making it an important research target.

Long-term aging is referred to as including aging periods of more than one day, up to several years. At this point of time, it is not clear what the approximate lifespan of a fingerprint is. However, the observations within this thesis show that even after three years, many fingerprints are still clearly visible. Fingerprint long-term aging is characterized by physical and chemical decomposition of less volatile substances, leading to a decrease of fingerprint material, contrast and contour at a slow speed. However, given the long time spans of such degradation, significant differences might still be observable, when long capturing periods are used.

In this work, the short- as well as the long-term aging is of interest, because fingerprints of both classes are often found at crime scenes [144]. Different degradation processes seem to occur during both time periods, potentially requiring or favoring different aging features. However, it seems particularly hard in the scope of this thesis to identify the causes of observed differences between the short- and long-term aging behavior (and even between different features of a single time period), because no information about the chemical composition of the traces is provided by the investigated capturing devices (selected in section 3.1.4).

3.1.3 The Concept of Time Series

The general concept of time series seems to be a straightforward approach for observing changes occurring during fingerprint aging. The main idea is to acquire a certain type of information in regular time intervals over a specific period of time. In the field of fingerprint age estimation, capturing a latent fingerprint in regular time intervals has barely been applied so far, because classical contact-based lifting techniques are usually invasive (e.g. by applying fingerprint developer or altering the print during the lifting process), preventing the repeated capture of unchanged prints. Because of this alterations, classical approaches are not able to observe the degradation behavior of a single fingerprint over time.

In some cases, techniques are used to overcome such restriction, e.g. by applying several fingerprints simultaneously to a certain substrate, followed by the lifting of a subset of these prints at each point of time [7]. Such technique is not regarded as being reliable, because the degradation behavior can vary greatly between prints and the simultaneous application of a huge number of prints will ultimately lead to different strengths of the prints, resulting in variations of their degradation behavior.

The here applied contactless and non-invasive devices do not alter a latent print during the capturing process. They therefore allow capturing time series of a single fingerprint over arbitrary time periods and in arbitrary capturing intervals. After a time series has been finalized, the fingerprint is still available in its original form for further time series or other investigation methods, such as DNA analysis. In the context of this thesis, only image guided techniques are used, producing intensity as well as topography image series of a selected fingerprint area.

3.1.4 Non-Invasive Devices Selected for Investigation

In the scope of this thesis, two selected non-invasive capturing devices are exemplary investigated. When selecting such devices, two main considerations are of great importance: precision and availability. Latent fingerprints contain comparatively small structures, which are determined in prior investigations to exhibit an average height of about 1 µm and a ridge thickness of approximately 300 µm. Prior own experiments have furthermore shown that without a microscopic objective lens, cameras usually deliver images of insufficient quality. A flat bed scanner has been investigated in own studies of [95], showing the general possibility of the device to capture changes during fingerprint aging. However, also severe limitations in terms of capturing precision and image quality are observed.

As a consequence of such findings, high-precision devices seem to be required for an adequate investigation of fingerprint aging behavior. Such devices are expensive, usually generating costs far beyond 100 000 €. Therefore, devices cannot be chosen freely, but are rather subject to the constraint of availability. In the scope of this thesis, the high precision surface measurement device FRT MicroProf200 CWL600 (CWL) [80] as well as the Keyence VK-X110 Confocal Laser Scanning Microscope (CLSM) [84] are available in the scope of the Digi-Dak [5] and DigiDak+ [145] projects. Both devices offer a high precision and allow for the computation of statistical as well as morphological aging features. The CWL device can achieve a maximum lateral dot distance of 2 µm and a longitudinal resolution of 20 nm, while the CLSM device can achieve even higher resolutions (lateral: 65 nm, longitudinal: 10 nm, using a 100x objective lens). Both devices are considered suitable for non-invasively studying the aging behavior of latent prints with a high precision, which is shown for the

CWL device in first studies of Leich et al. [28] and Hildebrandt et al. [29]. Additional information about the devices is provided in the general test setup of section 5.3.

Apart from the high cost as well as immobility, the main limitation of these devices can be seen in their inability to target specific substances or substance groups, which strongly reduces their feasibility for the investigation of changes in the qualitative and quantitative chemical composition of latent prints. Non-invasive, high-precision devices being able to capture such substance-specific properties do exist, yet are not available in the scope of this thesis. For example, FTIR spectroscopy seems to be an adequate technique for the non-invasive observation of chemical changes during aging, which remains a promising subject of future research.

3.1.5 The Used Indoor Application Scenario

In general, fingerprints might be found in different application scenarios. While prints are often lifted inside buildings [7], [9], [12], [14], [15], also outdoor scenarios are possible [9], [10], [11]. In some cases, fingerprints are even found submerged in certain substances, such as water [142], [146]. Because the experiments within this thesis are designed to exhibit a first qualitative evaluation of the potential of non-invasive capturing devices and digital processing techniques, the conducted experiments are exclusively focused on indoor crime scenes, namely office and laboratory environments. Influences (especially from the environment) are reduced here to a certain extent (e.g. precipitation is not present, temperature and humidity fluctuations are limited). Other influences, such as the sweat composition of the prints or the specific application process are not controlled and remain subject to a significant variation, similar to what can be expected from common crime scenes.

The influences in this indoor scenario are assumed to be small enough to achieve a certain age estimation performance. Such assumption is confirmed by the experimental results, which show that a certain accuracy can indeed be achieved for the used indoor locations. The age estimation performance might be iteratively improved in future work, by the identification of characteristic influences on the aging behavior in quantitative studies and a subsequent controlling or exclusion of these influences. A starting point for such investigations is provided by first qualitative studies of influences and the creation of a preliminary influence model in the scope of this thesis (sections 3.2.2.5 and 6.5). Additional application scenarios should be investigated in the future, such as the challenging outdoor crime scenes or fingerprints submerged under water.

3.1.6 The Digital Processing Pipeline

The creation of high-resolution, digitized fingerprint representations offers the possibility of designing a digital processing pipeline, which can be used for many different purposes, such as print enhancement, normalization, reduction of different distortions and influences, automated feature extraction as well as computer-based age estimation. Such digital processing pipeline has not been proposed so far for the age estimation of latent fingerprints. It is therefore a contribution of this thesis to design such digital processing pipeline and to evaluate selected aspects of it towards their feasibility for latent print age estimation. Digital processing can be seen as a major contribution of computer science to the field of crime scene forensics and is an emerging field of research for many forensic trace types, such as fingerprints, palm prints, ear prints, shoe prints, tool marks, fibers or bullets [4]. It is often referred to as computational forensics [19] or digitized forensics [20].

As the main methodological foundation of this thesis, a digital processing pipeline for the age estimation of latent fingerprints and its performance evaluation is designed. It is based on the general biometric processing pipeline, e.g. as described by Vielhauer in [147] and Pavesic in [148]. The detailed structure of its four main steps of data acquisition, preprocessing, feature extraction and classification is designed and implemented for the age estimation of latent fingerprints (Figure 3).

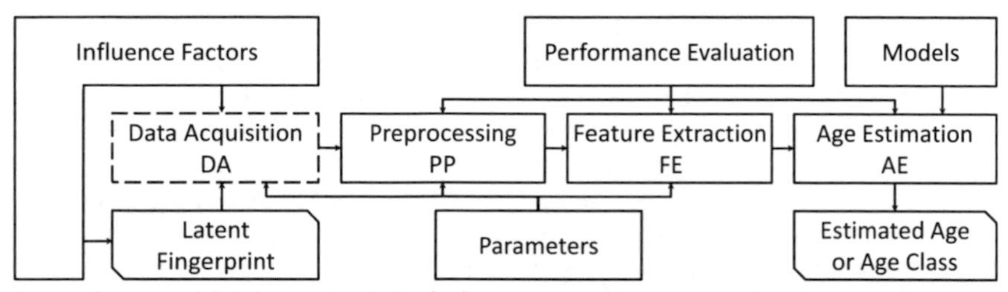

Figure 3: The proposed digital processing pipeline for fingerprint age estimation and performance evaluation, based on the general biometric processing pipeline, e.g. as described by Vielhauer in [147] and Pavesic in [148].

In the data acquisition step (DA), the designed processing pipeline captures a latent fingerprint image in regular time intervals, hence creating a digital time series. The print as well as the capturing process can be subject to different influences from the environment, the fingerprint application, the capturing device, sweat composition or surface material (section 2.4). Such influences can lead to certain distortions in the captured time series, which need to be addressed in later processing steps. In this thesis, data acquisition is not of interest beyond the identification of basic sensor settings as well as the identification of potential distortions, which later need to be reduced in the scope of image preprocessing. It is therefore marked in dashed lines in Figure 3. In future work, also optimizations of the hardware should be conducted, including researchers from the fields of sensor manufacturing.

After the capture of an aging time series is completed, all additional processing steps are conducted on the digitized data. The original fingerprint, which has not been altered by the capturing process, is not required anymore and can be used for other investigation methods, such as DNA analysis. For designing the three following processing steps of time series preprocessing (PP), feature extraction (FE) and age estimation (AE), existing processing techniques can be transferred and adapted from other application areas (such as algorithms for live fingerprints from biometrics or classification algorithms from machine-learning, section 2.3). However, also new methods need to be designed in many cases from basic image processing operations.

In the preprocessing step (PP) of the pipeline, images need to be transferred into a comparable form, by spatial alignment (long-term aging), reduction of various influences from the capturing devices and the environment as well as the segmentation of relevant fingerprint structures. The digital preprocessing of time series is regarded as especially challenging, because fluctuations of the capturing devices (e.g. changes of the overall image brightness over time) as well as changes from the environmental conditions (e.g. accumulating dust) need to be digitally reduced for the creation of reliable time series.

Various features are implemented and evaluated in the feature extraction step (FE), where different statistical, image-based features are applied for the first time to the challenge of age estimation. Furthermore, morphological features are designed and evaluated in the scope of long-term aging.

Features are computed from preprocessed print time series, which results in a sequence of feature values, representing the aging curve of such feature in respect to the investigated fingerprint. In the age estimation step (AE), formula-based age computation as well as machine-learning based classification into well defined age classes is investigated. For the first time, automated latent print age estimation is performed, resulting in objective performance measures, which might be used as a benchmark for future age estimation approaches.

The performance of the designed processing pipeline is evaluated in respect to selected exemplary aspects of the steps of data acquisition (DA), preprocessing (PP), feature extraction (FE) and age estimation (AE), to evaluate the potential and limitations of computer-based processing for digital fingerprint age estimation. Such evaluations include the selection of adequate sensor settings, fingerprint binarization methods as well as an optimization of different parameters. They are furthermore comprised of a feasibility assessment of the computed aging features as well as a performance examination and comparison of age estimation strategies. Also, influences on the captured fingerprint aging series are studied and combined to a first, preliminary influence model.

3.1.7 The Used Quality Measures

While the design of a digital pipeline for fingerprint processing and age estimation (chapter 4) is only one part of the contribution of this thesis, selected aspects of the proposed methods are exemplary investigated in the experimental evaluations (chapters 5 and 6). For this purpose, digital quality measures need to be defined, which can be used for an objective, automated computation of the performance of investigated methods. In the scope of this thesis, common quality measures are applied, which are well established in various scientific fields. These measures include regression [131] (e.g. as used in economics), the Pearson correlation coefficient [149] (e.g. as used in statistics) as well as Cohens Kappa [150] (e.g. as used in data mining). Furthermore, basic statistical metrics are applied, such as the mean, median or standard deviation.

3.2 Basic Assumptions and Research Objectives

When investigating the aging behavior of latent fingerprints, it seems appropriate to start examinations with the investigation of the aging process itself as well as potential influence factors impacting on it. Only after having gained sufficient understanding of the underlying processes, an evaluation of the accuracy of derived age estimation schemes seems to be reasonable. However, when researching the numerous possible influences on the fingerprint aging process and their complex interdependences, it can be concluded that the understanding of such process and its various influences will not be complete in the foreseeable future. Therefore, an alternative approach is taken in the limited scope of this thesis.

The experiments of this thesis are designed in a way to first show the general feasibility of the proposed age estimation approach using a limited indoor application scenario, exhibiting comparatively constant environmental conditions. Designing and evaluating different acquisition, preprocessing, feature extraction and age estimation techniques, first computer-based objective age estimation performance measures are derived. After experimentally verifying that the proposed approach can achieve a certain age estimation performance, it can then be iteratively revisited and improved by the identification and consideration of additional influences in future work. It can furthermore be extended to scenarios of stronger fluctuations of influences.

In the scope of this thesis, a basis is provided for such future iterative improvements, by conducting a variety of experiments on several influences and their combination into a first qualitative influence model. Such model might be used as a starting point for further investigations of fingerprint degradation characteristics, possibly leading to a continuous improvement of the age estimation performance in future decades. Furthermore, the here computed quality measures provide a potential benchmark for future age estimation schemes and can be used for rating the performance improvement achieved by future approaches.

In section 3.2.1, main assumptions are defined as the basis for the experiments conducted within this thesis. In practice, it cannot be assured that these assumptions are always met, requiring to revisit and refine them in regular time intervals in the future. After the introduction of main assumptions, the five research objectives of this thesis are presented in section 3.2.2, which are defined according to the earlier identified research gaps (section 1.2).

3.2.1 Assumptions

Similar to other scientific experiments, the investigations within this thesis are based on several assumptions, which form the necessary basis for the conducted evaluations. These assumptions are in particular:

- **Existence of suitable capturing devices, algorithms, features, age estimators and performance metrics:** One of the main assumptions of this thesis is that suitable non-invasive capturing devices as well as adequate procedures for the preprocessing, feature extraction, age estimation and quality evaluation do exist, including adequate aging features, age estimators and performance metrics. Only if such entities do exist, the proposed approach has the ability to estimate the age of latent prints to a certain extent.
- **Indoor application scenario:** All experiments conducted in the scope of this thesis are based on indoor application scenarios. This choice of application does not exclude environmental influences, but limits or controls them to a certain degree. Other influences from the sweat composition of a test subject or the specific application process are not limited by such scenario.
- **Representativeness of captured fingerprints:** The complete population of latent fingerprints, comprised of all latent prints left by humans (including all possible locations, times and different ethnic origins) cannot be investigated. Instead, a sample set of fingerprints from subjects in the local and temporal vicinity of this research are used as donors. It is assumed in the scope of this thesis that the sample set of acquired fingerprints is representative for the general population of latent prints. However, differences might be discovered in future work between certain fingerprint characteristics (e.g. sweat composition, size or shape) in respect to certain locations, ethnic origins or points of time, which could lead to different aging properties. In such case, more specific boundaries have to be defined for the application of the here presented findings.
- **Suitable capturing device parameterization:** In the scope of this thesis, the focus lies on the preprocessing, feature extraction and age estimation steps of the defined processing pipeline of Figure 3. The data acquisition step is only investigated as far as the selection of capturing parameters and the identification of introduced distortions is concerned. Parameters are chosen according to optimal values determined in first tests or as recommended by the sensor manufacturers. Such settings are assumed to be feasible for the studies conducted in the scope of this thesis. An optimization of these parameters needs to be performed in future work, including experts from

the area of sensor manufacturing, to provide enhanced functionality for the specific needs of age estimation schemes.

- **Constant capturing properties:** The capturing process itself is assumed to be constant in the time as well as the spatial domain. Light intensity, detector sensitivity and other properties should not differ between different capturing locations and times. As can be seen in section 4.1.2, this seems to not always be the case, and measures are taken to digitally reduce such distortions within the experiments of this thesis. However, this assumption seems to be important for a reliable observation of aging behavior and should be realized by the capturing hardware and software in the future.
- **Suitable frequency of captured fingerprint and background pixels:** For ensuring an adequate preprocessing and feature extraction of fingerprints, images are expected to be neither empty nor containing exclusively fingerprint pixels. Applied preprocessing techniques as well as computed aging features often rely on the presence of an adequate amount of pixels from each class. It is therefore assumed in the scope of this thesis that at least 10% of image pixels belong to each fingerprint and background. This assumption is ensured by manual inspection of fingerprint images before further processing.

While first restrictions of these assumptions are disclosed in the scope of this thesis (e.g. the non-constant capturing properties for certain cases), additional ones might be identified in future work. Therefore, the assumptions need to be reviewed regularly and additional measures have to be applied if necessary, to assure their validity. If an appropriate age estimation scheme can be proposed, some constraints might even be lifted (e.g. extending the application scenario from indoor scenes to other locations).

3.2.2 Research Objectives

In this section, the main research objectives O1 - O5 are introduced. They are defined according to the five main research gaps identified in section 1.2. Each objective is investigated separately in the remainder of this thesis, where adequate digital methods are designed in chapter 4 for addressing its specific aims. The research objectives are mapped to five corresponding test goals in section 5.1. Each test goal aims at investigating selected important aspects of its corresponding objective in the scope of the experimental evaluations of chapter 6.

3.2.2.1 Objective 1 (O1): Design of the Data Acquisition Process

Research Objective 1 (O1) represents the basic prerequisite for all other objectives. Its main aim is to extract basic settings of the two selected capturing devices, such as capturing resolution and measured area size. Furthermore, potential distortions introduced by the devices or other challenges to an accurate capture of comparable time series need to be identified. Once adequate sensor settings are chosen and challenges identified, the following objectives can be investigated based on the findings of O1.

3.2.2.2 Objective 2 (O2): Design of Digital Preprocessing Methods

The main aim of research objective 2 (O2) is the design and selection of adequate methods for the preprocessing of captured fingerprint time series, to allow for a successful extraction of aging features. Adequate methods are required for each capturing device (CWL vs. CLSM) and data type (intensity vs. topography) and need to be selected in respect to the short- and long-term aging. The aim of O2 is therefore to design adequate preprocessing methods for each investigated combination of capturing device, data type and time period, to provide time series in a comparable form for the following feature extraction.

Preprocessing sub-steps include spatial alignment, temporal normalization (to exclude changes of the overall image brightness over time), segmentation (segmentation targets need to be defined as an interface to the later conducted feature extraction), distortion correction and the enhancement of certain fingerprint particles used for the later performed feature calculation (e.g. pores, certain particles, fluid droplets, puddles or continuous ridges). Methods need to furthermore be designed in respect to the high resolution of the captured time series.

3.2.2.3 Objective 3 (O3): Design of Digital Aging Feature Sets

The aim of research objectives 3 (O3) is to provide a set of aging features, which can be used to estimate the age of latent fingerprints captured with both investigated devices. Such test set includes statistical aging features, which are a novel contribution of this thesis to the field. Furthermore, also features targeting morphological structures are designed, such as ridges, pores, dust, big and small droplets as well as puddles of liquid. Some features might be applicable for images of different capturing devices, while others are designed only for images of specific devices. Furthermore, some features might show a characteristic aging behavior for short-term aging periods, while others show a characteristic behavior for long-term periods (or both).

3.2.2.4 Objective 4 (O4): Design of Digital Age Estimation Strategies

In the scope of research objective 4 (O4), the main aim is to design adequate age estimation strategies for transferring computed aging features into data describing the age of a print. Based on the features designed in O3, formula-based age computation as well as a machine-learning based classification into specific age classes is investigated. In the scope of this thesis, only short-term aging periods are evaluated towards their age estimation performance, because a significant amount of samples is required for statistically reliable results, which cannot be produced for the long-term aging series, due to the limited capturing capabilities. Several hundred samples would have to be scanned here over several years for reliable results, which cannot be achieved in the given timeframe with the available devices.

For the design of age estimation strategies, different parameters need to be evaluated and optimized, such as the amount of required images, used feature representations, applied regression types, selected classifiers or required amount of training samples. Furthermore, the age estimation performance needs to be compared between different age classes, capturing devices, data types and features.

3.2.2.5 Objective 5 (O5): Modeling Influences on Fingerprint Aging and the Capturing Process

In the scope of research objective 5 (O5), the main aim is to obtain information about the nature of fingerprint degradation and to identify major influences on it. Here, also the impact of the capturing devices on an observed aging behavior is of interest. From the obtained results, the creation of a qualitative influence model is aimed at, which might be used as a base for future studies.

Fingerprints are investigated over short-term periods under systematic variations of exemplary selected influences. Different influences from the application process, used surface, sweat composition, environmental conditions and the capturing device are tested and compared. Such experiments are of qualitative and preliminary nature only, because the complex nature of the degradation and evaporation processes as well as the high number of involved substances can only be fully understood in the mid- or long-term future, including also other fingerprint properties, such as the qualitative and quantitative chemical composition of the residue.

4 Design of the Proposed Digital Processing Pipeline

Design Setup Results

In this chapter, the specific design of the proposed digital processing pipeline is introduced and formalized, addressing the research objectives O1 - O5. The objectives O1 - O4 refer directly to the four steps of the pipeline (data acquisition DA, preprocessing PP, feature extraction FE and age estimation AE). The proposed procedures for each objective are therefore designed to implement the corresponding sub-steps of each processing step (sections 4.1 to 4.4). Research objective O5 aims at conducting first qualitative studies of influences on the aging behavior of latent fingerprints as well as designing a first influence model from them. Therefore, its proposed procedures comprise all steps of the processing pipeline and are considered separately in section 4.5.

To allow for a comprehensive discussion with other experts in the field and to evaluate the general acceptability of the proposed approach in the scientific community, parts of the here presented concepts and results have been previously published, which is a common practice for a dissertation project in the field of computer science. The corresponding publications are:

2013

- [151] presented at the 1st ACM Workshop on Information Hiding and Multimedia Security (IH&MMSEC), Montpellier, France, June 17th - 19th, 2013

2012

- [152] presented at the 2012 IEEE International Workshop on Information Forensics and Security (WIFS), Tenerife, Spain, December 2nd - 5th, 2012
- [153] published in the journal Forensic Science International, Elsevier, Volume 222, Issues 1 - 3, pp. 52 - 70, October 10th, 2012
- [95] presented at the 14th ACM Workshop on Multimedia and Security (MMSEC), Coventry, UK, September 6th - 7th, 2012
- [154] presented at the SPIE conference on Optics, Photonics, and Digital Technologies for Multimedia Applications II, Brussels, Belgium, June 1st, 2012
- [79] presented at the SPIE conference on Three-Dimensional Image Processing (3DIP) and Applications II, Burlingame, California, USA, February 9th, 2012

2011

- [155] presented at the 2011 IEEE International Workshop on Information Forensics and Security (WIFS), Foz do Iguacu, Brazil, November 29th - December 2nd, 2011
- [156] presented at the 12th Joint IFIP TC6 and TC11 Conference on Communications and Multimedia Security (CMS), Ghent, Belgium, October 19th - 21st, 2011

- [157] presented at the SPIE conference on Optics and Photonics for Counterterrorism and Crime Fighting VII, Optical Materials in Defence Systems Technology VIII and Quantum-Physics-based Information Security, Prague, Czech Republic, October 13th, 2011
- [158] presented at the 13th ACM Workshop on Multimedia and Security (MMSEC), Niagara Falls, New York, USA, September 29th - 30th, 2011
- [159] presented at the 7th International Symposium on Image and Signal Processing and Analysis (ISPA 2011), Dubrovnik, Croatia, September, 4th - 6th, 2011

4.1 Design of the Data Acquisition Process (O1)

As discussed in chapter 3.1.6, the capturing process itself (data acquisition step) is considered as a given entity, which is not investigated in the scope of this thesis beyond the point of necessary prerequisites. Such prerequisites include the determination of basic sensor settings (section 4.1.1) as well as potential challenges from the used capturing devices (section 4.1.2), especially in regard to the earlier made assumptions of section 3.2.1. The first objective O1 comprises such first investigations of the used CWL and CLSM devices. Used quality measures are introduced in section 4.1.3.

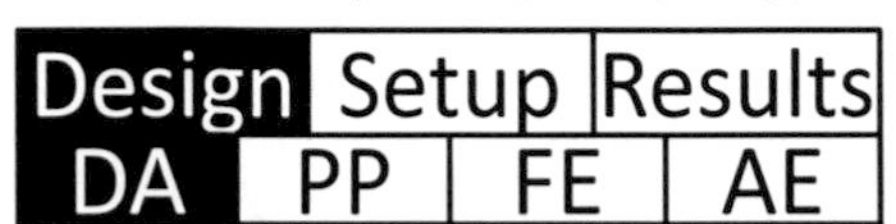

4.1.1 Basic Sensor Settings

Both capturing devices can be operated in different configurations, which are specified by certain sensor settings. This thesis is not aimed at a comprehensive optimization of such parameters. Merely basic settings are identified, which can be used to capture fingerprint time series of adequate quality. The measured area size as well as the used capturing resolution seem to be the most important of these parameters, for which adequate lower bounds need to be specified. Identifying such lower bound is required to allow for a maximum capturing speed, which is an important prerequisite for capturing sample sets of sufficient size. It is done by first experiments in the scope of this thesis as well as a theoretical discussion (sections 5.1.1, 5.4 and 6.1). Other parameters exist as well, such as the measurement frequency, longitudinal resolution, scan distance and brightness of the light source. Their value is chosen according to the recommendations of the sensor manufacturers to achieve an optimal capturing speed and precision.

4.1.2 Challenges From the Used Capturing Devices

Several challenges of capturing fingerprint time series with the used CWL and CLSM devices do arise, which have been identified in first studies of this thesis. This section gives an overview over such challenges. They are addressed in the scope of designing preprocessing sub-steps (sections 4.2.1 to 4.2.6) and are therefore referred to as preprocessing challenges PC1 - PC5. They are in some cases specific to the used capturing device and data type.

The first preprocessing challenge (PC1) is represented by the general spatial alignment of the consecutive images of a fingerprint time series. Because the short-term aging series are captured without removing the sample from the measurement table for both capturing devices, an alignment is not necessary, due to a sufficient precision of the measurement tables. However, long-term fingerprint time series are captured only once every week and therefore require the manual fixation of the sample to the measurement table for each temporal capture and its subsequent removal. This is realized by certain tools (e.g. fixation by screws or manual alignment at edges and fixation with plasticine, section 5.3). For such manual alignment, certain offsets cannot be avoided, creating the requirement for an automated alignment and cropping of the images in the scope of the preprocessing sub-steps (Figure 4).

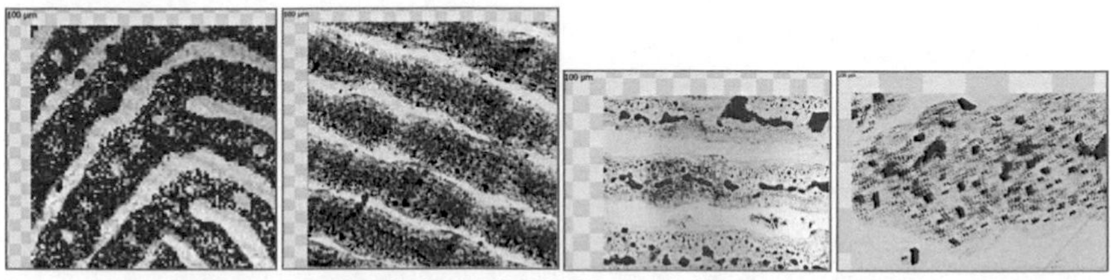

Figure 4: Consecutively captured and overlaid pairs of partial fingerprints (blue overlay of second image over black first image). From left to right: CWL short-term aging overlay; CWL long-term aging overlay; CLSM short-term aging overlay; CLSM long-term aging overlay (one box of the textured background represents 100 µm).

It can be seen from Figure 4 that for the long-term aging of both capturing devices (second and fourth image from the left), a certain offset between the consecutive, overlaid scans does exist, whereas for the short-term aging (first and third image), both scans are well aligned. The higher offset for long-term aging series is a result of the manual fixation of the samples to and their removal from the measurement table.

The second preprocessing challenge (PC2) can be seen in changes of the overall image brightness, which do occur during the course of a time series. As described in section 3.2.1, it is assumed for a proper age estimation that the overall image brightness remains constant throughout a time series. However, examinations in the scope of this thesis have shown that this is not always the case. In some cases, strong fluctuations in the overall image brightness seem to occur over time, which cannot be caused by fingerprint residue. Figure 5 and Figure 6 illustrate this phenomenon by visualizing exemplary time series of empty hard disk platters (ideal surfaces for the used capturing devices, see also section 5.3), displayed over a time period of 60 hours.

The CWL intensity images of Figure 5 (first row) exhibit a certain surface pattern of the hard disk platter substrate, which is not produced by the intensity images of the CLSM device (third row). This demonstrates the different nature of the various platters found in hard disks. However, in the scope of this thesis, platters exhibiting such pattern are avoided in the context of the experimental investigations, to not influence the measured aging characteristics. For the CLSM images (third and fourth row), round geometrical distortions (also called barrel distortions) can be observed, which are caused by the microscope objective lens. Such effect is sensor specific and can therefore not be avoided. However, because it is a systematic influence and fingerprint pixels can be assumed to be affected by such phenomenon with the same ratio than background pixels (because ridges and valleys are usually distributed approximately equal over the image), its influence on a measured aging characteristic is expected to be minimal.

When inspecting the overall brightness of the images of Figure 5 over time, significant changes can be observed for some time series. To visualize these changes in more detail, the mean image gray values of all four rows of Figure 5 is plotted over time in Figure 6 and the corresponding histograms are given for the four points in time t_1 = 0 h, t_2 = 20 h, t_3 = 40 h and t_4 = 60 h.

Figure 5: Exemplary images of empty hard disk platters captured with the CWL sensor as well as the CLSM device (first row: CWL intensity data; second row: CWL topography data; third row: CLSM intensity data; fourth row: CLSM topography data) over a total time period of 60 hours (from left to right: $t_1 = 0$ h, $t_2 = 20$ h, $t_3 = 40$ h, $t_4 = 60$ h). Images are not preprocessed except for histogram normalization for better visualization.

It is to note from Figure 6 that changes of the mean image gray value cannot be compared between the different data types (different rows in the image), because they have different ranges. However, the changes occurring for a single sample over time are of utmost importance in the scope of this thesis. It can clearly be seen from Figure 6, that the overall image brightness is not constant over time in most cases, which is regarded as a prerequisite for extracting aging properties (section 3.2.1). Some data types merely show a random-like fluctuation around a certain value (CWL intensity images - first row, CLSM topography images - fourth row). Such fluctuations might be interpreted as normal sensor noise of the capturing devices (which seems to be comparatively high in some cases). However, for the CWL topography as well as the CLSM intensity images (second and third row), a trend in the change of overall image brightness can be observed.

The exact reason for such behavior is difficult to determine in the scope of this thesis, because in-depth expertise of the physical properties of the capturing hard- and software are required, which are often proprietary. However, changes in the light source during long-term, uninterrupted service are a possible cause. Such changes might result from the laser diode degradation during its life. Furthermore, detector properties can change over time. Also, an increased temperature after starting a device might lead to changes of the overall image brightness. In the scope of this thesis, the aim is not to study such sensor behavior and its causes in detail. However, the observed changes need to be addressed for an adequate age estimation. In the future, it seems appropriate to expect these issues to be solved by sensor experts prior to a practical age estimation at crime scenes.

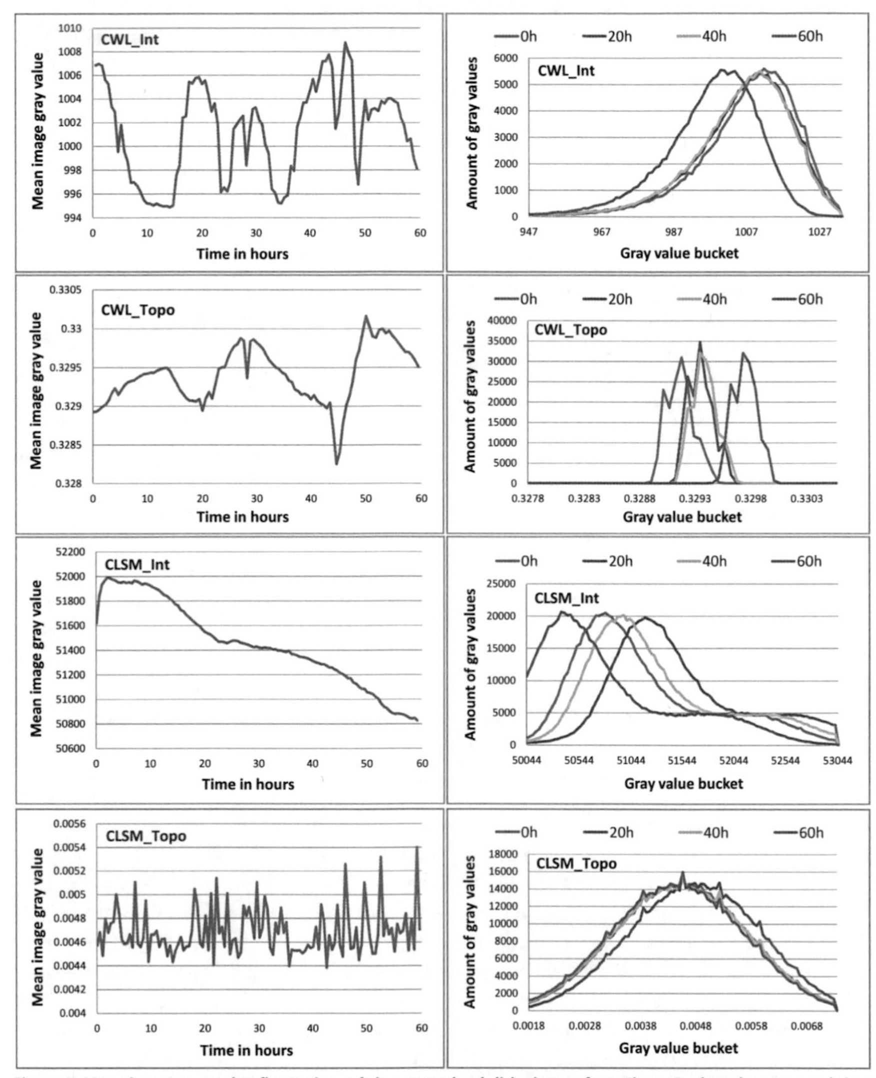

Figure 6: Mean image gray value fluctuations of the empty hard disk platters from Figure 5, plotted over a total time period of 60 hours (left images) and the corresponding image histograms of four selected points in time t_1 = 0 h, t_2 = 20 h, t_3 = 40 h and t_4 = 60 h (right images). Curves and histograms are depicted analogue to Figure 5 for CWL intensity (first row), CWL topography (second row), CLSM intensity (third row) and CLSM topography data (fourth row).

To address changes of the overall image brightness, the characteristics of their changes need to be analyzed. Fortunately, these changes seem to mostly affect the complete histogram, shifting it while leaving its shape intact (right images of Figure 6). Such finding is of great importance, because it allows to correct the occurring brightness changes by determining the shift of the histogram and performing a normalization. Excluding the temporal changes of the overall image brightness is regarded as a major challenge to the design of the preprocessing step. This challenge is addressed in the scope of this thesis by the creation of a background mask (separating fingerprint from background pixels) and a temporal normalization based on the image background specified by that mask (sections 4.2.2 and 4.2.3).

The third preprocessing challenge (PC3) can be seen in finding the right segmentation method. Different image components might be required to be segmented for different short- or long-term aging features. The main challenge can be seen in the task of finding the right fingerprint representation for each computed aging feature, defining which structures are considered relevant for a certain feature. Such relevant structure can be the fingerprint raw data or the correct separation of fingerprint pixels from background pixels in a binarized form. However, the representation of a fingerprint as a set of continuous, coherent ridge lines (rather than a collection of droplets and puddles) might also be required. Furthermore, certain structures (such as pores or particles) are of interest. Four different segmentation targets are therefore defined in the designed preprocessing step (section 4.2.4), in which captured fingerprint images need to be transferred.

The fourth preprocessing challenge (PC4) is given by the wide variety of different environmental influences. Such influences impact the aging behavior of fingerprints. However, they might additionally influence the capturing process. Some of these influences are studied in more detail in the scope of objective O5 (section 4.5), while others require extensive evaluations in future work. Some of them require to be addressed in the designed preprocessing step, to allow for a reliable feature extraction. First observations indicate temperature (CWL and CLSM), vibrations (CWL) as well as dust (CWL and CLSM) to be major sources of distortions, which need to be addressed before adequate aging features can be extracted. While changes in temperature as well as vibrations can induce changes in the optical system of a capturing device and therefore lead to a change in the overall image brightness as well as noise, dust might settle and even leave a fingerprint at arbitrary times, potentially distorting the computed aging features (Figure 32). Such issues therefore need to be addressed in the scope of fingerprint time series preprocessing.

The fifth preprocessing challenge (PC5) describes a phenomenon observed only for CLSM intensity images. In Figure 5 and Figure 6, it is shown that the changes of image brightness occurring on empty hard disk platters seem to mainly affect the complete histogram, shifting it in a certain direction, while leaving its shape unaltered. However, this is not always the case. For a significant amount of CLSM intensity time series, certain changes of the gray values of background pixels adjacent to the fingerprint residue can be observed, as depicted in Figure 7. Such changes are occurring only at background pixels (because no fingerprint material is detected in these regions at the beginning of a time series). After some time has passed, such artifacts become visible, characterized by a darkening of background (platter surface) pixels.

The temporal behavior of certain background areas depicted in Figure 7 is not desired for the calculation of aging features, because it potentially distorts them. However, it cannot be excluded by shifting the complete histogram (as is done for changes of the overall image brightness), because it affects only certain background areas. Because the phenomenon does not seem to be present in the first image of a time series, it can be accounted for by a general exclusion of background pixels in the feature extraction step. Using such technique, only the fingerprint pixels (and maybe their close vicinity) are used for the computation of aging properties, avoiding a distortion by such changes in background pixels (section 4.2.6).

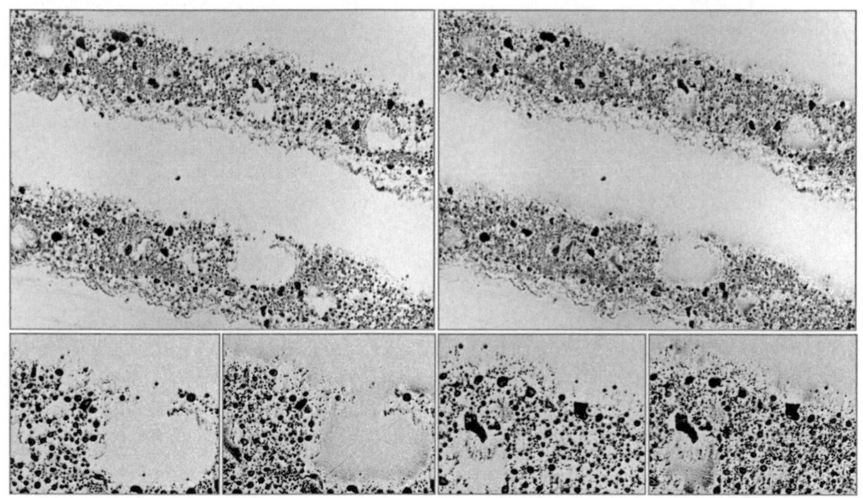

Figure 7: Exemplary CLSM fingerprint intensity image, captured directly after application (upper left image) and after 24 hours (upper right image). In the lower row, selected parts of the two images are enlarged. It can be seen that dark marks have formed near the edges of the fingerprint ridge and inside the pores.

All five identified preprocessing challenges PC1 - PC5 need to be addressed in the scope of fingerprint time series preprocessing, to allow for a reliable feature extraction. For that purpose, a sequence of adequate preprocessing sub-steps is designed in the scope of research objective O2 (section 4.2).

4.1.3 Quality Measures for the Data Acquisition Process

In the scope of objective O1, basic sensor settings are selected for the capturing process. In particular, the lower bound of the measured area size as well as the lateral resolution need to be chosen in respect to both capturing devices. This is achieved in the scope of test goal TG1 (section 5.1.1). The quality of a captured aging curve is measured by how well the curve can be approximated to a corresponding mathematical aging function using regression [131]. The Pearson correlation coefficient r between the captured curve and its corresponding function is therefore considered to be an adequate quality measure [149]. When determining the lower bound of capturing parameters, it is of interest which minimal settings still allow for a sufficiently high correlation coefficient and therefore an adequate quality of the captured aging curve. Further details about the quality measures used for selecting basic sensor settings techniques can be found in the experimental test setup (section 5.4.1).

4.2 Design of Digital Preprocessing Methods (O2)

The second objective O2 focuses on the preprocessing step of the overall processing pipeline described in Figure 3. For realizing the aim of O2, adequate preprocessing methods need to be designed. In first studies of this thesis using the CWL sensor, basic preprocessing techniques have been proposed and published in [154]. However, they have been shown to be of limited nature and do perform particularly bad on CLSM images (where a decreasing tendency of the overall image brightness was found to overlay the observed aging tendency in [152]). Therefore, new preprocessing techniques are designed here, including spatial alignment, temporal normalization of the overall image brightness, segmentation of relevant particles or areas needed for feature extraction as well as exclusion or reduction of undesired distortions. In short, the fingerprint time series need to be prepared for the following feature extraction step in a way to allow for a reliable extraction and comparison of aging features.

In section 4.1.2, five different challenges PC1 - PC5 have been introduced, which were identified during several examinations in the scope of this thesis and need to be addressed by adequate preprocessing methods to fulfill the defined aims. Different methods are therefore proposed and arranged into a sequence of preprocessing sub-steps, based on basic image processing operations known from the literature (section 2.3.1). The sub-steps include spatial alignment (PC1, section 4.2.1, only for long-term aging series), background mask creation and temporal normalization (PC2, sections 4.2.2 and 4.2.3), fingerprint segmentation (PC3, section 4.2.4), distortion correction (PC4, section 4.2.5) and valid pixels mask creation (PC5, section 4.2.6). They are visualized in Figure 8. Quality measures for the later conducted performance evaluation are given in section 4.2.7.

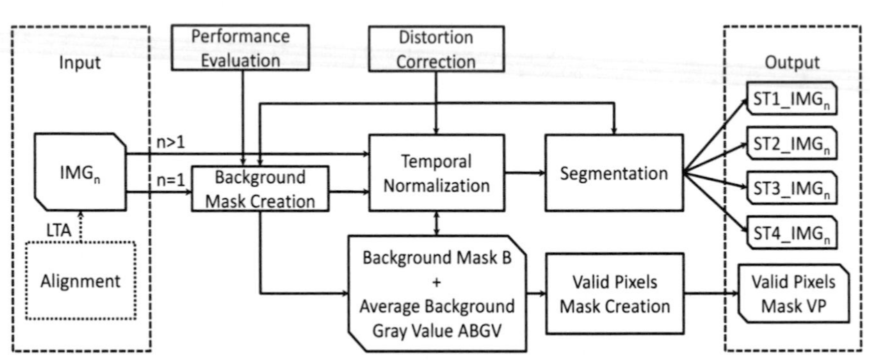

Figure 8: Detailed structure of the proposed preprocessing step of the overall processing pipeline.

As a part of the overall processing pipeline described in section 3.1.6 (Figure 3), the preprocessing step visualized in Figure 8 requires the definition of a well-specified interface to the capturing device (input) as well as to the feature extraction step (output). As input, an image IMG_n of an arbitrary time series is received. Such time series includes a total amount of n_max images from a certain capturing device (CWL vs. CLSM) and data type (intensity vs. topography), scanned over a fixed time period (e.g. short- vs. long-term aging). In case of long-term aging, the image has to be spatially aligned in respect to the other images of the corresponding time series (section 4.2.1). This is done prior to the time series preprocessing and is therefore marked in dotted lines in Figure 8.

If an input image IMG_n, n = {1,2,...,n_max} is the first of its time series (n = 1), it is preprocessed using the background mask creation sub-step (section 4.2.2), which creates and stores the background mask B as well as the average background gray value ABGV, later used for processing the complete time series. For all subsequent images of the time series (n > 1), the background mask creation step is skipped. The creation of B and ABGV from the first image of a time series is required for the following sub-step of temporal normalization (PC2).

After the background mask creation step has successfully been applied, the temporal normalization is performed (section 4.2.3). During this step, a given input image is transferred into a form comparable to other images of the same time series in terms of overall image brightness (PC2), which is an important prerequisite for the evaluation of aging behavior. The method of temporal normalization aligns the histogram of each image according to the gray value offset of its background region to the first image of the time series, using the background mask B and average background gray value ABGV.

Having been temporally normalized, fingerprint images are segmented (section 4.2.4). According to the specific requirements of a each aging feature, segmentation targets ST1 - ST4 are defined and each image is segmented according to the required target, extracting the desired fingerprint compounds (PC3).

During the three designed sub-steps of background mask creation, temporal normalization and image segmentation, different types of distortions are reduced (PC4). These distortions include image intensity fluctuations, measurement artifacts (null values and out of range images), temperature distortions, vibrations as well as measurement table propulsion artifacts (section 4.2.5). Most important, the influence of dust needs to be reduced as accurate as possible, to avoid its influence on the computed aging features.

Before a preprocessed time series is passed on to the feature extraction step of the overall processing pipeline, the background mask B is converted into a valid pixels mask VP (section 4.2.6). Such mask is used during feature computation to skip image pixels not belonging to the fingerprint residue from the feature extraction process. This seems to be a necessary step to avoid changes in the brightness of certain background areas to influence the computed aging behavior (PC5). After a fingerprint time series has successfully undergone all required preprocessing sub-steps, it is returned in the form of a sequence of output images, segmented according to a specified segmentation target ST. Also returned is its corresponding valid pixels mask VP (section 4.2.6).

In the following sections, the designed sub-steps for spatial alignment, background mask creation, temporal normalization, segmentation, distortion correction and valid mask creation are introduced and discussed in detail. Different methods can be required for certain capturing devices, data types or segmentation targets as well as combinations of them (e.g. a method might only be feasible for intensity images of the CWL and the CLSM device, while another method is feasible for all data types of all capturing devices). The methods are designed from basic image processing operations described in section 2.3.1.

4.2.1 Spatial Alignment (PC1)

Spatial alignment is only necessary for long-term aging series, because a significant offset is created by the manual fixation of the samples to and their removal from the measurement tables of the capturing devices (section 5.3). Because the intensity and topography images of each capturing device are always captured simultaneously and have therefore identical coordinates, only one data type needs to be aligned, which is chosen as the intensity images, having a much higher quality and visibility of the print (section 4.2.2). For alignment, all raw images R_IMG_n of a time series are first binarized using a certain threshold. Such threshold is determined by different binarization techniques, depending on the used capturing device, data type or other image properties (section 4.2.2). The same thresholding technique is applied to all images of the same time series. Binarized images B_IMG_n are then aligned using a basic auto-correlation measure AC designed in the scope of this thesis. For a chosen maximum offset off_max, each binarized image B_IMG_n is shifted pixel by pixel in x- and y-direction in respect to the first image B_IMG_1 of the corresponding time series:

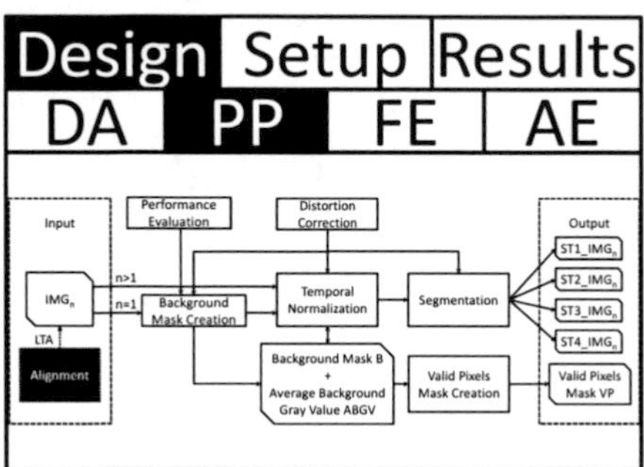


```
for(i=-off_max; i≤+off_max; i++)
   for(j=-off_max; j≤+off_max; j++)
      AC = 0
      for(x=off_max; x≤x_max-off_max; x++)
         for(y=off_max; y≤y_max-off_max; y++)
            AC = AC + computeAutoCorrelation(B_IMG1(x,y), B_IMGn(x+i,y+j))
         for end
      for end
   for end
for end
```

For each investigated offset (i,j), the auto correlation measure AC is computed, which is defined here as the equality of the pixel values $B\_IMG_1(x,y)$ and $B\_IMG_n(x+i,y+j)$, summed up over all pixels of the image B_IMG_1, cropped by an offset of 2·off_max:

$$AC(i,j) = \sum_{x=off\_max,y=off\_max}^{x\_max-off\_max,y\_max-off\_max} \begin{cases} 1, & B_{IMG\_1}(x,y) = B_{IMG\_n}(x+i,y+j) \\ 0, & else \end{cases} \tag{1}$$

$$x\_max = Rows(B_{IMG\_1}), y\_max = Cols(B_{IMG\_1}), 1 \leq off\_max < minimum\left(\frac{x\_max}{2}, \frac{y\_max}{2}\right)$$

For reducing the computational complexity of the alignment, off_max as well as the correlation area off_max ≤ x ≤ x_max-off_max and off_max ≤ y ≤ y_max-off_max can be decreased, depending on the specific properties of a time series. Furthermore, certain distortions (e.g. dust settling over time) can make certain regions of an image unfeasible for alignment, requiring a manual adaptation of the correlation area to regions of suitable quality. The correlation measure AC(i,j) is then compared between all investigated offsets (i,j) and the offset with the highest correlation is chosen as the final correction offset corr_off:

$$corr\_off = maximum\big(AC(i,j)\big), -off\_max \leq i \leq off\_max, -off\_max \leq j \leq off\_max \tag{2}$$

The correction offsets corr_off are computed in advance to the experimental evaluation and are stored for each long-term time series. In the final experiments, the corresponding images are shifted and cropped respectively. The cropping is performed in a way to create a similar image size for all time series of a certain test set, to ensure the comparability of the results.

4.2.2 Background Mask Creation (PC2)

In the scope of this thesis, a method is proposed to reduce the influence of changes of the overall image brightness. The method applies to the short- as well as the long-term aging in a similar way and is composed of the here introduced background mask creation in combination with the later performed temporal normalization. The main idea is to extract non-fingerprint pixels from the first image of a time series as accurate as possible, and to determine temporal changes of the overall image brightness from them (offset of the average background gray values ABGV), which are then corrected for the complete image. The aim of the background mask creation is therefore to assign each pixel of a captured image to either fingerprint pixels (foreground) or surface pixels (background) with the highest possible accuracy. This method is similarly applied to both data types and capturing devices as well as short- and long-term aging series.

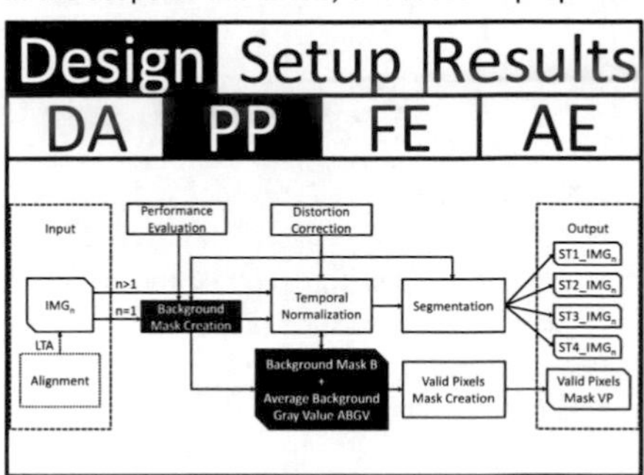

The background mask B(x,y) is calculated from the first image of each time series (n = 1) and then applied to all following images (n > 1). Its elements can assume values of zero and one, depending on its representation of fingerprint or background pixels. The device and data type specific background masks $B_{CWL\_Int}(x,y)$, $B_{CWL\_Topo}(x,y)$, $B_{CLSM\_Int}(x,y)$ and $B_{CLSM\_Topo}(x,y)$ are defined as the inverse matrix INV of the binarized fingerprint images $I_{CWL}(x,y)$, $T_{CWL}(x,y)$, $I_{CLSM}(x,y)$ and $T_{CLSM}(x,y)$, computed in the following sections 4.2.2.1 and 4.2.2.2:

$$B_{CWL\_Int}(x,y) = INV(I_{CWL}(x,y)) = 1 - I_{CWL}(x,y) \tag{3}$$

$$B_{CWL\_Topo}(x,y) = INV(T_{CWL}(x,y)) = 1 - T_{CWL}(x,y) \tag{4}$$

$$B_{CLSM\_Int}(x,y) = INV(I_{CLSM}(x,y)) = 1 - I_{CLSM}(x,y) \tag{5}$$

$$B_{CLSM\_Topo}(x,y) = INV(T_{CLSM}(x,y)) = 1 - T_{CLSM}(x,y) \tag{6}$$

However, when investigating fingerprint time series in the scope of aging feature design (O3) and age estimation strategies (O4), the spatial alignment of such images needs to be taken into account. Especially for long-term aging series, the spatial alignment of images (section 4.2.1) might not be perfectly accurate, leaving room for small offsets in its alignment. For that purpose, the background mask B is extended to the background mask B' for objectives O3 and O4. The mask B' is computed from B by the one-time application of the well known morphological dilation operation [103] (kernelsize: 3 x 3), to include also the background pixels adjacent to fingerprint pixels. Using such dilation, small shifts of the different images of a time series (up to one pixel) can be balanced. Furthermore, a one-time dilation also provides a certain robustness to errors of the background mask creation step, including fingerprint pixels falsely classified as background pixels, which are located in the close vicinity to the correctly classified residue:

$$B' = Dilate(B,1) \tag{7}$$

From the created mask B' (O3 and O4), the average background gray value ABGV is computed for the first image:

$$ABGV(R\_IMG_1) = \frac{1}{x\_max \cdot y\_max} \cdot \sum_{x=1,y=1}^{x\_max,y\_max} B'(x,y) \cdot R\_IMG_1(x,y) \quad (8)$$

To compute such average does only yield reliable results in case a normal distribution of background pixels is present, which is the case for the here investigated ideal hard disk platter surfaces. However, the distribution might have a longer tail on one side, in case not all fingerprint pixels are adequately selected by the binarization and background mask creation. To avoid such histogram tail influencing the correct computation of ABGV (mean of the distribution), the following iterative procedure is applied 20 times, which has been shown to terminate within 20 iterations for various samples:

```
----------------------------------------------------------------------------
i.   Calculate the mean μ and standard deviation σ of the hard disk platter back
     ground pixels specified by B'

ii.  Exclude all pixels outside the interval [μ-2.56·σ; μ+2.56·σ] from the next
     iteration
----------------------------------------------------------------------------
```

The mean value μ from the last iteration is then considered as the optimal ABGV for this time series. The interval [μ-2.56·σ; μ+2.56·σ] is exemplary chosen and contains most pixels of the normal distribution. Carrying out such procedure ensures that possible tails do not distort the computation of ABGV. The background mask B' is then stored together with the $ABGV(R\_IMG_1)$ of the first time series image and used for the temporal normalization of all following images (n > 1).

In the following sections 4.2.2.1 and 4.2.2.2, different state of the art as well as newly designed methods for the creation of binarized images $I_{CWL}(x,y)$, $T_{CWL}(x,y)$, $I_{CLSM}(x,y)$ and $T_{CLSM}(x,y)$ are introduced and discussed. It is of importance to select the best performing binarization method for each capturing device and data type, to ensure a maximum feature extraction and age estimation performance. Therefore, the selection of adequate binarization methods is the main aim of the conducted experimental evaluations in the scope of test goal TG2 (section 5.1.2). To investigate the performance of image binarization techniques, the temporal succession of images in a time series is not of importance, because B is computed from the first image of a time series only (and is applied to all following images of the series).

4.2.2.1 Binarization Techniques for Intensity Images of Both Capturing Devices

Threshold binarization is a straightforward technique for mapping image pixels into one of two classes: fingerprint residue or surface background. Such technique can be applied to the intensity images of short- and long-term series of both capturing devices in a similar way, where a decision is made for every pixel, independent of its neighboring pixels. Threshold binarization has been well studied in the literature and seems to work adequately for the intensity images I(x,y) of both capturing devices, in respect to the well reflecting hard disk platters primarily investigated in the scope of this thesis (section 5.3). Additional surfaces are discussed in the scope of objective O5. The selection of accurate binarization thresholds seems vital to a successful image preprocessing. Topography images of both devices are subject to severe distortions and noise and therefore require more complex preprocessing methods for a successful binarization and background mask creation, which are designed in section 4.2.2.2.

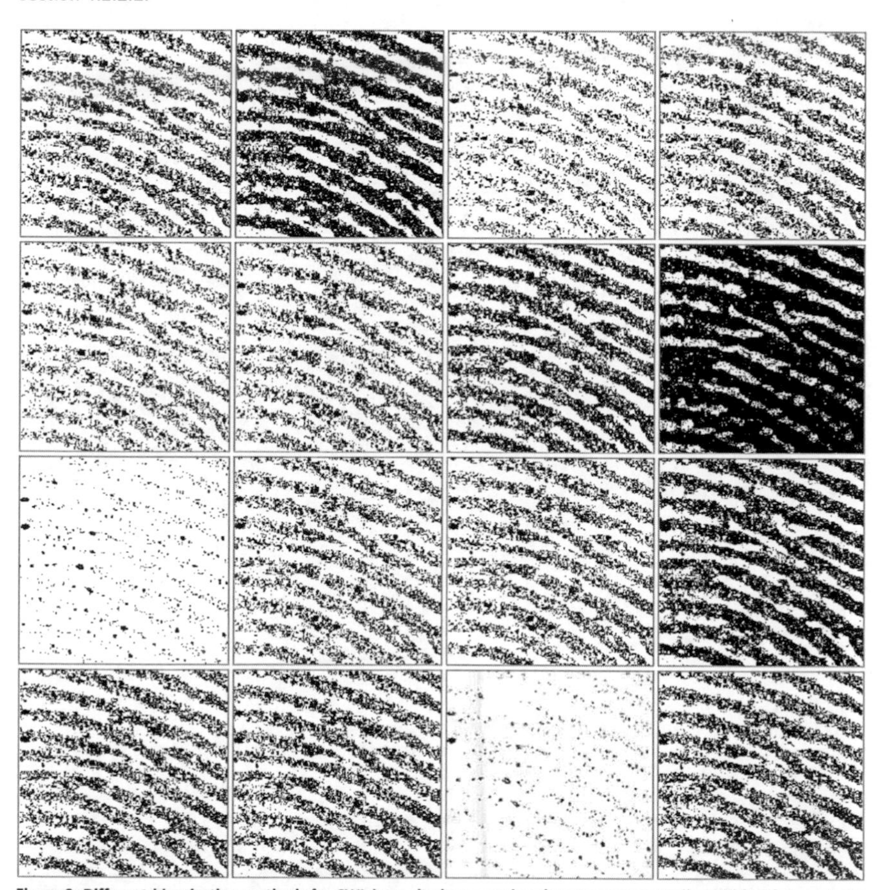

Figure 9: Different binarization methods for CWL intensity images using the open source toolbox Fiji [111]. From left to right and top to bottom: Default, Huang, Intermodes, IsoData, Li, MaxEntropy, Mean, MinError(I), Minimum, Moments, Otsu, Percentile, RenyiEntropy, Shanbhag, Triangle, Yen.

For binarizing CWL and CLSM intensity images, different threshold determination methods are investigated in the scope of this thesis. The following 16 state of the art methods are provided by the Fiji toolset [111]: Default (variation of the IsoData algorithm [111]), Huang (fuzzy thresholding based on Shannon's entropy function [112]), Intermodes (average of iteratively smoothed histogram [113]), IsoData (iterative threshold selection [114]), Li (minimum cross entropy thresholding [115]), MaxEntropy (maximum entropy based on Kapur-Sahoo-Wong method [116]), Mean (mean of pixel values [117]), MinError(I) (minimum error thresholding based on Kittler and Illingworth's method [118]), Minimum (minimum of iteratively smoothed histogram [113]), Moments (preservation of image moments based on Tsai's method [119]), Otsu (minimization of intra-class variance [120]), Percentile (uses 0.5 percentile [121]), RenyiEntropy (maximum entropy based on Renyi method [116]), Shanbhag (information measure as threshold [122]), Triangle (geometric method based on

Zack algorithm [123]) and Yen (Yen's threshold determination algorithm [124]). Exemplary intensity images of latent prints binarized with these methods are depicted in Figure 9 (CWL) and Figure 10 (CLSM).

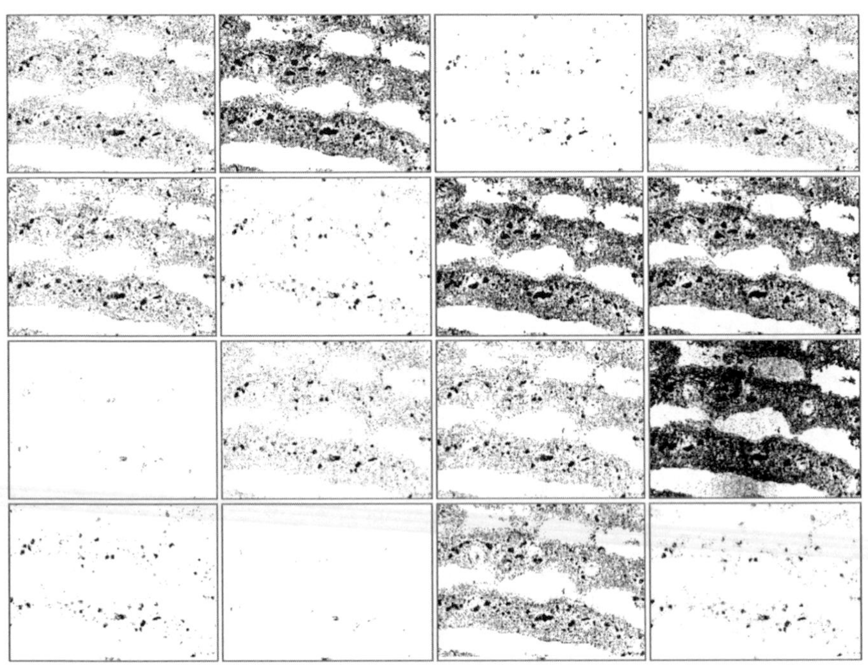

Figure 10: Different binarization methods for CLSM intensity images using the open source toolbox Fiji [111]. From left to right and top to bottom: Default, Huang, Intermodes, IsoData, Li, MaxEntropy, Mean, MinError(I), Minimum, Moments, Otsu, Percentile, RenyiEntropy, Shanbhag, Triangle, Yen.

Apart from these 16 methods, one additional algorithm RelRange is proposed here, reflecting the specific properties of the captured time series. In a first step, the method erases outliers from the image histogram. Such outliers are often introduced by pixel null values from the capturing process (e.g. where the light beam is refracted at droplets or reflected away from the detector, section 4.2.5.2). Also, external particles, such as dust or dirt, might contribute to outliers. To erase such outliers, one percent of the outermost pixel values are excluded from both sides of the histogram and the resulting minimum histogram bucket min(b) and maximum histogram bucket max(b) filled with nonzero values are determined. The threshold is then computed to be 80% of the resulting histogram range:

$$thresh_{RelRange} = min(b) + 0.8 \cdot (max(b) - min(b)) \tag{9}$$

Exemplary intensity images of latent prints binarized with the RelRange method are depicted in Figure 11 (CWL) and Figure 12 (CLSM). The figures furthermore present the original images as well as their histograms.

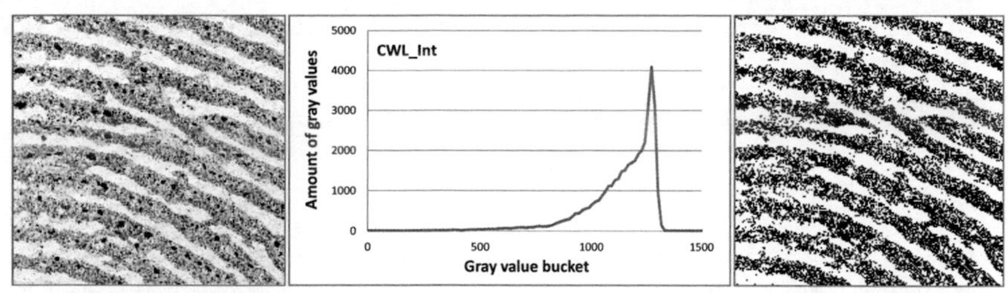

Figure 11: The designed RelRange binarization method for CWL intensity images. From left to right: original grayscale image; image histogram; image binarized with the RelRange method.

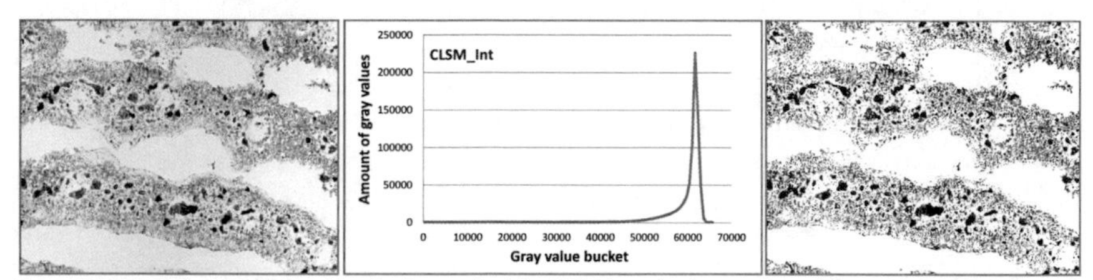

Figure 12: The designed RelRange binarization method for CLSM intensity images. From left to right: original grayscale image; image histogram; image binarized with the RelRange method.

From any computed threshold thresh (independent of the used technique), the binarized fingerprint image I_{CWL_CLSM} required for the background mask creation is computed using threshold binarization BIN:

$$I_{CWL\_CLSM} = BIN(I(x,y), thresh) \quad (10)$$

In general, it can be concluded from Figure 9 - Figure 12 that a variety of adequate methods exist for the binarization of intensity fingerprint images of both devices, from which the best-performing techniques are chosen in the later conducted experimental evaluations (test goal TG2, section 5.1.2).

4.2.2.2 Binarization Techniques for Topography Images of Both Capturing Devices

According to first studies in the scope of this thesis [153], topography images are in most cases of less quality than their intensity counterparts, because different capturing properties as well as environmental influences might lead to distorted images (Figure 13 - Figure 20). For example, emitted light is often scattered or refracted at droplets, which can lead to different amounts of light being captured by the detector, including null values (insufficient light reaches the detector, the captured pixel is set to zero). Furthermore, temperature changes, vibrations or other environmental influences lead to temporal changes in the captured images, represented by line wise distortions of gray values (especially for the CWL sensor, which traverses an object line by line) or other forms of noise. Although these influences might also apply to the intensity images of the capturing devices, their effect is significantly stronger for the topography images, because the tiny height differences of fingerprint particles (which are usually in the range of only 1 µm) are much more vulnerable to such kind of distortions. Consequently, for the successful separation of fingerprint pixels from surface pixels in topography images, fingerprint residue needs to be enhanced prior to a threshold binarization. Such enhancement is specific to the type of captured data and therefore differs between the capturing devices. However, it is similar for the investigation of short- and long-term aging series of the same device. In the scope of this thesis, three binarization methods are proposed for each the CWL as well as the CLSM topography data.

For binarizing topography images from any device, the image needs to first be planarized. Because the fingerprint particles are usually not higher than 1 µm, an angular position of the substrate (when fitted onto the measurement table) of only a fraction of a degree can lead to significantly different height information at different points of the surface. Therefore, a best-fit plane subtraction using least squares approximation [28], [105] is applied to all topography images as a first step, regardless of the capturing device. The resulting planarized topography images are formalized as T_P and their specific pixel values as $T_P(x,y)$. As a result of such subtraction, the substrate is represented by pixel values around zero and the fingerprint pixels take values significantly greater than zero. However, the light beam emitted by the sensor is often refracted and scattered at particles, which leads to pixels appearing to have a smaller height than the actual surface. Because this distortion does mainly occur at fingerprint pixels and not at the well reflecting hard disk platter substrate, these pixels are also considered as belonging to the residue. Therefore, after the best-fit plane subtraction, surface pixels take gray values close to zero, whereas fingerprint pixels are significantly different from zero, either in the positive or the negative direction. The image planarization process is exemplary depicted in Figure 13 and Figure 14.

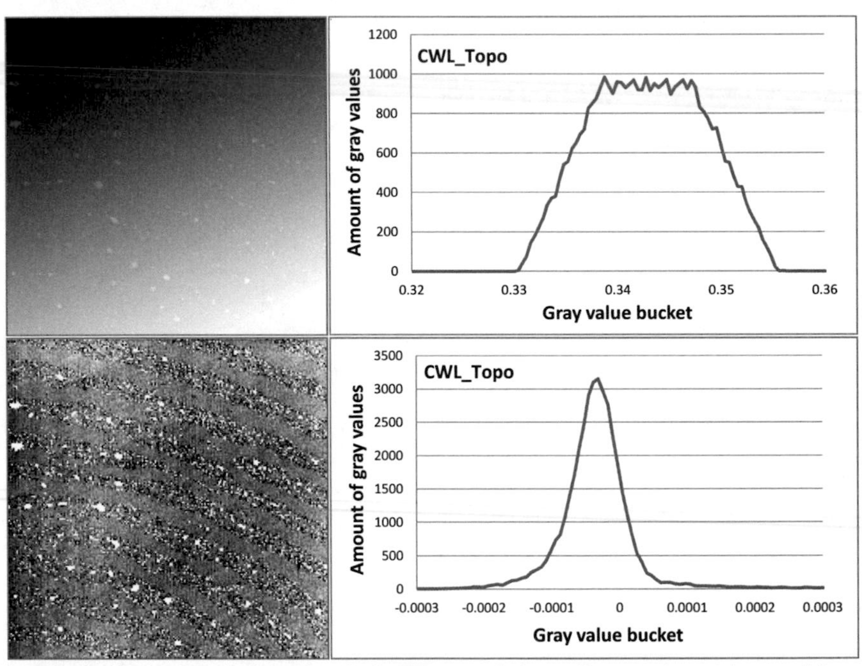

Figure 13: Visualization of best-fit plane subtraction for CWL topography images. Upper row: topography grayscale image (left); image histogram (right). Lower row: planarized topography grayscale image (left); image histogram (right).

For topography images of the CWL device, three different binarization methods CWL_DoG, CWL_Line and CWL_Var are designed in the scope of this thesis. The first method CWL_DoG utilizes a Difference of Gaussian (DoG) filter [103] to enhance the curved edges of fingerprint particles within the image. In particular, a planarized image T_P is convoluted with a 3 x 3 DoG filter kernel DoG(i,j), as depicted in Figure 15.

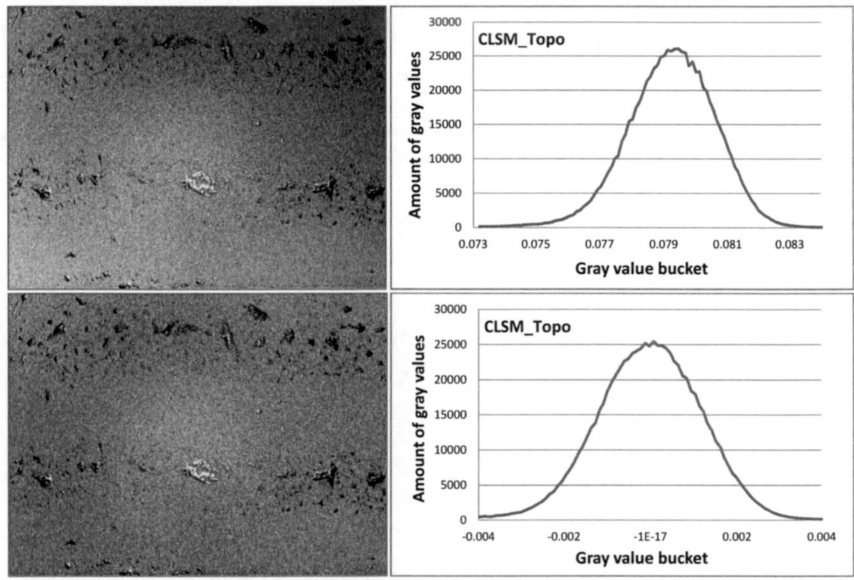

Figure 14: Visualization of best-fit plane subtraction for CLSM topography images. Upper row: topography grayscale image (left); image histogram (right). Lower row: planarized topography grayscale image (left); image histogram (right).

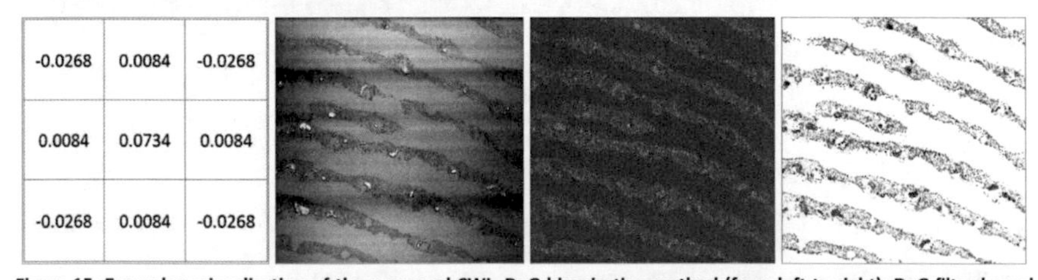

-0.0268	0.0084	-0.0268
0.0084	0.0734	0.0084
-0.0268	0.0084	-0.0268

Figure 15: Exemplary visualization of the proposed CWL_DoG binarization method (from left to right): DoG filter kernel used for the convolution; leveled topography fingerprint image before the preprocessing showing background distortions; image after application of the DoG filter; final fingerprint image T_{CWL_DoG} after two-sided threshold binarization.

In the filter result, fingerprint particles are significantly enhanced in contrast to the background distortions. Therefore, a threshold binarization is now possible. As described earlier, substrate pixels are scattered around the gray value of zero, whereas fingerprint pixels are significantly different from zero, in either the positive or the negative direction. Therefore, filtered topography images are binarized using a two-sided thresholding approach TS_BIN, where a fixed threshold ts_thresh is applied from both sides -ts_thresh and +ts_thresh for binarization:

$$T_{CWL\_DoG} = TS\_BIN(DoG(i,j) \cdot T_P(x,y), ts\_thresh = 5E^{-6}) \qquad (11)$$

The resulting binary image T_{CWL_DoG} represents the segmented fingerprint pixels using the CWL_DoG method. The specific value of ts_thresh = $5e^{-6}$ has been manually determined in prior investigations and seems to perform adequate. It is used here for all CWL topography images when binarizing with the CWL_DoG method.

The second proposed binarization approach for CWL topography images is referred to as the CWL_Line method. The approach is based on the observation that distortions often occur line wise, as can be seen in Figure 15 (second image from the left) and Figure 16 (first image from the left). Line wise distortions seem to be caused by the CWL sensor capturing prints line by line and changing environmental influences might affect only a certain amount of lines. In a first step, the Crop method applies a two-sided threshold ts_thresh = $3e^{-4}$ to the planarized topography images T_P, setting all pixels to zero which are smaller than $-3e^{-4}$ or greater than $3e^{-4}$, while leaving all other gray values unaltered. This is done to prevent outliers from influencing the following smoothening step. Afterwards, the image is convoluted with a line wise blurring filter Blur(i,j) [103] of kernel size 50 x 1 (based on the kernel mean value), to extract the characteristic form of the distortions. The resulting image T_{P_Line} is then subtracted from the original planarized image T_P, effectively removing line wise distortions. The enhanced image is binarized similar to the CWL_DoG method by application of a two-sided threshold ts_thresh:

$$T_{P\_Line}(x,y) = Blur(i,j) \cdot Crop(T_P(x,y), ts\_thresh = 3E^{-4}) \quad (12)$$

$$T_{CWL\_Line} = TS\_BIN\big(T_P(x,y) - T_{P\_Line}(x,y), ts\_thresh = 6E^{-5}\big) \quad (13)$$

The image T_{CWL_Line} is the final result of the CWL_Line binarization method. The optimal value of ts_thresh = $6e^{-5}$ is determined from prior tests and applied to all CWL topography images used with the approach. The different steps of the method are visualized in Figure 16.

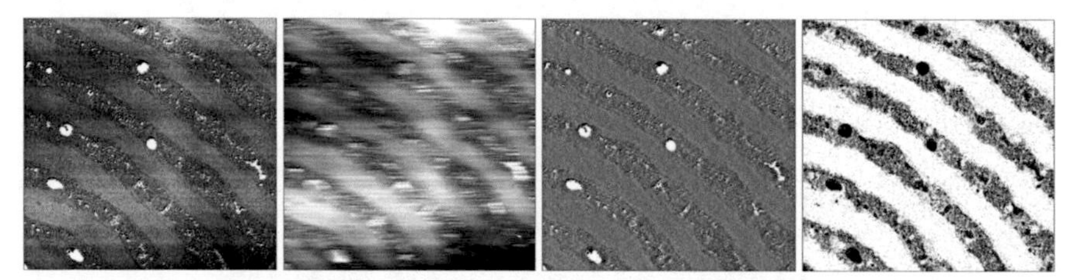

Figure 16: Exemplary visualization of the proposed CWL_Line binarization method (from left to right): best-fit plane subtracted topography image showing background distortions; cropped an blurred image after line wise smoothening; differential image after subtraction of blurred image from original image; final image T_{CWL_Line} after two-sided threshold binarization.

The CWL_Var method represents the third fingerprint binarization approach proposed for CWL topography images. Using such method, the 3 x 3 pixels block-based variance Var(i,j) of image gray values is calculated from a planarized image T_P. Because fingerprint residue is represented by many small particles and droplets, ridge areas lead to high local variances, whereas the variance of the smooth hard disk platter surface is comparatively low (Figure 17, second image from the left). The variance image has its absolute minimum at a gray value of zero (perfectly smooth surface area) and increased gray values represent and increased probability of belonging to fingerprint residue. Therefore, a threshold binarization BIN can be performed for creating the binary image. However, in the binarized image, print residue is represented by white pixels, whereas the substrate is represented by black pixels, which is an inverted representation in comparison to the earlier computed binary images. The final image T_{CWL_Var} is therefore defined as the inverse image INV(x,y) of the previously binarized image:

$$T_{CWL\_Var} = INV\big(BIN(Var(i,j) \cdot T_P(x,y), thresh = 1E^{-9})\big) \tag{14}$$

The binarization threshold thresh = $1e^{-9}$ is manually chosen from prior investigations and fixed for all images using the CWL_Var method. The different steps of the approach are visualized in Figure 17.

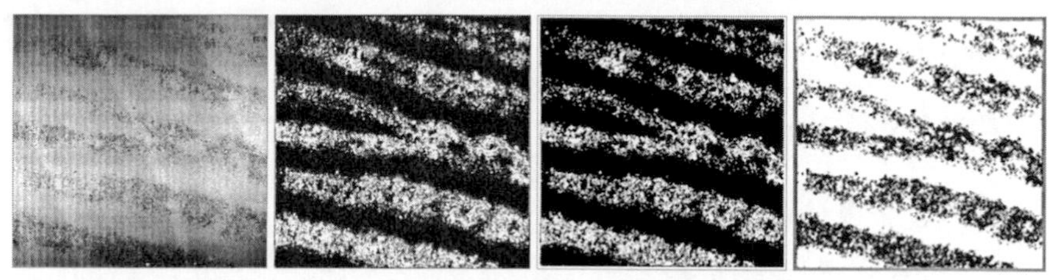

Figure 17: Exemplary visualization of the proposed CWL_Var binarization method (from left to right): planarized topography image showing background distortions; fingerprint image after block-based variance calculation; fingerprint image after threshold binarization; final image T_{CWL_Var} after inversion.

It can be observed from Figure 17 that the leveled topography image is distorted not only by horizontal lines, but also by regular vertical lines, possibly resulting from the propulsion of the measurement table. Also, diagonal distortions can be seen, crossing the image from the top left to the bottom right corner. Because the local variance of all such distortions is significantly smaller than the variance of the fingerprint droplets, these distortions can successfully be removed by using the local variance.

Apart from the three proposed methods for binarizing CWL topography images, three additional approaches CLSM_Thresh, CLSM_Blur and CLSM_Var are designed to segment fingerprints in CLSM topography images. Because the CLSM topography images seem to not be as distorted as their CWL counterparts, a simple threshold binarization approach is evaluated in the CLSM_Thresh preprocessing method, where a fixed threshold of thresh = $-2.7e^{-3}$ (manually chosen from prior tests) is applied to a planarized CLSM topography image T_P, to obtain the binarized fingerprint image T_{CLSM_Thresh}:

$$T_{CLSM\_Thresh} = BIN(T_P(x,y), thresh = -2.7E^{-3}) \tag{15}$$

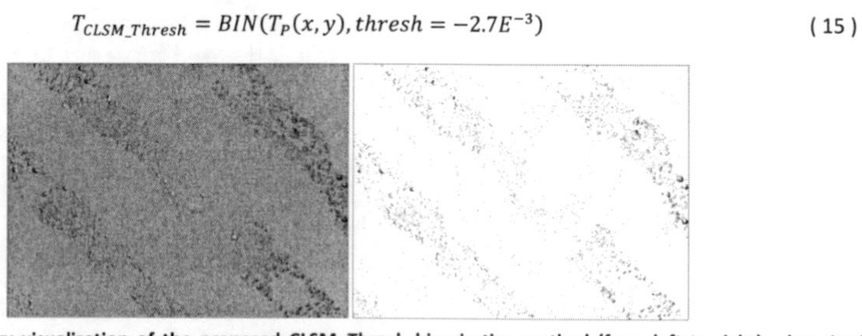

Figure 18: Exemplary visualization of the proposed CLSM_Thresh binarization method (from left to right): planarized topography image showing background noise; fingerprint image T_{CLSM_Thresh} after threshold binarization.

The CLSM_Thresh approach is visualized in Figure 18. It can be seen from the images that a certain amount of fingerprint pixels take similar values than the background noise, leading to only a fraction of the original fingerprint remaining after binarization. Increasing the threshold value would lead to more fingerprint pixels being selected, however, would also select a lot of background noise as fingerprint. Especially in the area of the earlier described barrel distortions (section 4.1.2), many fingerprint pixels seem to be affected during the acquisition process and cannot be segmented using the here proposed binarization approach. However, a significant amount of fingerprint pixels from other image regions can successfully be selected, which seems to be sufficient for a further evaluation of topography-based images.

For the successful reduction of background noise, the CLSM_Blur approach is furthermore proposed and investigated. The method smoothes a planarized CLSM topography image T_P with a standard Gaussian filter kernel Blur(i,j) [103] with a kernelsize of 5 x 5 pixels (sigma = 10). Such approach seems to effectively widen the difference between fingerprint pixels and background pixels in the histogram, because the blurring effect removes noise pixels of small density (background), whereas more homogeneous pixels of high density (fingerprint area) are only slightly changed. In contrast to the earlier investigated CWL images, the filter kernel is of increased size (5 x 5 pixels in comparison to 3 x 3 pixels for the CWL sensor), because the CLSM images are of significantly higher resolution. After image blurring, a fixed threshold of thresh = $-6.5e^{-4}$ is applied to create the binary image T_{CLSM_Blur}:

$$T_{CLSM\_Blur} = BIN(Blur(i,j) \cdot T_P(x,y), thresh = -6.5E^{-4}) \qquad (16)$$

The used binarization threshold is manually determined from prior tests. A visualization of the method is depicted in Figure 19, showing that indeed many of the observable fingerprint pixels seem to be present in the final binarized image, whereas the background noise can successfully be removed. Barrel distortions can again be detected by a lack of selected fingerprint pixels in the center of each image. However, sufficient fingerprint pixels seem to have been selected for further investigation.

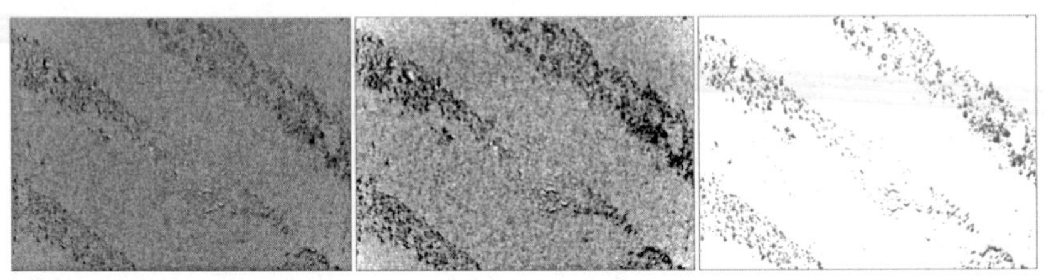

Figure 19: Exemplary visualization of the proposed CLSM_Blur binarization method (from left to right): planarized topography image showing background noise; fingerprint image after block-based smoothening; fingerprint image T_{CLSM_Blur} after threshold binarization.

Another proposed fingerprint binarization method is the CLSM_Var approach, which is similar to the earlier introduced CWL_Var technique. The method is based on the assumption that the block-based local variance of image pixels is much higher at fingerprint particles (especially at their edges) than that of the background area. From a planarized CLSM topography image T_P, the method calculates the block-based local variance Var(i,j) of 5 x 5 image blocks. Again the used kernelsize is bigger than that of the corresponding CWL approach, due to the higher capturing resolution. The obtained image of local variances is binarized using the fixed threshold of thresh = $1.8e^{-6}$ (manually selected from prior investigations). Similar to the CWL_Var approach, the resulting binary image is inverted for a consistent representation of fingerprint residue as black pixels throughout the experiments:

$$T_{CLSM\_Var} = INV\big(BIN(Var(i,j) \cdot T_P(x,y), thresh = 1.8E^{-6})\big) \qquad (17)$$

The method is exemplary visualized in Figure 20. The resulting binarized topography image T_{CLSM_Var} seems to represent the fingerprint pixels quite well. However, also a certain amount of noise remains in the print valleys. Such noise highlights the difficulty of finding an optimal threshold for binarization: if the threshold is selected too small, many fingerprint pixels are removed together with the noise. If the threshold is selected too big, a lot of noise remains in the binarized image.

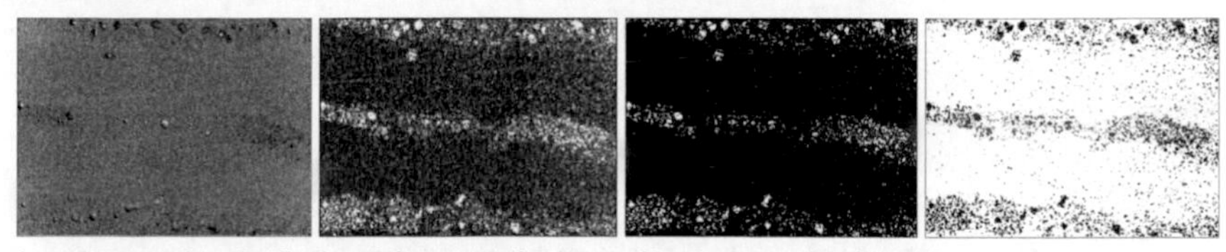

Figure 20: Exemplary visualization of the proposed CLSM_Var binarization method (from left to right): planarized topography image showing background noise; fingerprint image after block-based variance calculation; fingerprint image after threshold binarization; final image T_{CLSM_Var} after inversion.

From the different binarization techniques, the best performing one is selected for each capturing device and data type in the scope of the experimental evaluations of test goal TG2 (section 5.1.2). Such best performing technique is then used for the background mask creation in all further conducted experiments of this thesis (test goals TG3 - TG5).

4.2.3 Temporal Normalization (PC2)

Normalization is a tool widely used for transferring data into a form in which it can be compared. In the scope of age estimation, the primary goal is not to compare different fingerprints (which is of interest in other contexts), but rather to compare the consecutive temporal images of a specific print over time. Temporal normalization is therefore designed in this thesis as a means of making images comparable in respect to their changes over time. In particular, changes in the pixel values of the print residue are of interest. Therefore, all other changes affecting the image gray values should be excluded. This especially applies to changes of the overall brightness of a captured image over time, which might be subject to significant fluctuations. Therefore, a temporal normalization sub-step is designed here to remove non aging-related changes of the overall image brightness over time. Temporal normalization is applied to all capturing devices, data types as well the short- and long-term aging in a similar manner.

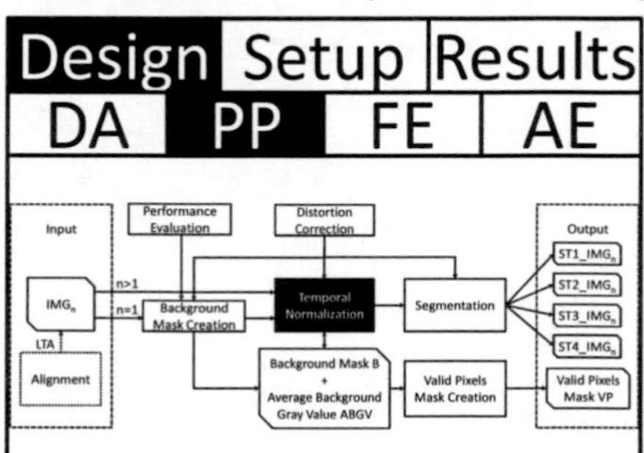

In the scope of temporal normalization, the background mask B'(x,y) is used, which has been computed from the first raw image R_IMG_1 of a time series and stored together with the average background gray value $ABGV(R\_IMG_1)$ in section 4.2.2. For all subsequent raw images R_IMG_n of the time series, the average background gray value $ABGV(R\_IMG_n)$ is computed (x_max and y_max describe the total amount of background pixels according to B'(x,y)):

$$ABGV(R\_IMG_n) = \frac{1}{x\_max \cdot y\_max} \cdot \sum_{x=1,y=1}^{x\_max,y\_max} B'(x,y) \cdot R\_IMG_n(x,y) \qquad (18)$$

All images R_IMG_n of a time series are then transferred into their temporally normalized form TN_IMG_n, by adding the difference of $ABGV(R\_IMG_1)$ and $ABGV(R\_IMG_n)$ to each image pixel:

$$TN\_IMG_n(x,y) = R\_IMG_n(x,y) + ABGV(R\_IMG_1) - ABGV(R\_IMG_n) \qquad (19)$$

The focus of temporal normalization lies on the exclusion of non aging-related changes of the overall image brightness over time. This can effectively be measured by observing the ABGV and the background histogram (according to the background mask B') of an image. If the AVBG is approximately constant or if the background histograms are correctly aligned, changes in the image histogram can be attributed to the fingerprint aging targeted in this thesis. Figure 21 depicts selected images of an exemplary CLSM series and its corresponding histograms before and after temporal normalization.

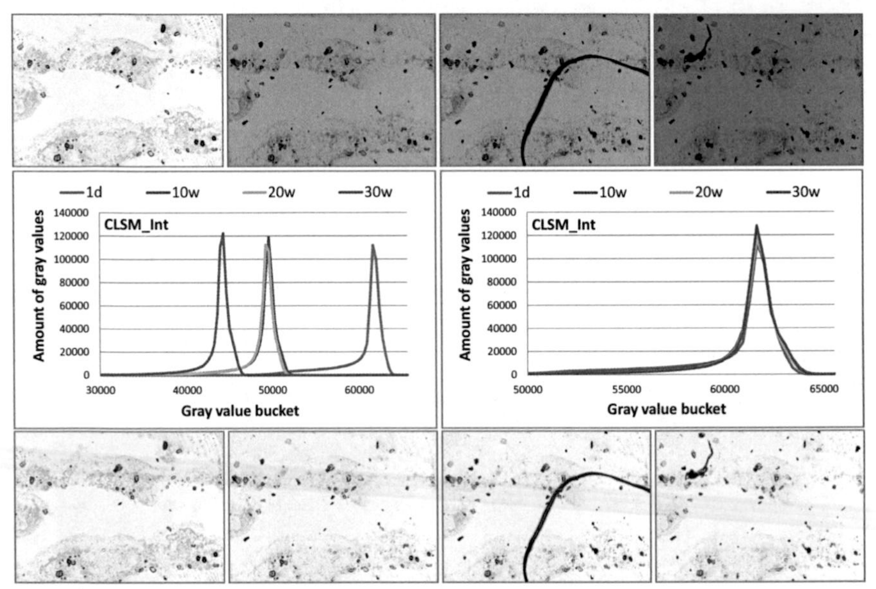

Figure 21: The proposed temporal normalization visualized at the example of a CLSM intensity long-term aging series. First row: Grayscale images at points in time t_1 = 1 d, t_2 = 10 w, t_3 = 20 w, t_4 = 30 w (at t_3, a hair has settled on the sample, which has disappeared at t_4). Second row: corresponding image histograms before (left image) and after (right image) temporal normalization. Third row: corresponding temporally normalized gray scale images.

After spatial alignment (section 4.2.1) as well as background mask creation (section 4.2.2) and temporal normalization (section 4.2.3), the captured fingerprint time series are prepared for the segmentation of relevant structures.

4.2.4 Segmentation (PC3)

Segmentation is used to separate relevant structures of the fingerprint residue from the remaining image. This sub-step is part of the proposed preprocessing step of Figure 3 and succeeds the temporal normalization. In this section, different segmentation methods are designed in respect to different segmentation targets for the short- and long-term aging. While in some cases, methods from the literature can be re-used with certain adaptations, other segmentation targets require novel methods. They need to be designed in a way to act as a general interface between the preprocessing step and the follow-

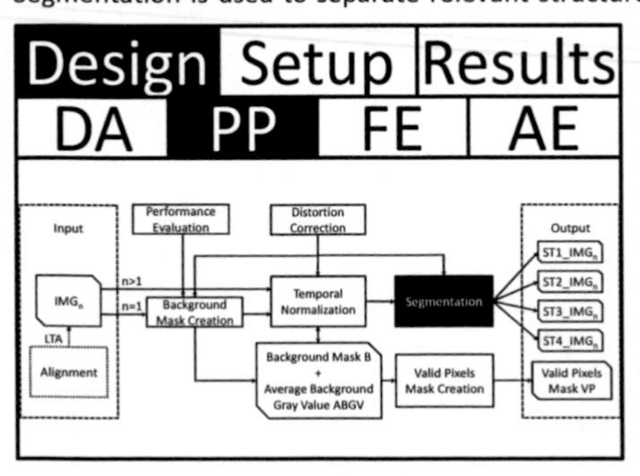

ing feature extraction step of the proposed processing pipeline. In the scope of this thesis, the following four segmentation targets are defined, from which all investigated aging features can be computed:

- **Segmentation Target 1 (ST1)**: For the first segmentation target, a captured fingerprint image is present in its plain form, without any segmentation. This target is utilized for features, which need to be calculated directly on the image raw data.
- **Segmentation Target 2 (ST2)**: For this target, a captured fingerprint image is transferred into a binary image, where for each pixel a decision is made on whether it represents fingerprint residue or background surface. Resulting images usually consist of a collection of particles, droplets and puddles, which form the fingerprint ridges.
- **Segmentation Target 3 (ST3)**: For the third target, fingerprint ridges are segmented as continuous, cohesive lines, rather than a collection of droplets, puddles and particles. The segmentation process is therefore much more complex than for target ST2.
- **Segmentation Target 4 (ST4)**: Target ST4 requires specific structures or particle types to be segmented from the captured fingerprint images, such as pores, dust, big and small particles as well as puddles. All other pixels are discarded.

The four defined segmentation targets are designed in the scope of the segmentation sub-step. Such step can naturally be applied only after a successful temporal normalization, excluding changes of the overall image brightness. The different segmentation techniques for the four segmentation targets are introduced in the following sections (4.2.4.1 to 4.2.4.4).

4.2.4.1 Fingerprint Raw Data (ST1)

Certain statistical features do not require any prior segmentation of fingerprint components. They rather operate on the image raw data. However, this does not necessarily indicate that no preprocessing is required at all. The earlier described spatial alignment (for long-term aging only), best-fit plane subtraction (for topography images), background mask creation and temporal normalization are still necessary. However, in the segmentation sub-step, no additional preprocessing needs to be performed:

$$ST1\_IMG_n = TN\_IMG_n \tag{20}$$

Exemplary images of each data type segmented according to ST1 are depicted in Figure 22.

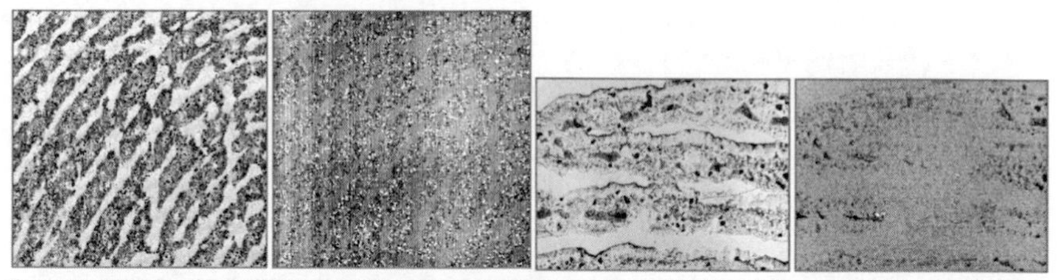

Figure 22: Exemplary fingerprint images $ST1\_IMG_n$ segmented according to ST1 (from left to right): CWL intensity image; CWL topography image; CLSM intensity image; CLSM topography image.

The images of Figure 22 exhibit a comparatively good quality for intensity data (first and third image from the left). The topography images still exhibit certain distortions, such as vertical lines (second image) or barrel distortions (fourth image). Such distortions are addressed by further preprocessing methods when using ST2 (in the scope of image binarization). However, when using ST1 for the later conducted feature extraction, these distortions limit the quality of the computed aging features.

4.2.4.2 Binarized Fingerprint Data (ST2)

For this segmentation target, captured images are segmented pixel-wise into fingerprint pixels (foreground) and substrate pixels (background). As a result, fingerprint residue is represented in the form of small droplets, particles and puddles. The segmentation process is similar to the fingerprint binarization introduced in sections 4.2.2.1 and 4.2.2.2. For all images of short- and long-term series of both capturing devices, the corresponding binarization procedure is applied. For the segmentation of intensity time series, the threshold thresh computed in sections 4.2.2.1 from the first image of a time series R_IMG_1 is applied to all temporally normalized images TN_IMG_n, leading to the images $ST2\_Int\_IMG_n$, which are segmented according to the specifications of ST2:

$$ST2\_Int\_IMG_n = I_{CWL\_CLSM}(TN\_IMG_n) = BIN(TN\_IMG_n, thresh) \quad (21)$$

For the time series of topography data, the binarization procedures of section 4.2.2.2 are applied to each temporally normalized image TN_IMG_n:

$$ST2\_Topo\_IMG_n = T_{CWL\_CLSM}(TN\_IMG_n) \quad (22)$$

After segmentation, intensity and topography images are not further distinguished and therefore referred to as $ST2\_IMG_n$. Exemplary images of each data type segmented according to ST2 are depicted in Figure 23.

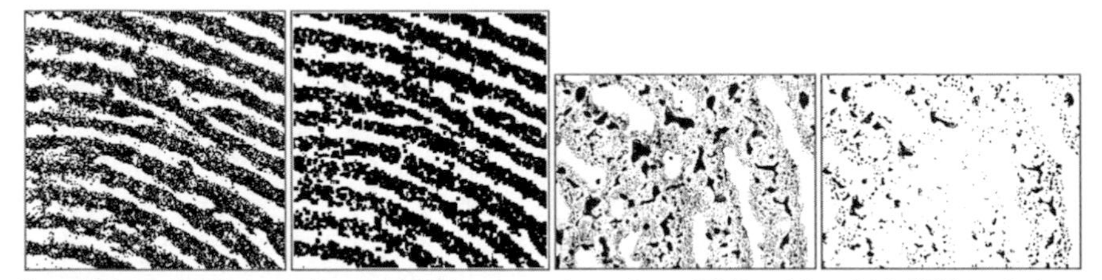

Figure 23: Exemplary fingerprint images $ST2\_IMG_n$ segmented according to ST2 (from left to right): CWL intensity image (using the Huang method); CWL topography image (using the CWL_Var method); CLSM intensity image (using the Triangle method); CLSM topography image (using the CLSM_Blur method).

The images of Figure 23 are of adequate quality for the later conducted feature extraction. During the binarization, the distortions remaining in ST1 (Figure 22, second and fourth image from the left) have been largely addressed. A few artifacts from the barrel distortions of the CLSM device might remain (fourth image of Figure 23), where less fingerprint pixels are segmented in the image region affected by barrel distortions.

4.2.4.3 Enhanced Ridge Segmentation (ST3)

Segmentation target ST3 creates binary images of consecutive, smooth ridges without gaps. They are used to compute and investigate changes in fingerprint ridge thickness over time. Such feature is only of interest for the long-term aging, because short periods of up to 24 hours are not expected to induce significant changes to the ridge thickness. Furthermore, only CWL intensity images are considered, because the CLSM device captures (as a result of its high resolution) only a comparatively small area of the fingerprint, which might be too small for a reliable print ridge thickness extraction. Also, such high-resolution CLSM image cannot be adequately enhanced using the adapted algorithm. Topography images seem to be of insufficient quality for ridge segmentation. In the scope of ST3, fingerprint ridges are enhanced to smooth and uninterrupted lines, where all gaps have been filled:

$$ST3\_IMG_n = EnhanceRidges(TN\_IMG_n) \quad (23)$$

To achieve such segmentation, the approach proposed by Hong et al. in [125] is used. The approach is based on the computation of block-based orientation and frequency fields of a fingerprint image, followed by the application of a Gabor filter, enhancing the specific orientation and frequency of each block. The method of Hong et al. is optimized to work with exemplary fingerprints of a resolution of 500 dpi (50.8 µm) and is not publicly available. However, a Matlab implementation of the method is provided by Kovesi et al. in [126]. The method is re-implemented in C++/OpenCV in the scope of this thesis [151], incorporating the functionality of Hong et al. and extending it to work with the here investigated, high-resolution latent fingerprint images. The consecutive steps of the method are depicted in Figure 24, where the original components are visualized in solid lines and the extended components are marked in dashed lines.

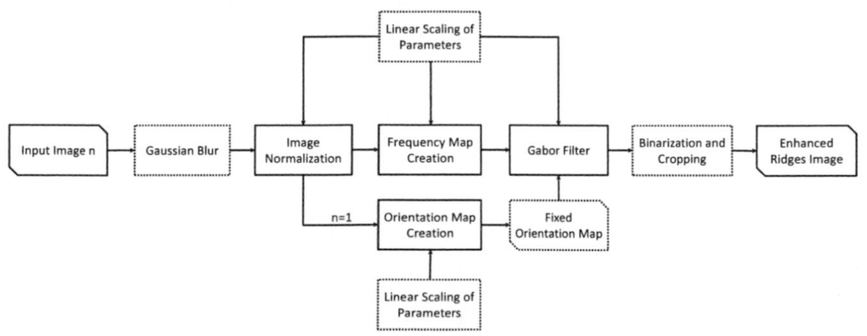

Figure 24: Proposed segmentation sub-steps for the enhancement of latent fingerprint ridges according to ST3, based on the approach of Hong et al. [125] and its Matlab implementation of Kovesi et al. [126], adapted for the investigation of contactless captured, high-resolution latent fingerprints (original components are visualized in solid lines; extended components are marked in dashed lines).

For a detailed description of the processing components of the method of Hong et al., the reader may refer to [125]. One specific adaptation in the scope of this thesis can be seen in the application of a Gaussian Blur filter (kernelsize = dpi/100, sigma = dpi/500) prior to the image processing, to avoid ridges being split at pores, which does sometimes occur when using high-resolution images. The major adaptation can be seen in the linear scaling of the filter kernels and other parameters, to work with resolutions of 2540 dpi (10 µm) from the CWL sensor. For the CLSM device, much higher resolutions exist, requiring additional enhancement filters in future work.

While the method of Hong et al. computes a separate frequency and orientation map for each image, an adaptation is made to the orientation map usage in the scope of this thesis. Because dust has a major influence on the investigated time series, especially in the scope of long-term aging, the increasing accumulation of dust would significantly distort the orientation map creation. This phenomenon is also discussed in section 4.2.5.6. The influence of dust is avoided by computing the orientation map only from the first image of a time series (n = 1) and applying this map to all subsequent images (n > 1). Using such method does successfully exclude the influence of dust from the investigated temporal images, while at the same time does not influence the ridge thickness, which is determined by the frequency of ridge lines and not dependent on the orientation. In practice, this approach can be used to investigate if changes in the average ridge thickness do occur over time, as is the aim in the conducted experiments of test goal TG3. For a practical age estimation, such approach cannot be applied, because a fingerprint sample might have already been distorted by dust before even the first scan is conducted. However, this is only of relevance if the ridge thickness can be identified as being subject to characteristic changes when aging (therefore representing a suitable feature for age estimation), which is shown in section 6.3.3.1 to not be the case.

Certain components of the original Hong method are removed, because they are no longer valid or necessary. The segmentation of fingerprint from non-fingerprint areas proposed by Hong et al. is not of interest for the here conducted studies, because only areas within a fingerprint are investigated. Furthermore, the reliability map proposed by Hong et al., identifying blocks with a high probability of false enhancement based on differences in the orientation of adjacent blocks are removed. A major reason for this action can be seen in the fact that latent fingerprints are investigated here, in contrast to exemplary prints of much higher quality, for which the original algorithm was designed. In combination with the significantly increased image resolution, many image regions would be marked as unreliable, which is in contrast to a manual inspection of this region. The reliability functionality is therefore disabled in the scope of this investigation. Also, the confidence interval for computed wavelengths given by Hong et al. is removed, because a greater variety of wavelengths does exist due to the significantly increased kernel sizes. As proposed by Kovesi at al. in their Matlab implementation [126], the median frequency of all image blocks is selected instead of individual block frequencies, achieving better results. After the successful enhancement of the fingerprint ridges, the resulting images are binarized using a fixed global threshold of thresh = 0.85. The binarization is followed by a cropping of 120 pixels at each image edge, to exclude artifacts potentially introduced by the ridge enhancement. Exemplary images visualizing the different enhancement sub-steps are given in Figure 25.

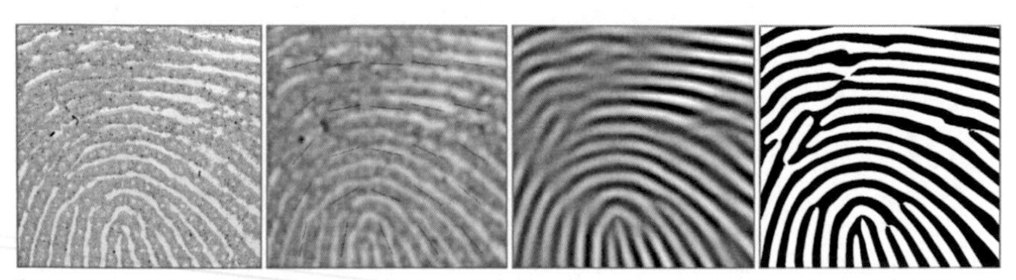

Figure 25: Visualization of the fingerprint ridge segmentation according to ST3, exemplary demonstrated using a CWL intensity image (from left to right): original grayscale image; computed orientation map; image after block-based Gabor-filtering; final segmented image $ST3\_IMG_n$.

It can be seen from Figure 25 that the computed ridge enhancement is not optimal and shows certain distortions, especially when ridges merge with their neighboring ridges (compared between leftmost and rightmost image of Figure 25). The example is deliberately chosen to visualize that the method does not work well for latent fingerprints, because it was originally designed by Hong et al. for the enhancement of exemplary prints, which are of much better quality. However, the algorithm seems to perform sufficiently accurate for representing the mean thickness of fingerprint ridges, investigated in the scope of this thesis.

4.2.4.4 Pores and Specific Particle Types (ST4)

The aim of the fourth segmentation target ST4 is to segment pores as well as certain types of particles from a fingerprint image, to separately investigate their aging behavior. Such different structures are only investigated for the long-term aging, because first manual inspections suggest that they only change over long time periods. Furthermore, topography images do not seem to exhibit sufficient quality for an accurate segmentation of most of these particle types, which is therefore limited to intensity images. In particular, structures of interest are pores ST4_PORE, dust particles ST4_DUST, big fingerprint particles (e.g. big droplets) ST4_BIG, small fingerprint particles (e.g. small droplets) ST4_SMALL as well as puddles (coherent, comparatively bright areas of liquid, forming puddle-like structures) ST4_PUDDLE. Examples of these structures are depicted in Figure 26 - Figure 30.

Because pores are comparatively big and can be well extracted from the CWL images, ST4_PORE is segmented from CWL images only. ST4_DUST, ST4_BIG, ST4_SMALL and ST4_PUDDLE can only be successfully extracted at higher image resolutions, therefore requiring the examination of CLSM images. All investigations of certain types of particles in the scope of this thesis are based on masks. From the first image of a time series, a particle mask PM is created in respect to a certain particle type PT = {PORE, DUST, BIG, SMALL, PUDDLE}, segmenting only particles of this type and its close surroundings. This mask is then applied to all consecutive, temporally normalized images TN_IMG_n of a time series, excluding all other parts of the residue and the background. The masked images represent the segmentation target ST4.

ST4_PORE: For the extraction of pores, a Mexican Hat filter MH(x,y) is used as proposed by Jain et al. in [127]. The scale factor of sigma = 1.32 suggested for live scans (1000 dpi) is adapted here to sigma = 6 for the used high-resolution CWL intensity images (2540 dpi). The Mexican Hat filter is applied to the first image TN_IMG_1 of each temporally normalized time series, followed by a manual processing due to an insufficient accuracy of the method on the investigated latent prints. During such manual processing, each filter response is binarized, selected non-pore artifacts are removed and additional, not-selected pores are added. The resulting pore mask is referred to as PM_PORE_{TEMP}:

$$PM\_PORE_{TEMP} = ManualEnhancement(MH(x,y) \cdot TN\_IMG_1) \qquad (24)$$

The decisive factor for deleting an artifact or adding a pore is considered the human observer, which seems to be the possibly most reliable measure in this matter, when well reflecting, ideal surfaces are used. However, not all latent fingerprints exhibit pores, even when captured with an adequately high resolution. This is due to the fact that the quality of latent prints is often much worse than that of exemplary prints and depends on the application conditions. Fingers might only be slightly tipped to a surface, pressed hard or can even be smeared.

After the pore mask PM_PORE_{TEMP} is manually enhanced to adequately represent pores, further preprocessing steps are required. Dust needs to be removed as it can significantly distort pores, especially when considering the investigated long-term aging periods of over two years. For an effective exclusion of dust, the cumulative dust mask CD of a time series is used. The creation of such mask is described in section 4.2.5.6 and contains each pixel of an image, which is covered by dust for at least one point of time during the capturing period. Such dust mask is dilated eight times, to also prevent the broader surroundings of dust particles from being selected as pores (eight times dilation is required for the investigated area around pores, which is selected by increasing the pore regions by six dilations and should not contain dust within this region). The dilated dust mask CD is then applied to the pore mask PM_PORE_{TEMP}, to mark all pixels in PM_PORE_{TEMP} as invalid, which are marked in CD as containing dust. Furthermore, the filter result is processed by a size filter MinSize, which removes all pores being smaller than an exemplary selected minimal size of 20 pixels from the mask. This step is performed in order to remove pores being partially cropped by the exclusion of dust:

$$PM\_PORE = MinSize(Dilate(CD, 8) \cdot PM\_PORE_{TEMP}, 20) \quad (25)$$

The final segmentation target $ST4\_PORE\_IMG_n$ is computed by dilating the final pores mask PM_PORE six times (to include also the broader surroundings of the pores). The dilated mask is then applied to each temporal image TN_IMG_n of the time series by marking all image pixels as invalid, which do not belong to a pores region according to PM_PORE:

$$ST4\_PORE\_IMG_n = Dilate(PM\_PORE, 6) \cdot TN\_IMG_n \quad (26)$$

The final masked images $ST4\_PORE\_IMG_n$ of each time series are used in the scope of the experiments to investigate changes in the pores themselves as well as their surrounding regions. The creation of $ST4\_PORE\_IMG_n$ is exemplary visualized in Figure 26:

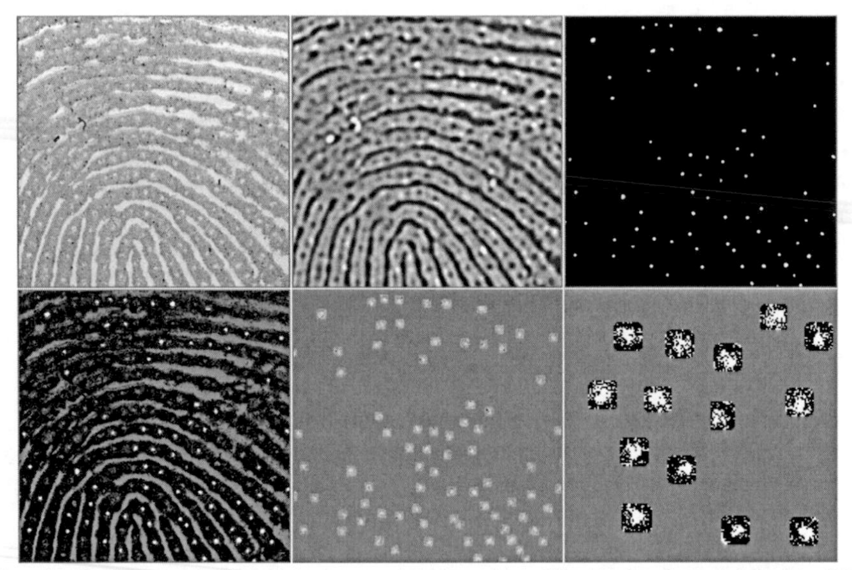

Figure 26: Visualization of the fingerprint pore segmentation according to ST4, exemplary demonstrated using a CWL intensity image (from left to right and top to bottom): original grayscale image; Mexican Hat filter response; pores mask PM_PORE (after manual preprocessing, binarization, dust exclusion and size filtering); overlay of pores mask PM_PORE with the original grayscale image (only for visualizing the segmented pores); grayscale image $ST4\_PORE\_IMG_n$ after masking of non-pore region (masked values are marked gray for better visualization); enlarged region of $ST4\_PORE\_IMG_n$ (binarized for better visualization, invalid values are marked in gray).

It can be seen from Figure 26 that the Mexican Hat filter response enhances pores quite well, yet not in all cases (upper middle image). A manual correction is therefore performed, excluding falsely segmented pores and adding missing ones. Such manual preprocessing in combination with a binarization, dust exclusion and size filtering seems to deliver adequately extracted pores for further processing (Figure 26, upper right and lower left image). Not all pores need to be detected, a subset of correctly detected pores is regarded as being sufficient for the evaluation of pores aging behavior.

The middle lower image of Figure 26 depicts the original grayscale image overlaid with the pores mask PM_PORE, providing a sufficient segmentation performance of $ST4\_PORE\_IMG_n$. For a better visualization, the original image is binarized and enlarged in the rightmost lower image of Figure 26.

ST4_DUST: Dust can best be observed over time using CLSM intensity images, because they provide the required resolution and precision. Dust is often composed of skin flakes and hair. In case dust is overlaying fingerprint residue, it might be hard to differentiate a dust particle from fingerprint particles, especially because residue does often also contain skin flakes, which look similar to those of dust. To avoid such issue, changes of dust over time are measured exclusively for the background (surface) region of a fingerprint image, which can be assumed to be dust free at the beginning of a time series (because the surfaces have been thoroughly cleaned prior to the experiments). The background mask B computed in section 4.2.2 separates fingerprint pixels from background pixels. Such mask is therefore used to investigate the changes in dust over time:

$$PM\_DUST = B \tag{27}$$

To also exclude the border region between fingerprint residue and background to provide robustness against small offsets due to inaccurate alignment of the temporal images, a dilate operation is applied two times to the PM_DUST mask. The result is applied to each temporal image TN_IMG_n of a time series, to mark all pixels as invalid not belonging to the investigated region:

$$ST4\_DUST\_IMG_n = Dilate(PM\_DUST, 2) \cdot TN\_IMG_n \tag{28}$$

The masked background region $ST4\_DUST\_IMG_n$ is considered as being free from fingerprint particles. Over the complete capturing period, dust settles in this region, which is investigated as a potential feature in the here conducted experiments. The creation process of $ST4\_DUST\_IMG_n$ is exemplary depicted in Figure 27.

Figure 27: Exemplary visualization of the dust segmentation according to ST4_DUST, shown for an aged CLSM intensity image (t = 30 w). From left to right: aged grayscale fingerprint image (including settled dust); segmented area PM_DUST for dust investigation (dilated two times, black: masked fingerprint region, white: background region used for dust observation); segmented dust according to $ST4\_DUST\_IMG_n$ (binarized for better visualization).

The rightmost image of Figure 27 visualizes the dust particles of the image background area and is used for the later conducted evaluation of dust. If a dust particle covers both ridge and background area at the same time, only the part covering the background area is considered. The rightmost image of Figure 27 has been binarized already for a better visualization, a step which is performed in section 4.3.2.3 for the computation of the dust feature F12.

ST4_BIG: Apart from dust, three main types of particles seem to be present within a fingerprint image: big particles, small particles and puddles. In the scope of ST4, adequate segmentation methods are designed for each of them. Big particles often appear as big droplets of liquid or crystalline structures, whereas small particles usually represented tiny droplets. Puddles seem to be mainly caused by evenly distributed areas of liquid, where several droplets seem to have merged to a puddle-like structure. An adequate separation between these three types of particles seems to be only possible using the increased resolution of the CLSM device. Therefore, only CLSM intensity images are used for the segmentation of these particles. CLSM topography images should also be studied in future work.

To segment big particles for ST4_BIG, the first image of a series TN_IMG_1 is binarized similar to ST2. However, the computed threshold thresh is reduced to thresh = 0.85 · thresh. This reduction is necessary to separate big particles from puddles. Puddles are usually comparatively bright, whereas big droplets are much darker (or at least have a dark outline around them, caused by sensor artifacts created at the droplet edges). Therefore, a decreased value of thresh leads to most puddles being selected as background (white), whereas most droplets are selected as belonging to the residue (black). After binarization, dust is removed from the mask by usage of the cumulative dust mask CD, similar to the earlier introduced segmentation of ST4_Pore (please refer to section 4.2.5.6 for information about the creation of CD). The cumulative dust mask is dilated two times to create robustness against small offsets between the temporal images (due to insufficient alignment) and to compensate small shifts of the dust particles. Afterwards, a method MinSize is applied, which selects only particles with an exemplary chosen size of more than 50 pixels. For some big particles, only the outlines are selected using such method. Therefore, a manual enhancement is necessary, closing incomplete outlines and filling outlined particles if necessary. After the manual enhancement, the big particles mask PM_BIG is obtained:

$$PM\_BIG = ManualEnhancement(MinSize(Dilate(CD,2) \cdot BIN(TN\_IMG_1, 0.85 \cdot thresh),50)) \quad (29)$$

The big particles mask does not necessarily contain all big particles present in the image. However, for investigating the aging behavior of these particles, a subset of the present big particles is sufficient. The mask PM_BIG is then applied to all images TN_IMG_n of a time series, marking all pixels as invalid, which do not belong to this particle class. In this scope, PM_BIG is dilated two times, to compensate for potential small offsets due to insufficient alignment:

$$ST4\_BIG\_IMG_n = Dilate(PM\_BIG, 2) \cdot TN\_IMG_n \quad (30)$$

The complete process of big particles selection is exemplary visualized in Figure 28.

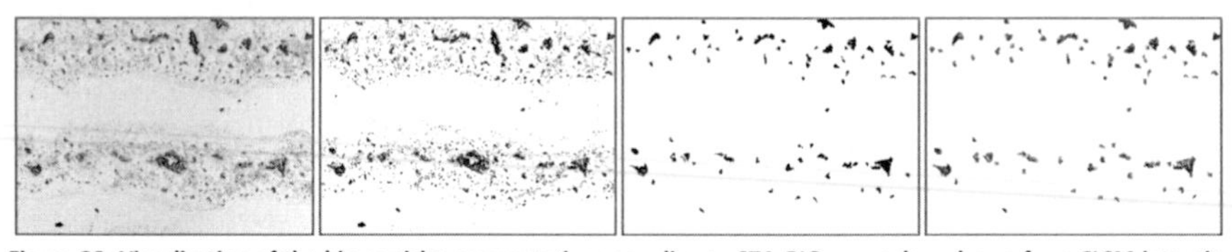

Figure 28: Visualization of the big particles segmentation according to ST4_BIG, exemplary shown for a CLSM intensity image (t = 1 d). From left to right: grayscale fingerprint image; binarized fingerprint image (using the reduced threshold thresh = 0.85 · thresh); big particles mask PM_BIG after dust removal, MinSize filtering and manual enhancement; final image $ST4\_BIG\_IMG_n$ with segmented big particles.

Figure 28 visualizes the segmentation of big particles. The rightmost image depicts the original grayscale values of the filtered particles. It cannot be said without additional chemical analysis, which particles such class consists of. In Figure 28, the particles have comparatively sharp edges, possibly representing crystallized salt or similar substances, while for other samples, round droplet-like shapes have also been observed. However, even when observing droplet-like shapes, it cannot be determined without chemical analysis, if the droplets consists of water or fatty components.

ST4_SMALL: Analogue to ST4_BIG, the segmentation target ST4_SMALL selects small particles within a fingerprint time series. It is computed on CLSM intensity images only. The designed algorithm first binarizes an image TN_IMG_1 of a time series according to ST2, using the threshold thresh computed in the background mask creation step. Here, thresh does not need to be decreased (because such measure is only necessary to separate big particles from puddles). To the binarized image, the cumulative dust mask CD as well as the big particles mask PM_BIG are applied. The cumulative dust mask CD is introduced in section 4.2.5.6 and includes each pixel, which is covered by dust for at least one point of time during the capturing period. The big particles mask PM_BIG represents big particles and has been introduced earlier. Both masks are dilated two times before application, to create robustness against small offsets caused by insufficient alignment or small shifts of dust particles. The resulting image contains the binarized residue without big particles and dust pixels, which have been marked as invalid. In a last step, particles bigger than 50 pixels are removed by a MaxSize method, to exclude any remaining non-small particles:

$$PM\_SMALL = MaxSize(Dilate(PM\_BIG, 2) \cdot Dilate(CD, 2) \cdot BIN(TN\_IMG_1, thresh), 50) \qquad (31)$$

The final small particles mask PM_SMALL is then applied to all images TN_IMG_n of a time series to obtain segmentation target $ST4\_SMALL\_IMG_n$. Analogue to the previously discussed segmentation targets, the mask is dilated two times to provide robustness against small offsets between the images or small shifts of the dust particles:

$$ST4\_SMALL\_IMG_n = Dilate(PM\_SMALL, 2) \cdot TN\_IMG_n \qquad (32)$$

The complete process of small particles selection is exemplary visualized in Figure 29.

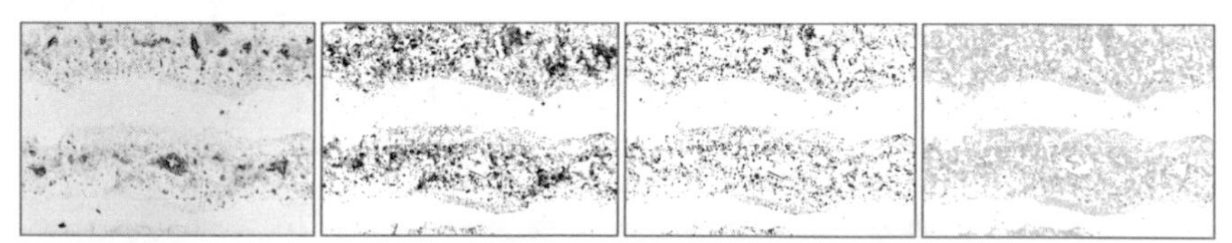

Figure 29: Visualization of the small particles segmentation according to ST4_SMALL, exemplary shown for a CLSM intensity image (t = 1 d). From left to right: grayscale fingerprint image; image after binarization and exclusion of dust (using mask CD) and big particles (using mask PM_BIG); particle mask PM_SMALL after MaxSize filtering; final segmented small particles image $ST4\_SMALL\_IMG_n$.

It can be seen from Figure 29 that small particles comprise most parts of a latent fingerprint. However, they are also overlaid with bright puddles in some cases, which tend to split up during segmentation. A completely accurate separation of small particles from puddles is therefore not possible.

ST4_PUDDLE: The designed algorithm for ST4_PUDDLE segments bright puddles of liquid from the binarized first image TN_IMG_1 of a CLSM intensity time series. The binarization is performed according to ST2 using the threshold thresh computed in the scope of background mask creation. The binarized image is filtered by application of the cumulative dust mask CD, the big particles mask PM_BIG as well as the small particles mask PM_SMALL. The creation of the cumulative dust mask CD is introduced in section 4.2.5.6, while the creation of PM_BIG and PM_SMALL has been introduced earlier. The masks are dilated two times prior to application (to remove small offsets caused by inaccurate alignment or small shifts of dust particles):

$$PM\_PUDDLE = Dilate(PM\_SMALL, 2) \cdot Dilate(PM\_BIG, 2) \cdot Dilate(CD, 2) \cdot BIN(TN\_IMG_1, thresh) \quad (33)$$

The resulting particle mask PM_PUDDLE represents a certain amount of puddles segmented in the images of the current time series. Analogue to the other particle types, it cannot be guaranteed that all puddles are included into the mask. However, this is not of importance, because a subset of particles of a certain type seems sufficient for the evaluation of its aging characteristics. To create the segmentation target $ST4\_PUDDLE\_IMG_n$, the mask PM_PUDDLE is dilated two times analogue to the earlier described particles and applied to each image of a time series:

$$ST4\_PUDDLE\_IMG_n = Dilate(PM\_PUDDLE, 2) \cdot TN\_IMG_n \quad (34)$$

Figure 30 visualizes exemplary images of the different puddle selection steps.

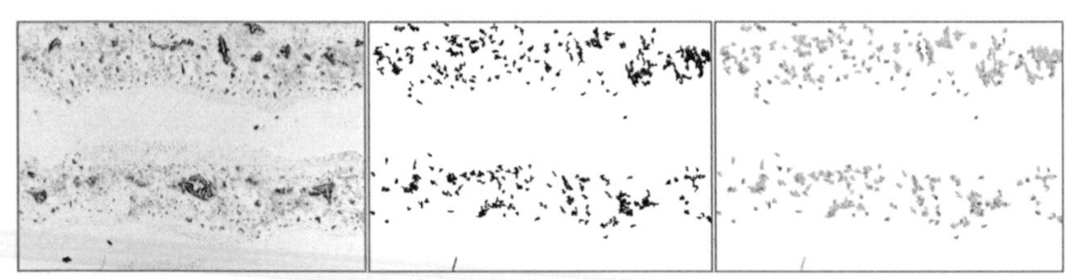

Figure 30: Visualization of the puddles segmentation according to ST4_PUDDLE, exemplary shown for a CLSM intensity image (t = 1 d). From left to right: grayscale fingerprint image; puddles mask PM_PUDDLE after binarization and exclusion of dust (using mask CD), small particles (using mask PM_SMALL) and big particles (using mask PM_BIG); final segmented puddle image $ST4\_PUDDLE\_IMG_n$.

It can be observed from Figure 30 that puddles cover continuous areas of different size and shape. Such structures are possibly composed of certain liquid substances, such as water or fat. However, a comprehensive analysis of such structures can only be obtained using chemical methods.

4.2.5 Distortion Correction (PC4)

Different influences are impacting the estimation of a fingerprints age. Many of these influences alter the observed aging behavior of the print residue, e.g. the speed of degradation. These influences are investigated in the scope of research objective O5. However, some influences lead to distortions or even inconsistencies of the captured time series. A correction of such distortions is therefore regarded as a prerequisite for the creation of adequate time series. Possible causes of such distortions are plentiful and cannot be investigated in detail in the scope of this thesis. They include fluctuations in the intensity

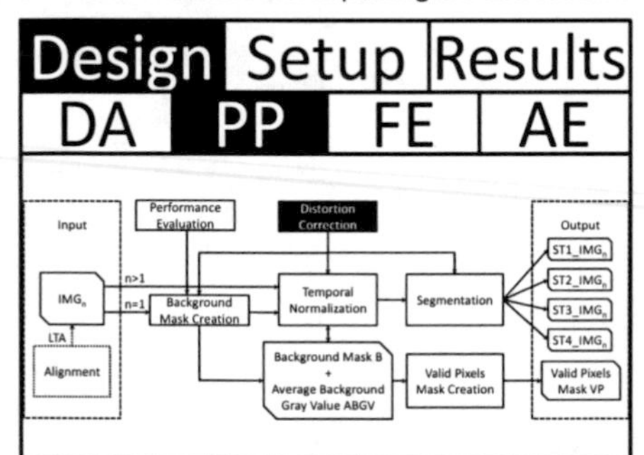

of emitted radiation by the light source, sensor noise, aging of the light source and the detector, limitations of the measurement interval, refraction at droplets, changes in ambient temperature, vibrations, the measurement table propulsion as well as dust.

In the scope of the here designed distortion correction sub-step, image brightness fluctuations (section 4.2.5.1), null value distortions (section 4.2.5.2), out of range images (section 4.2.5.3), temperature distortions (section 4.2.5.4), vibrations and measurement table propulsion distortions (section 4.2.5.5) as well as dust (section 4.2.5.6) are discussed. They are reduced or even corrected in the scope of the different preprocessing sub-steps.

4.2.5.1 Reduction of Fluctuations in Image Brightness

Fluctuations in the overall image brightness of captured time series are a major preprocessing challenge identified as PC2 (section 4.1.2). They are addressed by the earlier introduced background mask creation (section 4.2.2) and temporal normalization (section 4.2.3) sub-steps, performing a histogram alignment based on the background of an image. Local fluctuations in the image brightness have also been identified (PC5) (section 4.1.2). They are addressed by the later introduced valid pixels mask VP creation (section 4.2.6).

For the captured topography images, a major distortion of the image brightness can be seen in sensor noise. While intensity images seem to be only marginally effected, the small heights of fingerprint particles captured by the topography images seem to be subject to a significant amount of distortions (Figure 13 - Figure 20). Such noise distortions are addressed in the fingerprint binarization sub-step (section 4.2.2.2), where special methods are designed to exclude noise when binarizing topography images of both devices. Furthermore, barrel distortions as identified in Figure 5 overlay certain regions of the topography images. They can only partially be addressed by the binarization techniques of section 4.2.2.2.

4.2.5.2 Exclusion of Null Values

In the scope of CWL images, null values do occur. Capturing a null value means that an insufficient amount of light is received by the detector. Hence, the pixel value is set to zero. Null values do mainly occur at droplets or sharp edges, because light is refracted or reflected away from the detector. Because the CLSM device is operated with monochromatic light, no such differentiation is performed. Here, refraction or reflection artifacts are represented as noise and cannot be differentiated from the general sensor noise.

Null values sometimes occur in the scope of long-term aging, as a product of degrading structures and increasing dust particles. Therefore, a cumulative null value mask CN is designed for CWL long-term aging series, to allow for an exclusion of such pixels. It is set to a value of one, if at least for one point of time n the function IsNull($TN\_IMG_n(x,y)$) returns true:

$$CN(x,y) = BIN\left(\sum_{n=1}^{n\_max} \begin{cases} 1, & IsNull\big(TN\_IMG_n(x,y)\big) = true \\ 0, & else \end{cases}, 0.5\right) \quad (35)$$

The function IsNull computes for each image pixel $TN\_IMG_n(x,y)$ of a time series image TN_IMG_n, if the corresponding pixel value is zero:

$$IsNull\big(TN\_IMG_n(x,y)\big) = \begin{cases} true, & TN\_IMG_n(x,y) = 0 \\ false, & else \end{cases} \quad (36)$$

Exemplary images visualizing the cumulative null value mask creation are given in Figure 31.

Figure 31: Visualization of the cumulative null value mask creation, exemplary shown for a CWL intensity image (t = 26 w). From left to right: grayscale fingerprint image; null values of the image; cumulative null value mask CN for the complete time series; cumulative null values highlighted in original grayscale image (for visualization purposes only).

As can be observed from Figure 31, null values are of minor importance for single images (second image form the left). However, they might accumulate over time (third image from the left), in which case they have to be excluded using the mask CN.

4.2.5.3 Exclusion of Out of Range Images

In the scope of capturing long-term aging series, fingerprints need to be fixed onto the measurement table of the capturing devices for each acquisition. In some cases, the hard disk platters might not be applied totally planar, resulting in small angles between the investigated surface and the measurement table. Such small angles do not have a significant influence on the CWL sensor, due to an automated height adjustment of the sensor for each captured image. However, the CLSM device can only be focused once in the beginning of a scan session and can run out of focus when traversing large areas of a platter. In practice, such phenomenon occurs rarely and can be recognized by visual inspection. In such case, images are manually removed from a time series, effectively excluding such distortion.

4.2.5.4 Reduction of Temperature Distortions

The influence of temperature fluctuations on the aging behavior of latent fingerprints is evaluated in the scope of research objective O5. Here, the impact of temperature fluctuations on the capturing process is briefly discussed. Changes in the ambient temperature can lead to changes in the optical system of the capturing devices (e.g. the objective lenses), which might be a reason for the observed fluctuations in overall image brightness (PC2, section 4.1.2). Furthermore, starting a capturing device can lead to an initial increase of temperature, until the working temperature of the device is reached.

Whether such temperature influences are indeed causing the observed fluctuations in the overall image brightness can only be presumed and needs to be studied in the scope of future research. However, in case an influence on the overall image brightness does exist, it is reduced by the designed background mask creation (section 4.2.2) and temporal normalization (section 4.2.3) substeps. Temperature fluctuations might also add to sensor noise, which is a challenge especially for topography images, as has been discussed earlier (section 4.2.5.1). The proposed methods for noise reduction in the scope of fingerprint binarization seem to exhibit a feasible approach for reducing such noise.

4.2.5.5 Reduction of Vibrations and Measurement Table Propulsion Distortions

Vibrations and measurement table propulsion distortions seem to be mainly a challenge to CWL topography images. Because the CLSM device captures images tile-wise and therefore has a fixed position of the measurement table when capturing a single tile, no distortions caused by the movement of the measurement table can be observed. However, because the CWL device captures images line-wise and each line is captured by a continuous movement of the measurement table underneath the sensor, vibrations as well as measurement table propulsion distortions do occur.

Experts passing by the device and even trams passing by outside the laboratory can produce line-wise distortions of the captured topography images, where a certain amount of lines exhibit darker or brighter pixel values than the rest of the image. Furthermore, the propulsion of the measurement table leads to line- as well as row-wise distortions. However, it cannot be clearly distinguished from the resulting images, if a certain distortion is caused by the earlier discussed sensor properties, temperature fluctuations, vibrations or even additional, yet undiscovered influences. Independent of the cause of a distortion, the different methods designed for topography image binarization (section 4.2.2.2) are used to reduce such distortions.

4.2.5.6 Reduction of Dust

Dust seems to be a major source of distortion. Dust particles can assume different forms, such as grains, dust fluffs or even hair (Figure 32) and can appear or disappear at random. Grains do barely seem to influence the observed aging behavior, due to several reasons. First of all, they are usually small and occupy only tiny areas of a fingerprint. Moreover, grains are represented by dark gray values in an image histogram (due to their diffuse reflection properties) and are therefore located at the lower end of the histogram, barely influencing the print binarization.

For the short-term aging, dust plays only a minor role, because only a limited amount of dust particles settles on the captured fingerprints during the first 24 hours. However, when studying the long-term aging of prints, including time periods of over two years, significant amounts of dust can settle onto a print, representing a major source of distortion. Therefore, mechanisms need to be designed to remove or at least reduce the influence of dust on the long-term aging. In some cases, this influence is already considered to a certain extent by the different preprocessing sub-steps, e.g. when excluding the surface background from the feature calculation using the later introduced valid pixels mask VP (section 4.2.6). Furthermore, the computation of the ridge thickness reduces the influence of dust by applying the orientation field of the first image of a time series to all remaining images of the series.

However, for all other features, dust needs to be considered separately. In the scope of this thesis, the influence of dust is considered by the design of a cumulative dust mask CD of a time series. Using such approach, dust is detected in each temporal image TN_IMG_n of the series and the results are accumulated to the final cumulative dust mask CD(x,y), which marks every image pixel as dust pixel, which has been covered by dust for at least one time during the complete examination period. In this case, the function $IsDust(TN\_IMG_n(x,y))$ returns true for at least one capturing point n:

$$CD(x,y) = BIN\left(\sum_{n=0}^{n\_max} \begin{cases} 1, & IsDust\big(TN\_IMG_n(x,y)\big) = true \\ 0, & else \end{cases}, 0.5 \right) \tag{37}$$

If CD(x,y) takes a value of one, the pixel at the coordinates (x,y) has been covered with dust for at least one time during the examination period. It is therefore considered as a dust pixel. In case CD(x,y) takes a value of zero, the pixel has at no point of time been covered by a dust particle and is therefore considered as valid for further examinations. The cumulative dust mask CD(x,y) can be directly used in the earlier introduced preprocessing sub-steps as well as for the later introduced valid pixels mask VP creation (section 4.2.6), to exclude regions from the evaluation, in which dust will eventually settle at some point of the examination period.

The function IsDust decides for a given temporal image coordinate $TN\_IMG_n(x,y)$, if a dust particle is present in the specific image at the specific coordinate. Because intensity and topography images of both devices are captured simultaneously and therefore have similar coordinates, it is sufficient to compute IsDust from either the intensity or the topography data of a sensor and to use the resulting mask CD for both data types. Because the quality of the intensity images is much higher for both devices, the intensity images are used here for the computation of IsDust. Dust accumulates in most cases at the lowest end of a histogram, therefore allowing to be extracted by threshold binarization. For an accurate differentiation between dust and fingerprint pixels, such binarization is conducted using a manually determined threshold man_thresh, which can vary for the different capturing devices and time series:

$$IsDust\big(TN\_IMG_n(x,y)\big) = \begin{cases} true, & TN\_IMG_n(x,y) < man\_thresh \\ false, & else \end{cases} \tag{38}$$

For creating the final dust mask CD, only dust particles of a certain minimum size are considered, to avoid an inaccurate segmentation. A MinSize filter is used for the CWL device, to exclude dust particles smaller than 50 pixels from the final CD mask. For the CLSM Device, a median blur filter (kernelsize = 5) is applied directly to each temporal image, before IsDust is computed. The different methods perform optimal for each capturing device in separating dust particles from fingerprint particles. Exemplary images visualizing the dust mask creation are given in Figure 32.

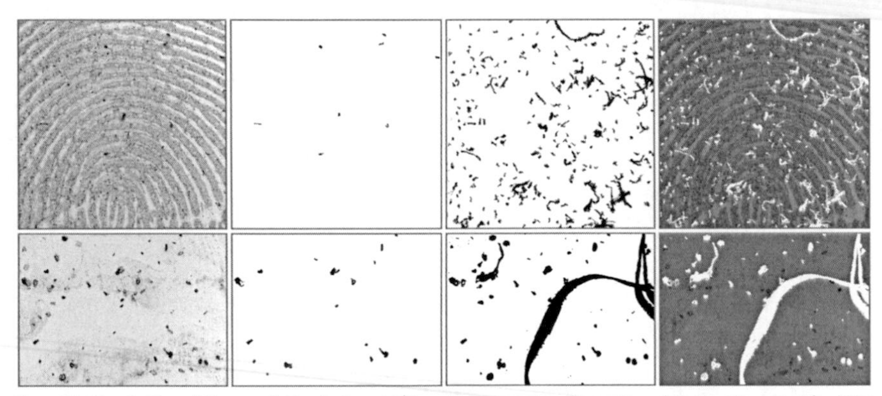

Figure 32: Visualization of the cumulative dust mask CD creation, exemplary shown for a CWL intensity image (t = 26 w, upper row) and a CLSM intensity image (t = 26 w, lower row). From left to right: grayscale fingerprint image; dust mask of the single image; cumulative dust mask CD for the complete time series (CWL: t_{max} = 2.3 y, CLSM: t_{max} = 30 w); cumulative dust values highlighted in original grayscale image (for visualization purposes only).

Figure 30 depicts for an exemplary CWL as well as CLSM time series the amount of dust which can accumulate during aging. The time period of the CWL series is much higher (t_{max} = 2.3 years), leading to a much higher accumulation of dust. However, the much shorter CLSM series (t_{max} = 30 weeks) includes images of a significantly higher magnification, leading to at least similarly big areas being covered by dust. Furthermore, a hair settles eventually on the CLSM sample, rendering a significant amount of the image infeasible for aging feature computation.

4.2.6 Valid Pixels Mask Creation (PC5)

The valid pixels mask VP is proposed for the later performed feature extraction (O3), to exclude non-

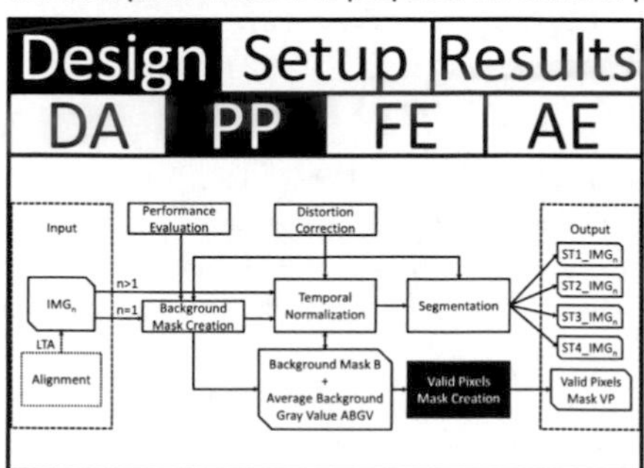

fingerprint pixels from the feature calculation. For the short-term aging, VP is defined as the inverse of the background mask B' of a time series. It is to note that the VP mask specifies all pixels as valid, which are identified as belonging to the fingerprint in the preprocessing step as well as their direct neighboring pixels (due to the one dilation operation performed on the residue pixels for the creation of B' in section 4.2.2). Therefore, small alignment offsets of the temporal images of a time series are balanced:

$$VP = 1 - B' \quad (39)$$

In the scope of long-term aging, VP fulfills some additional functionality. Because dust acts as a major influence factor over long time periods (section 4.2.5.6), VP does not only mark background pixels as invalid, but also dust pixels, defined in the cumulative dust mask CD (sections 4.2.5.6). Furthermore, both masks are dilated two times, to allow for and increased robustness against small offsets caused by incorrect alignment and small shifts of dust particles, which do especially occur during long-term aging. For CWL long-term aging series, the null values are furthermore excluded, using the cumulative null value mask CN (section 4.2.5.2). This mask is not dilated, because null values in CN refer to specific pixel values, which are computed for each image and are not subject to shifts or movement:

$$VP = 1 - CN \cdot Dilate(CD, 2) \cdot Dilate(B, 2) \quad (40)$$

The mask VP is passed on together with the preprocessed images of a fingerprint time series to the feature extraction step of the introduced processing pipeline. In the scope of the later designed feature extraction, VP defines the valid pixels used for the short- as well as long-term aging feature computation. The different specifications of VP given for the short- and long-term aging in formulae (39) and (40) are automatically applied in dependence on the chosen examination period.

4.2.7 Quality Measures For Digital Preprocessing Methods

For addressing the research objective O2, several preprocessing methods have been designed in re-

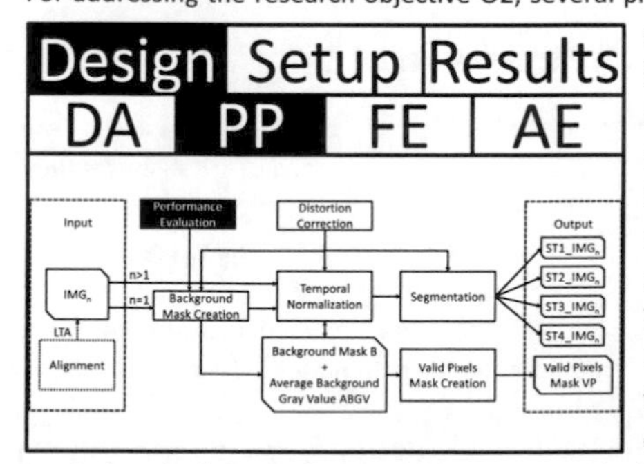

spect to the five identified preprocessing challenges PC1 - PC5. Fingerprint binarization conducted in the scope of background mask creation (PC2, section 4.2.2) is regarded as the most fundamental of these methods, because most other designed methods rely on a correct separation of fingerprint and background pixels. Therefore, the performance of the binarization methods designed in section 4.2 is experimentally investigated in the scope of the conducted experiments (test goal TG2, section 5.1.2). In particular, the best performing binarization method is selected in respect to each capturing device and data type and is applied to all following experiments (test goals TG3 - TG5). Other evaluations remain subject to future research, such as investigating the performance of the other designed preprocessing methods as well as to which extend a certain error of the image binarization is propagated, enhanced or decreased during the remaining processing steps.

For evaluating the correctness of a binarization method, a binary ground truth image is required, accurately specifying which image pixels belong to the fingerprint and which belong to the background. In general, it is not possible to create perfect ground truth, e.g. to correctly decide for each pixel of a captured image if it belongs to the fingerprint or to the background. However, the subjective assessment of a human observer seems to be adequately close to a perfect ground truth, in case ideal, well reflecting substrates are used, which is the case for the hard disk platter surfaces of the later performed experimental investigations. The field of dactyloscopy entirely relies on the judgment of dactyloscopic experts, who make the final decision on whether a fingerprint matches another one. Therefore, it is considered adequate in the scope of this thesis to regard the judgment of a human observer as ground truth of sufficient accuracy.

To compare an automatically binarized image IMG_{AU} with its corresponding ground truth image IMG_{GT}, the binarization error b_err is defined, providing an objective measure of their difference. For computing b_err, both images are compared pixel-wise. If a pixel of the automatically binarized image $IMG_{AU}(x,y)$ is not equal to its corresponding pixel from the ground truth image $IMG_{GT}(x,y)$, a counter is increased. The final value of the counter is then divided by the total amount of image pixels, leading to the final error rate b_err:

$$b\_err = \frac{1}{x\_max \cdot y\_max} \cdot \sum_{x=1,y=1}^{x\_max,y\_max} \begin{cases} 1, & IMG_{GT}(x,y) \neq IMG_{AU}(x,y) \\ 0, & else \end{cases} \quad (41)$$

The binarization error b_err is regarded as a feasible measure for describing how close an automatically binarized fingerprint image is to its optimal (manually determined) ground truth image. A certain deviation between both images can barely be avoided, because even between the judgments of different human observers a certain deviation of the determined ground truth would occur. However, such fluctuations are assumed to be negligible. In the remainder of this thesis, b_err is presented interchangeably either by its computed relative value or in percentage points. Further details about the quality evaluation of automated binarization techniques can be found in the experimental test setup (section 5.5.1).

4.3 Design of Digital Aging Feature Sets (O3)

After a captured fingerprint time series has been preprocessed into the required segmentation targets ST1 - ST4 in the scope of O2, aging features are designed in the feature extraction step of the proposed processing pipeline (objective O3). The designed feature extraction step is depicted in Figure 33. It consists of only one sub-step, which is the feature extraction. An input image STx_IMG_n is required, which is the nth image of a time series and has been preprocessed according to the segmentation target STx (x = {1,2,3,4}). Together with the preprocessed image, the valid pixels mask VP of the corresponding time series is provided. In the feature extraction sub-step, different

features Fk are computed, depending on the segmentation target STx. After feature calculation, one or more computed features Fk (k = {1,2,...,k_max}) are provided as output, representing the interface to the following age estimation step.

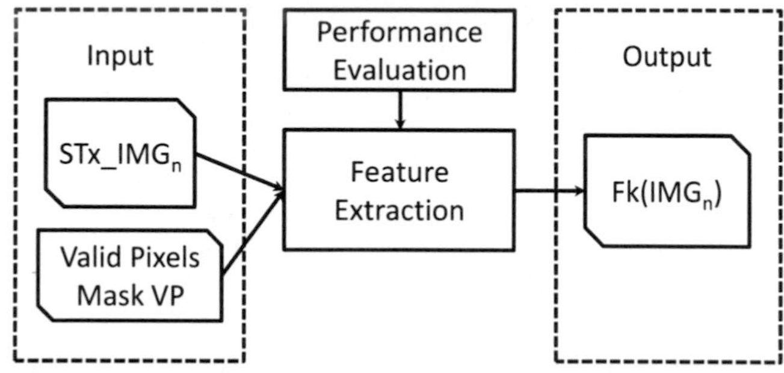

Figure 33: Detailed structure of the proposed feature extraction step of the overall processing pipeline.

As introduced earlier, features can be divided into morphological, statistical and substance-specific ones. Statistical features have not been explored so far for the age estimation of fingerprints. They are therefore designed here for the first time. Morphological features (ridge thickness and pores) have been targeted for the long-term aging in an earlier non-invasive approach of Popa et al. [75], presumably based on a manual evaluation. Therefore, methods are designed in the scope of O3 for the digital and automated computation of the ridge thickness and pore properties, to provide a means of comparison and to elaborate on the feasibility of such kind of features. Furthermore, additional morphological features targeting dust, big and small particles as well as puddles are designed. Morphological features are considered to be only feasible in the scope of long-term aging, because the comparatively slow changes in their values seem to manifest themselves only over long aging periods. Contactless captured, substance-specific features are also of interest. However, they require additional non-invasive capturing devices, which can acquire high-resolution, substance-specific fingerprint data and are therefore subject of future research.

In section 4.3.1, seven statistical age estimation features F1 - F7 for the investigation of short- and long-term aging periods are exemplary proposed, followed by eight morphological features F8 - F15 for the long-term aging (section 4.3.2). Section 4.3.3 defines quality measures for an objective evaluation of the feature performance in the scope of the conducted experiments. Exemplary fingerprint time series for each segmentation target ST1 - ST4 used as input for the designed feature sets are visualized in Figure 34 - Figure 36 and Figure 38 - Figure 42 for exemplary chosen capturing devices, data types and aging periods.

4.3.1 Statistical Feature Set for Short- and Long-Term Aging Series

In this section, seven statistical features for the short- and long-term aging are exemplary proposed. They are applied in the later performed experimental investigations to both capturing devices and data types in respect to the short- and long-term aging. A differentiation between capturing devices and data types is not necessary, because the earlier defined segmentation targets ST1 - ST4 act as universal interfaces between the sensor and data type specific preprocessing (section 4.2.4) and the here designed feature extraction. The features target statistical properties of captured fingerprint images and therefore measure changes in the overall appearance of a print, e.g. the loss of contrast.

The proposed statistical, image-based feature set has been published in the scope of this thesis in [153] using obsolete preprocessing methods. In contrast to [153], the features are extended here by adequate preprocessing methods in the scope of O2 and the application to valid fingerprint pixels only, which are specified according to the earlier created mask VP (section 4.2.6). The features are introduced in more detail in the following sections. Because they are of statistical nature, they are comparatively resilient to small alignment offsets, which can occur in the scope of long-term aging.

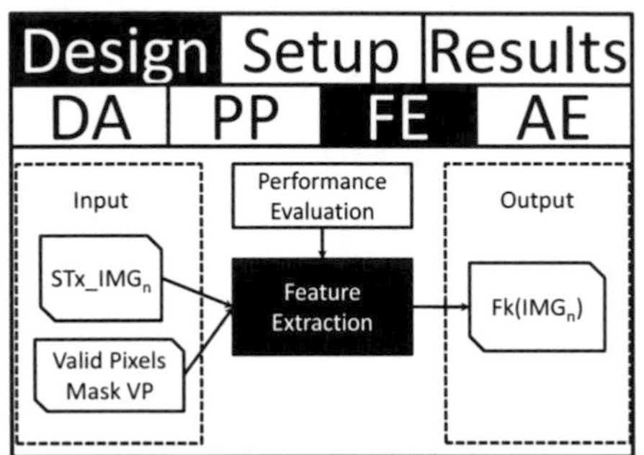

4.3.1.1 *Binary Pixels (F1)*

The feature F1 is specifically designed in the scope of this thesis to measure the loss in image contrast during aging. It is referred to as the Binary Pixels feature and shows a characteristic, promising aging tendency. It has been extensively studied and published in the scope of this thesis [151], [153], [156], [159]. Because volatile substances are evaporating (e.g. water or short-chained fatty acids) and degrading, fingerprint pixels seem to disappear over time. Such degradation happens very fast in the beginning until the print has dried up and then continuous with a decreasing speed. It seems to be a very effective feature. Time series need to be segmented according to segmentation target ST2, deciding for each single pixel of the binarized image if it belongs to the fingerprint (foreground, represented as black pixels with a gray value of zero) or the substrate (background, represented by white pixels with a gray value of one). A segmented image IMG_{n_ST2} (which can be of either four types I_{CWL}, I_{CLSM}, T_{CWL}, T_{CLSM}) is used for feature calculation. Exemplary time series of segmentation target ST2 are given in Figure 34. It is to note that for the long-term aging samples, a print is captured for the first time after 24 hours, to reduce the short-term aging effects caused by the drying of the fingerprint.

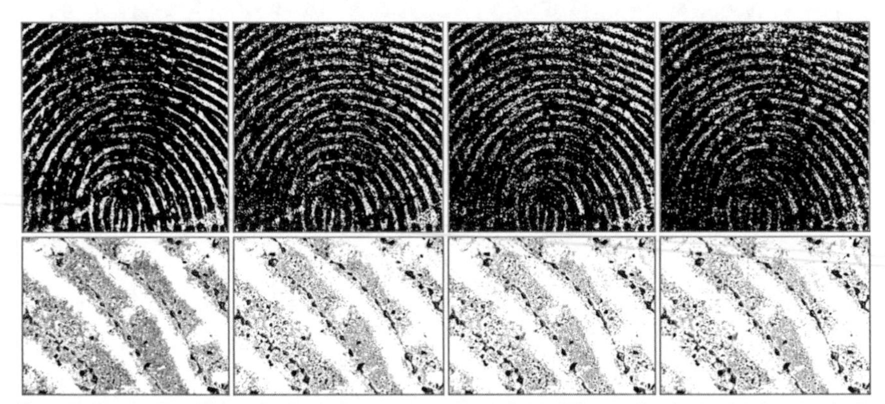

Figure 34: Visualization of fingerprint time series preprocessed according to ST2, exemplary depicted at the example of a CWL topography long-term aging series (upper row, t_1 = 1 d, t_2 = 35 w, t_3 = 79 w, t_4 = 120 w) as well as a CLSM intensity short-term aging series (lower row, t_1 = 0.5 h, t_2 = 9 h, t_3 = 17 h, t_4 = 24 h).

A slight decrease in image brightness over time can be observed in Figure 34 (from left to right), especially between the first and the second image, indicating that most changes happen in the beginning of an aging period. Feature F1 is calculated from ST2 by determining the relative frequency of (white) background pixels of the binary image IMG_{n_ST2} in respect to the valid pixels mask VP, where the coordinates x and y describe the coordinates of all valid fingerprint pixels specified in VP:

$$F1 = \frac{1}{x\_max \cdot y\_max} \cdot \sum_{x=1,y=1}^{x\_max,y\_max} VP(x,y) \cdot IMG_{n\_ST2} \qquad (42)$$

The Binary Pixels feature F1 is regarded as a promising aging feature, because it measures the loss in overall image contrast, especially targeting the border region between fingerprint and background pixels in the histogram (by designing adequate thresholds), where fingerprint pixels gradually turn into background pixels during aging.

4.3.1.2 Mean Image Gray Value (F2)

Calculating the global mean of image gray values represents another promising aging feature, which is referred to as F2. In comparison to the earlier introduced Binary Pixels feature F1, the calculation of the image mean does not only measure the disappearance of fingerprint pixels, but also a potential increase in pixel gray values. Certain print compounds do not disappear instantly, but rather exhibit a steady increase of their brightness and only eventually disappear. Calculating the mean of image gray values might be an effective way to measure such changes. Especially for topography images, where pixel gray values represent fingerprint height information, the supposedly decreasing height of particles might manifest themselves in constantly changing image pixel gray values. However, such feature targets the complete histogram and is therefore vulnerable to certain distortions, even if they impact only gray values at the far end of the histogram.

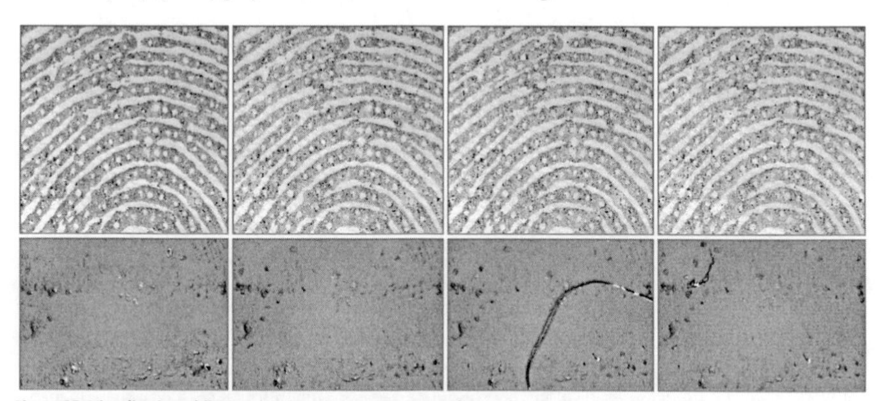

Figure 35: Visualization of fingerprint time series preprocessed according to ST1, exemplary depicted at the example of a CWL intensity short-term aging series (upper row, $t_1 = 0.5$ h, $t_2 = 8$ h, $t_3 = 16$ h, $t_4 = 24$ h) as well as a CLSM topography long-term aging series (lower row, $t_1 = 1$ d, $t_2 = 10$ w, $t_3 = 20$ w, $t_4 = 30$ w).

The calculation of F2 is similar to that of the Binary Pixels feature. However, images are not binarized beforehand. A temporally normalized grayscale image IMG_{n_ST1} is used, as is computed in the scope of ST1 (Figure 35). The exemplary depicted grayscale images of Figure 35 show a slight decrease of the image contrast during aging (from left to right). Feature F2 is computed according to the following formula, where x and y describe the coordinates of all valid fingerprint pixels according to the mask VP:

$$F2 = \frac{1}{x\_max \cdot y\_max} \cdot \sum_{x=1,y=1}^{x\_max,y\_max} VP(x,y) \cdot IMG_{n\_ST1} \qquad (43)$$

Computing the mean image gray value is based on the same general characteristics than the Binary Pixels feature F1, which is the change of fingerprint pixels from darker into brighter gray values. However, while F1 only measures these changes at a well defined threshold, feature F2 averages them over the complete histogram. It might therefore be subject to an increased performance (because all pixels are included) or a decreased performance (because changes occurring in other histogram areas are also captured, possibly increasing the overall vulnerability to distorting influences).

4.3.1.3 Global Standard Deviation (F3)

Standard deviation is a well known statistical measure to compute the general deviation of the elements of a normally distributed set from their mean value. It is computed as feature F3 from the segmentation target ST1 for all valid fingerprint pixels according to VP:

$$F3 = \sqrt{\frac{1}{x\_max \cdot y\_max - 1} \cdot \sum_{x=1,y=1}^{x\_max,y\_max} \left(VP(x,y) \cdot IMG_{n\_ST1} - \overline{VP(x,y) \cdot IMG_{n\_ST1}}\right)^2} \qquad (44)$$

In the scope of fingerprint aging, image pixels are expected to become brighter over time and eventually take the value of background pixels. Therefore, pixels potentially become more similar, leading to a decrease of the standard deviation. However, if particles are degrading, they might also collapse into smaller particles, which could lead to an increase of the standard deviation.

4.3.1.4 Mean Local Variance (F4)

The local variance of image blocks Var(i,j) describes the variability of image pixels within a block. Such measure is characteristic for detecting edges and has been used earlier to detect fingerprint residue in topography images (section 4.2.2.2). The used block size is specified according to the capturing device (CWL: 3 x 3 pixels, CLSM 5 x 5 pixels, due to the different resolutions of both devices). The mean local variance feature F4 of an image IMG_{n_ST1} is computed for all valid fingerprint pixels VP, based on segmentation target ST1:

$$F4 = \frac{1}{x\_max \cdot y\_max} \cdot \sum_{x=1,y=1}^{x\_max,y\_max} VP(x,y) \cdot Var(i,j) \cdot IMG_{n\_ST1}(x,y) \qquad (45)$$

During aging of a fingerprint, the edges of particles and droplets deteriorate, leading to so far unknown changes of the mean local variance. While such variance might increase due to the collapse of structures into smaller structures, it could also decrease due to an increased amount of pixels turning into background pixels over time.

4.3.1.5 Mean Gradient (F5)

Gradients are a widely used metric to determine edges within images. The mean gradient value is therefore investigated as feature F5, based on segmentation target ST1. To calculate the gradients, a Sobel filter [103] $Sobel_x(i,j)$ is applied to the image blocks (CWL: 3 x 3 pixels, CLSM: 5 x 5 pixels, due to the different resolutions of both capturing devices). The filter can be applied to either x- or y-direction as well as in both directions, leading to the gradient image $G_{IMG\_n\_ST1\_X}$, $G_{IMG\_n\_ST1\_Y}$ or $G_{IMG\_n\_ST1\_XY}$:

$$G_{IMG\_n\_ST1\_X}(x,y) = Sobel_X(i,j) \cdot IMG_{n\_ST1}(x,y) \quad (46)$$

$$G_{IMG\_n\_ST1\_Y}(x,y) = Sobel_Y(i,j) \cdot IMG_{n\_ST1}(x,y) \quad (47)$$

$$G_{IMG\_n\_ST1\_XY}(x,y) = Sobel_{XY}(i,j) \cdot IMG_{n\_ST1}(x,y) \quad (48)$$

In the scope of feature F5, the gradients are calculated in both directions $G_{IMG\_n\_ST1\_XY}$, because the specific orientation of fingerprint particles is arbitrary (gradients of single directions are also of use for the later introduced Coherence feature F7). The absolute values of the computed gradients are averaged, forming the final feature value F5 for all valid pixels according to VP:

$$F5 = \frac{1}{x\_max \cdot y\_max} \cdot \sum_{x=1,y=1}^{x\_max,y\_max} VP(x,y) \cdot \left|G_{IMG\_n\_ST1\_XY}(x,y)\right| \quad (49)$$

Similar to the local variance, the deterioration of fingerprint components might lead to either an increase or a decrease in the gradients of an image, which could be characteristic for the aging of latent fingerprints.

4.3.1.6 Mean Surface Roughness (F6)

Surface roughness is an established measure to determine the characteristics of a certain substrate. The ISO norm 4287 [128] describes different measures of roughness, one of which is exemplary investigated in the scope of feature F6. The mean surface roughness is computed for all valid pixels according to VP, based on segmentation target ST1:

$$F6 = \frac{1}{x\_max \cdot y\_max} \cdot \sum_{x=1,y=1}^{x\_max,y\_max} \left|VP(x,y) \cdot IMG_{n\_ST1}(x,y) - \overline{VP(x,y) \cdot IMG_{n\_ST1}}\right| \quad (50)$$

During aging of a print, the surface roughness potentially decreases, because print particles, which introduce a certain amount of roughness, deteriorate and finally disappear. However, the breaking up of contours of certain compounds might also lead to a medium-term increase in such roughness.

4.3.1.7 Coherence (F7)

Coherence is another statistical measure, which has been proposed in [129] for the computation of fingerprint features. It is based on the orientation of print ridges and could be of interest for the age estimation. It is therefore investigated in the scope of feature F7 and computed according to [129] for all valid pixels of VP, based on segmentation target ST1:

$$p_1 = \left(\overline{VP \cdot (G_{IMG\_n\_ST1\_X} \cdot G_{IMG\_n\_ST1\_X})} - \overline{VP \cdot (G_{IMG\_n\_ST1\_Y} \cdot G_{IMG\_n\_ST1\_Y})}\right)^2 \quad (51)$$

$$p_2 = \left(\overline{VP \cdot (G_{IMG\_n\_ST1\_X} \cdot G_{IMG\_n\_ST1\_Y})}\right)^2 \quad (52)$$

$$p_3 = \overline{VP \cdot (G_{IMG\_n\_ST1\_X} \cdot G_{IMG\_n\_ST1\_X})} + \overline{VP \cdot (G_{IMG\_n\_ST1\_Y} \cdot G_{IMG\_n\_ST1\_Y})} \quad (53)$$

$$F7 = \frac{\sqrt{p_1 + 4 \cdot p_2}}{p_3} \quad (54)$$

It is unclear at this point whether the coherence metric has the potential to represent characteristic changes during fingerprint degradation and if an increasing or decreasing trend is dominant. However, it is of interest to evaluate this feature towards its potential for age estimation.

4.3.2 Morphological Feature Set for Long-Term Aging Series

Apart from the earlier introduced statistical feature set, the long-term aging offers the potential to investigate certain morphological features, some of which might only marginally change in the scope of short aging periods, but could exhibit more characteristic changes during long-term aging. Such features are here designed and include the ridge thickness, pores, dust, big and small particles as well as puddles. They are based on the segmentation targets ST3 or ST4 and are designed for intensity images of selected capturing devices. Topography images are regarded as exhibiting an insufficient quality for a reliable extraction of morphological structures and are therefore discarded.

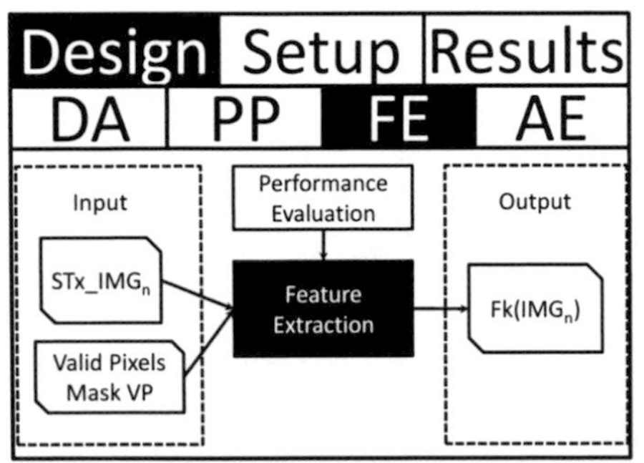

In contrast to the earlier defined statistical feature set, the valid pixels mask VP is not necessary for the computation of morphological features, because for the segmentation targets ST3 and ST4, distorting influences are reduced in the scope of the segmentation sub-step. Distortion reduction is realized for feature F8 by using a similar orientation map for all images of a time series (section 4.2.4.3) and for features F9 - F15 by the use of specific masks for particle extraction, which exclude all image pixels not belonging to the selected particle type as well as dust (section 4.2.4.4). The long-term aging features are introduced in more detail in the following sections. Features F8 - F11 have been published in the scope of this thesis in [151], based on obsolete preprocessing methods. They are redesigned here using the advanced preprocessing methods introduced in the scope of O2.

4.3.2.1 Mean Ridge Thickness (F8)

The mean ridge thickness feature (F8) of a fingerprint image is exemplary investigated in the scope of this thesis for CWL intensity images. Topography images seem to be of less quality, while CLMS intensity images provide a very high resolution, which limits the size of the measured area and leads to ridges being broken up during enhancement. The feature F8 is based on the earlier introduced segmentation target ST3 (section 4.2.4.3), which depicts fingerprint ridges as continuous, smooth lines without gaps (Figure 36).

Figure 36 exemplary visualizes fingerprint segmentation according to ST3 for a time period of over two years. It can be seen that the quality of the segmented ridges decreases slightly over time (images from left to right), even if measures have been taken to limit the influence of dust (section 4.2.4.3). The comparatively low quality of latent fingerprints in general and the significant amount of dust settling during aging are major sources of such distortions. However, for investigating the average fingerprint ridge thickness in the scope of feature F8, the distortions seem to be acceptable.

Figure 36: Visualization of fingerprint time series preprocessed according to ST3, depicted for an exemplary chosen CWL intensity long-term aging series (from left to right: t_1 = 1 d, t_2 = 35 w, t_3 = 79 w, t_4 = 120 w).

Feature F8 is calculated using the method of skeletons described in [130]. A fingerprint skeleton in the scope of this thesis consists of all center points c of valid circles val_cir(c) present along the ridges of an image IMG_{n_ST3}. A valid circle is defined as having exactly one intersecting point ip(c) with each edge of a ridge. It furthermore lies inside the ridge and its diameter d(c) describes the distance between two opposite intersecting points ip1(c) and ip2(c):

$$val\_cir(c) = \begin{cases} true, & |ip1(c)| = 1 \; and \; |ip2(c)| = 1 \; and \; \left|\overrightarrow{ip1(c)ip2(c)}\right| = d(c) \\ false, & else \end{cases} \tag{55}$$

In a valid circle val_cir(c), the diameter d(c) is always equal to the ridge thickness at the center point c, being the distance between the two edges of the ridge. Circles not having their center in the middle of a ridge or having a diameter different from the ridge width would ultimately lead to having more or less than one intersection with each edge of the ridge and therefore being an invalid circle. At certain minutiae, such as bifurcations, the distance between the ip1(s) and ip2(s) is different to the diameter of the circle, leading to an exclusion of the corresponding center points from the skeleton. Exemplary skeletonized fingerprint images are depicted in Figure 37.

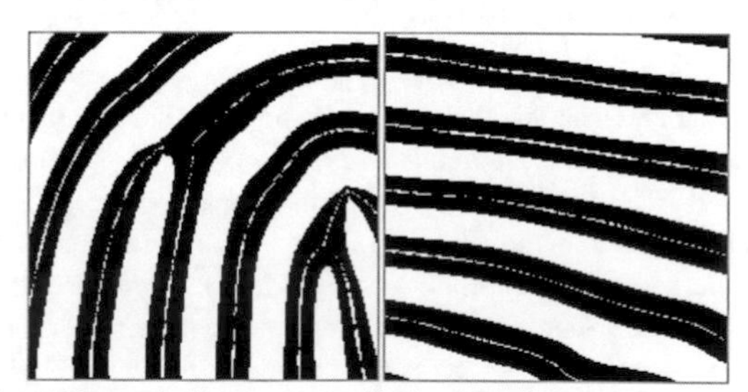

Figure 37: Skeletonized partial fingerprint images computed in the scope of feature F8.

The mean ridge thickness feature F8 is computed from all valid center points c found in a fingerprint image IMG_{n_ST3}, by determining the mean value of the diameters d(c) of their corresponding circles c:

$$F8 = \frac{1}{x\_max} \cdot \sum_{x=1}^{x\_max} d(c_x) \tag{56}$$

The designed feature F8 provides information about changes in the ridge thickness of a latent print during aging, not only at specific locations of an image, but averaged over the complete image. It is therefore considered as a robust means of capturing aging-related alterations.

4.3.2.2 *Area, Amount and Average Size of Pores (F9-F11)*

The investigation of pores (F9-F11) is based on CWL intensity images, because they provide an adequate amount of pores (in contrast to the CLSM device) as well as sufficient image quality (in contrast to topography images). Segmentation target $ST4\_PORE\_IMG_n$ is required as input (Figure 38, upper row). Images of this type consist of temporally normalized images TN_IMG_n, in which all pixels but those belonging to pores and their surroundings (a six times dilation) have been marked as invalid (marked gray in Figure 38).

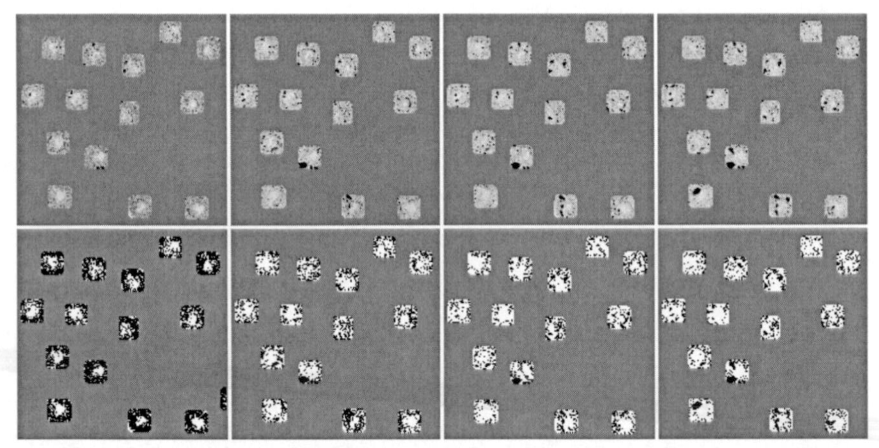

Figure 38: Visualization of fingerprint time series preprocessed according to ST4_PORE (upper row) and additionally binarized (second row, binarized for better visualization only). The depicted time series is exemplary chosen from the CWL intensity long-term aging series (from left to right: t_1 = 1 d, t_2 = 35 w, t_3 = 79 w, t_4 = 120 w). Images are enlarged; non-pore pixels according to PM_PORE are marked in gray.

From Figure 38 it can be observed that pores are adequately segmented using the proposed approach. It can furthermore be seen that the regions around pores become significantly brighter over time. Therefore, changes in the relative frequency of background (surface) pixels in the vicinity of pores are measured in the scope of feature F9. Such feature is similar to the earlier introduced Binary Pixels feature F1, but applied to pore regions only. For that purpose, $ST4\_PORE\_IMG_n$ is binarized according to the earlier determined threshold (section 4.2.2.1). In the resulting image, valid pixels can have one of two possible values: a value of zero represents fingerprint residue (material around pores). A value of one represents either pores (round gaps in the residue) or fingerprint material around pores, which has disappeared during aging (including a few natural gaps). All computations are applied only to pixel values marked as belonging to pore regions in $ST4\_PORE\_IMG_n$. Therefore, the variables x, y, x_max and y_max refer to such pixels only:

$$F9 = \frac{1}{x\_max \cdot y\_max} \cdot \sum_{x=1,y=1}^{x\_max,y\_max} BIN(ST4\_PORE\_IMG_n, thresh) \qquad (57)$$

Feature F10 computes the amount of pores present in a fingerprint image $ST4\_PORE\_IMG_n$, investigating a potential decrease of pores. Feature F11 calculates the average size of a pore. For the computation of both features, $ST4\_PORE\_IMG_n$ is binarized using the earlier determined threshold (section 4.2.2.1). Afterwards, a Mexican Hat filter MH is applied as proposed in [127] to enhance the pores (using a value of sigma = 6 for the 2540 dpi high-resolution images). To the Mexican Hat filter results, an adaptive binarization method Adapt_Bin_Mean is applied. The method computes for each pore region (pore and its surroundings as marked in ST4_PORE_IMG$_n$) a separate mean value, which is used as an adaptive threshold for this single pore region. The resulting binary pore areas contain the detected pores and in some cases a few additional filtering artifacts in their vicinities. To remove also such filtering artifacts, only structures greater than 50 pixels are selected by a MinSize function. The resulting image contains only the segmented pores and is referred to as P_IMG$_n$:

$$P\_IMG_n = MinSize(Adapt\_Bin\_Mean(MH \cdot BIN(ST4\_PORE\_IMG_n, thresh)),50) \tag{58}$$

From the pores image P_IMG$_n$, feature F10 is computed by determining the amount of binary objects present in the image:

$$F10 = Binary\_Objects\_Count(P\_IMG_n) \tag{59}$$

Feature F11 computes the mean pore size in pixels by dividing the absolute amount of white (pore) pixels from P_IMG$_n$ by the total amount of pores (F10):

$$F11 = \frac{1}{F10} \cdot \sum_{x=1,y=1}^{x\_max,y\_max} P\_IMG_n(x,y) \tag{60}$$

Using features F9 - F11, changes in the direct vicinity of pores as well as in the pores themselves can be observed over time, to identify characteristic alterations during the degradation of latent prints.

4.3.2.3 Dust, Particles and Puddles (F12-F15)

The analysis of dust, particles and puddles requires a comparatively high resolution and good image quality. It is therefore limited to the intensity images of the CLSM device. As input, the segmentation targets ST4_DUST_IMG$_n$, ST4_BIG_IMG$_n$, ST4_SMALL_IMG$_n$ and ST4_PUDDLE_IMG$_n$ are provided, of which exemplary time series are depicted in Figure 39 - Figure 42. Within each image, all pixels not representing the respective structure (or its vicinity) are masked. Some of the features are based on a student work conducted in the scope of this thesis [160], which are redesigned here.

Feature F12 calculates the relative amount of dust per investigated pixel over time. As discussed in section 4.2.4.4, only the background region of a fingerprint is included in ST4_DUST_IMG$_n$, for a reliable investigation of dust. For computing feature F12, the input image ST4_DUST_IMG$_n$ is binarized using the earlier determined threshold (section 4.2.2.1), followed by a MinSize filter, to exclude all dust particles smaller than 50 pixels:

$$BD\_IMG_n = MinSize(BIN(ST4\_DUST\_IMG_n, thresh),50) \tag{61}$$

The resulting background dust image BD_IMG$_n$ contains only dust particles with a size greater 50 pixels. This size is chosen in order to capture only clearly visible dust particles and to exclude sensor and surface artifacts, which sometimes manifest themselves in the form of small dots. The final feature value F12 is computed by determining the relative frequency of dust particles (represented by a pixel value of one) in respect to the overall background pixels according to ST4_DUST_IMG$_n$ (represented by a pixel value of zero):

$$F12 = \frac{1}{x\_max \cdot y\_max} \cdot \sum_{x=1,y=1}^{x\_max,y\_max} BD\_IMG_n(x,y) \tag{62}$$

An exemplary time series for feature F12 is given in Figure 39. It visualizes the increased amount of dust settling on the surface over a total time period of 30 weeks (from left to right).

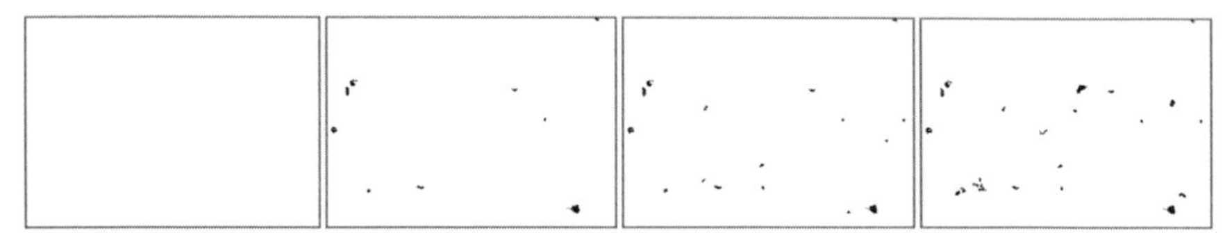

Figure 39: Visualization of fingerprint time series preprocessed according to ST4_DUST, depicted for an exemplary chosen CLSM intensity long-term aging series (from left to right: $t_1 = 1$ d, $t_2 = 11$ w, $t_3 = 20$ w, $t_4 = 30$ w). Pixels not belonging to the dust area according to PM_DUST are marked in white. For better visualization, images are already binarized and MinSize filtered according to F12 (formula (61)).

Features F13 - 15 are computed in a similar way to represent the relative amount of big particles (F14), small particles (F15) and puddles (F16) per investigated pixel over time. The images $ST4\_BIG\_IMG_n$, $ST4\_SMALL\_IMG_n$ and $ST4\_PUDDLE\_IMG_n$ are binarized according to the earlier determined threshold (section 4.2.2.1). The final feature values are computed by calculating the amount of pixels belonging to the respective particle type (taking a value of one) and dividing them by the total amount of investigated pixels (taking a value of zero):

$$F13 = \frac{1}{x\_max \cdot y\_max} \cdot \sum_{x=1,y=1}^{x\_max,y\_max} BIN(ST4\_BIG\_IMG_n, thresh) \qquad (63)$$

$$F14 = \frac{1}{x\_max \cdot y\_max} \cdot \sum_{x=1,y=1}^{x\_max,y\_max} BIN(ST4\_SMALL\_IMG_n, thresh) \qquad (64)$$

$$F15 = \frac{1}{x\_max \cdot y\_max} \cdot \sum_{x=1,y=1}^{x\_max,y\_max} BIN(ST4\_PUDLLE\_IMG_n, thresh) \qquad (65)$$

Exemplary fingerprint time series of features F13 - F15 are given in Figure 40 - Figure 42. They are segmented according to ST4_BIG, ST4_SMALL and ST4_PUDDLE.

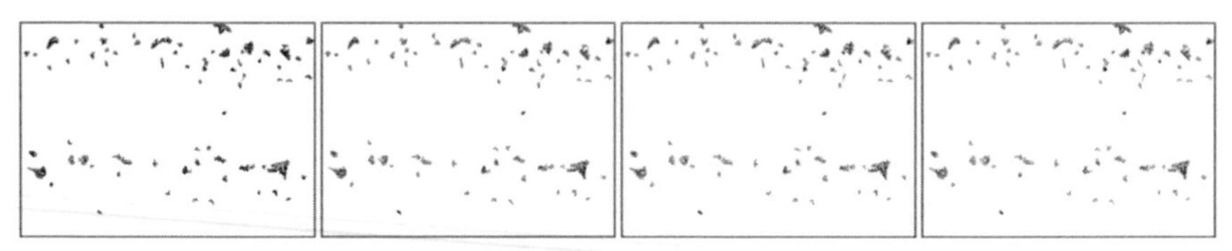

Figure 40: Visualization of fingerprint time series preprocessed according to ST4_BIG, depicted for an exemplary chosen CLSM intensity long-term aging series (from left to right: $t_1 = 1$ d, $t_2 = 11$ w, $t_3 = 20$ w, $t_4 = 30$ w). Pixels not belonging to big particles according to PM_BIG are marked in white.

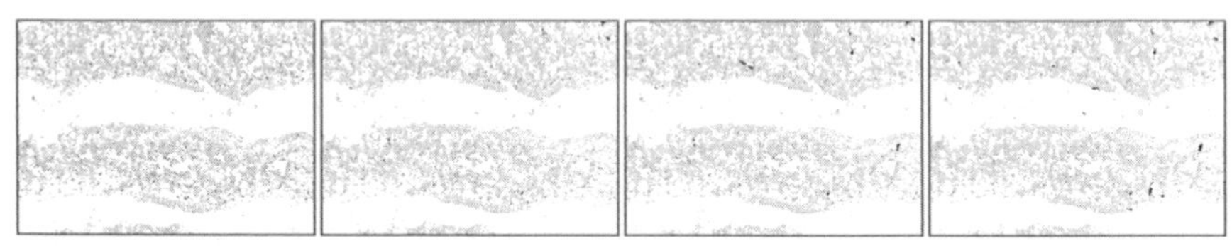

Figure 41: Visualization of fingerprint time series preprocessed according to ST4_SMALL, depicted for an exemplary chosen CLSM intensity long-term aging series (from left to right: $t_1 = 1$ d, $t_2 = 11$ w, $t_3 = 20$ w, $t_4 = 30$ w). Pixels not belonging to small particles according to PM_SMALL are marked in white.

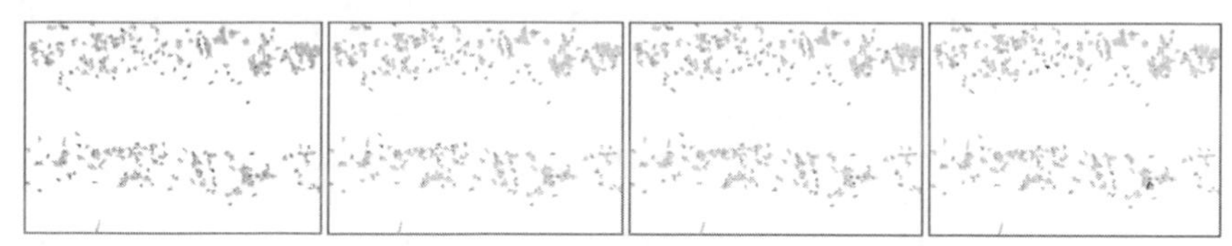

Figure 42: Visualization of fingerprint time series preprocessed according to ST4_PUDDLE, depicted for an exemplary chosen CLSM intensity long-term aging series (from left to right: t_1 = 1 d, t_2 = 11 w, t_3 = 20 w, t_4 = 30 w). Pixels not belonging to puddles according to PM_PUDDLE are marked in white.

The fingerprint time series of features F13 - F15 depicted in Figure 40 - Figure 42 seem to exhibit an increase in image brightness over time, possibly caused by particle pixels becoming brighter. However, a specific decrease in particle size cannot be observed. Therefore, pixels seem to disappear randomly within a structure, rather than at its edges.

4.3.3 Quality Measures for Digital Aging Feature Sets

All aging features designed in the scope of objective O3 for the short- as well as the long-term aging are subject to a quality evaluation in the later conducted experiments (test goal TG3, section 5.1.3). To evaluate the quality of a designed aging feature, objective quality measures need to be established. In the scope of this thesis, it is of main interest to investigate to which extent the computed aging curves of different features exhibit a characteristic, reproducible trend during aging. To successfully evaluate such property, it seems feasible to use regression [131]. The approximation of mathematical aging functions and the correlation of the experimental aging curves to such functions is considered an adequate, objective measure. To this extent, different characteristic aging trends should be considered. In this thesis, a linear, logarithmic and exponential trend seems to be appropriate for first studies. An aging property Pr = {lin,log,exp} is therefore defined as the type of aging curve which is approximated. Other shapes are also possible in future work, such as polynomial, moving average or power.

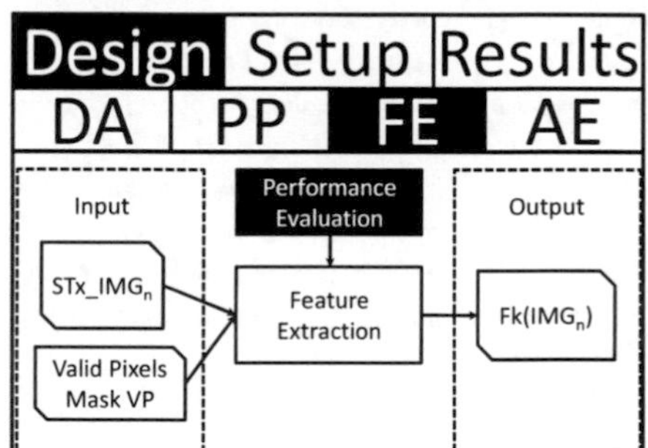

When using regression to approximate mathematical aging functions, three main variables seem to be of interest, which are the slope (parameter 'a') and offset (parameter 'b') of the approximated mathematical aging function $Fk_M(t)$ as well as the Pearson correlation coefficient r [149], describing the correlation between the experimental aging curve $Fk_E(t)$ and the approximated mathematical aging function $Fk_M(t)$. The approximated aging functions $Fk_M(t)$ have the following form, where Fk represents an investigated feature, M refers to a mathematical function (rather than an experimental aging curve E) and $Fk_M(t)$ describes the feature value at a certain point of time t:

$$Linear\ approximation: Fk_M(t) = a \cdot t + b \quad (66)$$

$$Logarithmic\ approximation: Fk_M(t) = a \cdot ln\,(t) + b \quad (67)$$

$$Exponential\ approximation: Fk_M(t) = Fk_M(0) \cdot e^{-a \cdot t} \quad (68)$$

For the quality evaluation of an experimental aging curve $Fk_E(t)$, its best-fitting mathematical aging function $Fk_M(t)$ is approximated using regression and the sum of squared errors. The approximation is done in respect to a certain aging property Pr and the correlation coefficient r describes how well $Fk_E(t)$ fits $Fk_M(t)$. In case of a high correlation, the approximated parameters 'a' and 'b' of the mathematical function $Fk_M(t)$ are characteristic for the aging behavior of the series. The parameter 'a' describes the slope of such function, i.e. the speed of fingerprint pixels becoming background pixels over time, also referred to as aging speed As. The parameter 'b' describes the y-offset of the function, e.g. indirectly representing the amount of fingerprint pixels present in the image at or shortly after deposition for certain features.

The Pearson correlation coefficient r is an adequate measure to describe how well an experimental aging curve $Fk_E(t)$ correlates to its approximated mathematical aging function $Fk_M(t)$. It can assume values in the interval [-1;1], where -1 represents a perfect correlation to a monotonic decreasing curve, 1 represents a perfect correlation to a monotonic increasing curve and 0 represents no correlation at all. Correlations are always given in respect to a certain aging property Pr. In the scope of this thesis, correlation coefficients of $r \geq 0.8$ and $r \leq -0.8$ are considered as exhibiting a strong increasing or decreasing correlation. The amount of experimental aging curves showing such strong correlation r in respect to a certain aging property Pr is regarded as an adequate quality measure for evaluating the quality of a feature. Only a high reproducibility of a characteristic aging property Pr seems to be a reliable indicator for feasible aging features. The average correlation coefficient r_avrg of experimental aging curves to a mathematical function $FK_M(t)$ as well as the relative frequency of experimental aging curves exhibiting a strong correlation $h(r \geq 0.8)$ or $h(r \leq -0.8)$ towards such function are therefore used as objective criteria for the quality evaluation and comparison of features. Further details about the quality evaluation of aging feature sets can be found in the experimental test setup (section 5.6.3).

4.4 Design of Digital Age Estimation Strategies (O4)

In the scope of research objective O4, digital age estimation strategies are designed. They can be applied to sequences of feature values from arbitrary devices, data types and aging periods. The detailed structure of sub-steps proposed for the age estimation step of the overall processing pipeline is depicted in Figure 43. As input, a sequence of consecutive feature values $Fk(IMG_n)$ is provided, which have been computed in the previous feature extraction step. The required amount of consecutive values is defined by the variable Ti and can start at any Image IMG_n of a time series ($n \leq n\_max - Ti + 1$). In addition to such sequence, the equidistant time offset Δt between two consecutive values is also provided. From the Ti feature values, an age estimation feature AEF is computed in the age estimation feature computation sub-step. This can be achieved by passing on a single unaltered value or by computing the changes occurring within the provided sequence, depending on the used parameter specification of this sub-step.

Design | Setup | Results
DA | PP | FE | AE

Prior to the computation of age estimation features AEF, a normalization of the experimental aging curves should be considered, because certain variations are expected for the slope (parameter 'a') of the curves as well as their y-offset (parameter 'b', section 4.3.3). However, in the scope of this thesis, a normalization of the experimental aging curves is not performed, due to the following three identified reasons:

- In a practical age estimation scheme, it might not be possible or reasonable to normalize fingerprint time series, because for an effective normalization, the first value of a time series is required. Such image is not present at a crime scene, because an arbitrary amount of time can have passed between fingerprint application and its age estimation by a forensic expert. Merely a normalization in respect to the first captured fingerprint image can be performed in practice, which seems to be rather unreliable considering the unknown state of the prints degradation.
- A wide variety of different influences on the aging process exist, such as the sweat composition, environmental conditions or surface properties. Changes in the speed of fingerprint degradation seem to be mainly caused by differences in such influences, rather than different starting values. Therefore, normalization would shift such differences rather than eliminating them. Preliminary investigations seem to confirm such argument, by not exhibiting an increased performance of the proposed age estimation schemes, even if normalized time series are used.
- While the y-offset (parameter 'b') seems to be subject to significant variations, the computation of the slope (parameter 'a') of an experimental curve excludes this influence and therefore the variation of the y-offset, effectively serving as a form of normalization.

In the scope of research objective O4, two main strategies for age estimation are designed, which are the formula-based age estimation as well as the machine-learning based classification into predefined age classes. Both strategies take one or more age estimation features AEF as input. The formula-based approach processes features AEF according to its specified formula and parameters, resulting in estimates of the absolute fingerprint ages t_est. The machine-learning based approach classifies age estimation features AEF into age classes Ac_est, based on a model trained prior to the age estimation (marked with dotted lines in Figure 43). The resulting estimates of both approaches are later examined towards their performance for the short-term aging (test goal TG4, section 5.1.4). The age estimates are evaluated in respect to the ground truth age t_real(IMG_n) of an AEF input sequence, which is defined as the time t_real(IMG_n) of the first feature value Fk(IMG_n) of the sequence.

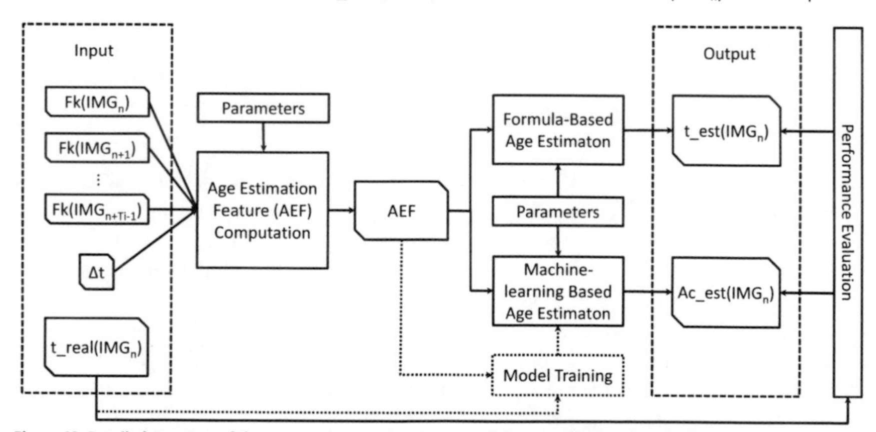

Figure 43: Detailed structure of the proposed age estimation step of the overall processing pipeline.

In section 4.4.1, four formula-based age estimation strategies are designed, followed by the proposal of a machine-learning based strategy in section 4.4.2, including a first concept for systematic parameter optimization and statistical significance determination. Quality measures for performance evaluation and comparison are introduced in section 4.4.3.

4.4.1 Formula-Based Age Estimation

A first approach for estimating the age of a latent fingerprint is considered to be a formula-based one. Because mathematical aging functions can be approximated from experimental aging curves using regression, such approach seems promising. The general idea of using a formula-based technique for age estimation has been published in the scope of this thesis in [79], proposing a first approach without presentation of specific experimental results of estimated ages. The idea is here extended to four different approaches and practically evaluated in the later conducted experiments of test goal TG4 (section 5.1.4). In the scope of formula-based age estimation, a sequence of age estimation features AEF is computed according to Figure 43, followed by the estimation of its absolute age t_est in the formula-based age estimation sub-step. The specific mathematical function to be used as well as its constant parameters 'a' and 'b' are provided by a parameter specification.

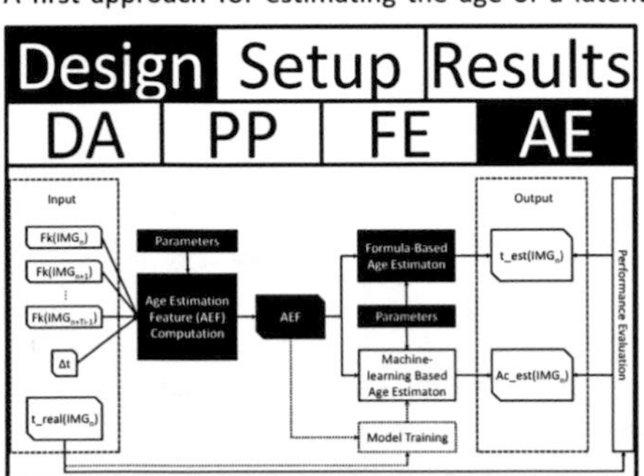

Formula-based age estimation based on prior knowledge of aging characteristics (Pk): For determining a fingerprints age using a formula-based approach, the earlier introduced aging property Pr = {lin,log,exp} defines the type of the used age computation formula. As is reported in the scope of the experimental results of test goal TG3 (section 6.3.1), a logarithmic short-term aging property seems to be dominant for most features of both devices and data types and is therefore used in the scope of O4 for the formula-based age estimation. An exponential property does not seem to adequately represent the occurring fingerprint degradation behavior, performing similar to a linear property (section 6.3.1). Using a linear aging property would furthermore exhibit the disadvantage of its first derivative being a constant and its second derivative being zero, preventing the use of derivatives for the age estimation.

Therefore, the aging property Pr = log is selected for designing formula-based age estimation schemes in the scope of O4. Using such property offers three main possibilities (referred to as feature representations Fr), similar to the later investigated machine-learning based classification: first, computed feature values can be used in their original form (Fr = 0); second, the slope of the experimental aging curve at the point of interest can be approximated and considered as the first derivative of the corresponding aging function at this point (Fr = 1); third, the slope of the slope values at a certain point of interest might be computed and can be considered as the second derivative of the corresponding aging function, representing the rate of change in the slope of the aging curve (Fr = 2). The corresponding mathematical functions can be formulated as follows:

$$Fr = 0:\ f(t) = a \cdot ln(t) + b \qquad (69)$$

$$Fr = 1\colon\ f'(t) = (a \cdot ln(t) + b)' = \frac{a}{t} \tag{70}$$

$$Fr = 2\colon\ f''(t) = \left(\frac{a}{t}\right)' = -\frac{a}{t^2} \tag{71}$$

When designing formula-based age estimation schemes, the values f(t), f'(t) and f''(t) of the mathematical aging functions are substituted with the computed age estimation features AEF. For Fr = 0, AEF is identical to the input feature value $Fk(IMG_n)$ and Ti = 1:

$$AEF(Fr = 0, n) = Fk(IMG_n) \tag{72}$$

For Fr = 1 and Fr = 2, AEF is be approximated from a certain amount Ti of consecutive feature values (using linear regression), to enhance the accuracy of the results and reduce the influence of outliers. Values of Ti = 3 and Ti = 5 are exemplary chosen for the here designed two feature representations, leading to the following computed AEFs (Reg refers to the linear regression method and outputs the slope of the approximated mathematical aging function):

$$AEF(Fr = 1, n) = Reg(FK\_IMG_n, FK\_IMG_{n+1}, FK\_IMG_{n+2}) \tag{73}$$

$$AEF(Fr = 2, n) = Reg(AEF(Fr = 1, n), AEF(Fr = 1, n + 1), AEF(Fr = 1, n + 2)) \tag{74}$$

As a result, the three designed schemes return the absolute estimated age $t\_est(IMG_n)$ of a fingerprint, whose accuracy can then be evaluated by a comparison to its real age $t\_real(IMG_n)$, provided as input.

A major limitation of formula-based age estimation using the formulae (69) - (71) can be seen in the requirement of knowing the parameter 'a' of the aging curves. In case of f(t), even the parameter 'b' is required. In the scope of the later conducted experiments (TG4, section 5.7.1), both parameters are averaged from all time series of a certain test set and used for the formulae (69) - (71). However, because the parameters 'a' and 'b' exhibit a significant variation due to the natural variability of print characteristics and aging behavior, using average values might deliver comparatively inaccurate results.

Formula-based age estimation without prior knowledge of aging characteristics (Npk): To avoid the limitation of using average values for the parameters 'a' and 'b' of the mathematical logarithmic aging function, such parameters can be computed directly from a given sequence of feature values. In this case, only a specific time series is considered, potentially leading to parameters of much greater accuracy. For that purpose, the following system of equations can be used for consecutive feature values f(t), f(t+Δt), f(t+2Δt):

$$f(t) = a \cdot ln(t) + b \tag{75}$$

$$f(t + \Delta t) = a \cdot ln(t + \Delta t) + b \tag{76}$$

$$f(t + 2 \cdot \Delta t) = a \cdot ln(t + 2 \cdot \Delta t) + b \tag{77}$$

The system of equations can be merged to the following form:

$$\frac{f(t + \Delta t) - f(t)}{f(t + 2 \cdot \Delta t) - f(t + \Delta t)} = \frac{ln\left(1 + \frac{\Delta t}{t}\right)}{ln\left(\frac{t + 2 \cdot \Delta t}{t + \Delta t}\right)} \tag{78}$$

It can be seen from formula (78) that when using three consecutive feature values (based on images captured with an equidistant time difference Δt), the parameters 'a' and 'b' are removed from the equation. The influence of an averaged parameter 'a' or 'b' is therefore excluded. Formula (78) can furthermore be rewritten as:

$$\frac{f(t+\Delta t)-f(t)}{f(t+2\cdot\Delta t)-f(t+\Delta t)}=\frac{\frac{f(t+\Delta t)-f(t)}{\Delta t}}{\frac{f(t+2\cdot\Delta t)-f(t+\Delta t)}{\Delta t}} \qquad (79)$$

The transformation of formula (78) into formula (79) shows that the given fraction can be interpreted as the ratio of two difference quotients and therefore as the approximated ratio of the two slopes f'(t) and f'(t+Δt):

$$\frac{f'(t)}{f'(t+\Delta t)}\approx\frac{ln\left(1+\frac{\Delta t}{t}\right)}{ln\left(\frac{t+2\cdot\Delta t}{t+\Delta t}\right)} \qquad (80)$$

Based on formula (80), the slope of an experimental aging curve at two consecutive points in time can be used for a time series specific age estimation. The values of f'(t) and f'(t+Δt) are substituted with the age estimation features AEF(Fr = 1,n) and AEF(Fr = 1,n+1), which are computed according to formula (73) using linear regression Reg (Ti = 3 is exemplary used). Applying such parameters, formula (80) is solved computationally by a systematic variation of the estimated time t_est in steps of Δt = 0.01 h between t_est_1 = 0.01 h and t_est_{max} = 1000 h. If the curve does not produce a valid result within this time period, the computation is aborted. If terminated within the given range of t_est, the algorithm returns the estimated absolute age $t\_est(IMG_n)$ of a fingerprint, whose accuracy can be evaluated using its real age $t\_real(IMG_n)$.

The main limitation of such approach can be seen in the nature of a logarithmic function, where small changes in the feature values or derivatives might lead to significant changes in the estimated age. This applies especially to prints of higher ages, because small changes in the later progression of a logarithmic curve lead to significant changes in its estimated earlier progression. Therefore, the older a fingerprint sample is, the more unreliable its approximated age seems to be.

4.4.2 Machine-Learning Based Age Estimation

In this section, a method for fingerprint age estimation using machine-learning based classification is designed. Furthermore, a first concept for parameter optimization as well as statistical significance determination is proposed. Machine-learning based classification is a widely used pattern recognition technique, which can be utilized using different methods (please refer to [132] for further reading). First studies on this issue have been published in the scope of this thesis in [79] and [153], based on obsolete pre-processing and age estimation techniques and without performing a parameter optimization

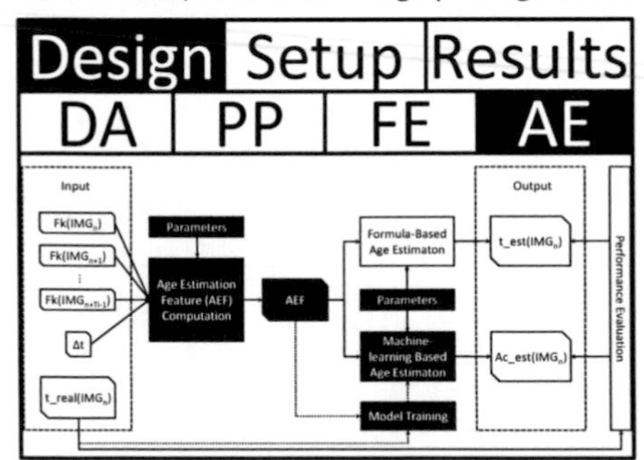

and statistical significance determination. An extended approach is therefore designed here for a better performance and higher reliability of the scheme.

In the scope of machine-learning based age estimation, a training phase is conducted (model training sub-step of Figure 43), in which a model is trained from a set of age estimation features AEF and their corresponding ground truth ages t_real(IMG_n), which are mapped to well defined age classes Ac. In the scope of this thesis, only two-class problems are considered, specifying if a fingerprint is younger as or older than a certain time threshold Tt. Furthermore, a classifier Cl is defined, describing the methods used for model creation and classification. The model training phase is depicted in dotted lines in Figure 43, because it is conducted prior to the main age estimation.

After the model is sufficiently trained, age estimation can be performed. Age estimation features AEF are computed according to Figure 43 and are provided to the machine-learning based age estimation sub-step. Based on the trained model, they are classified into one of the earlier defined age classes Ac. The correctness of the assignment of a fingerprint image IMG_n to an age class Ac_est(IMG_n) can be evaluated by a comparison to its ground truth age t_real(IMG_n) in the performance evaluation sub-step (Figure 43).

In the scope of machine-learning based age estimation, several parameters need to be specified and optimized, for the age estimation feature AEF computation as well as the model training sub-step. In the scope of this thesis, the following six parameter are of interest, which need to be optimized in respect to both investigated capturing devices and data types as well as different features Fk:

- **Amount of temporal images required for an optimal age estimation (Ti):** In the scope of the age estimation feature AEF computation sub-step (Figure 43), it is of importance to determine the amount of temporal images Ti (supporting points) required for an optimal computation of AEF (Ti = 1..n_max). An increasing amount of supporting points might help to exclude outliers caused by noise. However, if too many supporting points are used, the resulting AEFs lose their characteristic differences. The amount of temporally captured images Ti directly determines the capturing time required for a practical age estimation scheme.
- **Feature Representation (Fr):** The feature representation Fr = {0,1,2} is a parameter used in the scope of the age estimation feature AEF computation sub-step (Figure 43). Age estimation features AEF can be computed by merely copying the feature value Fk of an image (Fr = 0; Ti = 1) or by computing the slope of the regression curve formed by a sequence of feature values (Fr = 1; approximating the first derivative of the partial aging curve created by the sequence of feature values; Ti = 2...n_max). Even the change of this regression line can be computed, by performing an additional regression on the regression line (Fr = 2; approximating the second derivative of the partial aging curve represented by the sequence of feature values; Ti = 3...n_max). However, the second derivative can only be applied if a logarithmic approximation is used, because it is always zero for a linear approximation (because the slope of a linear curve is similar at all points in time).
- **Linear vs. logarithmic regression (Re):** In the scope of the age estimation feature AEF computation sub-step (Figure 43), different regression types Re can be used for the slope approximation, such as linear or logarithmic regression (Re = {lin,log}). An exponential approximation is not investigated here, because such property does not seem to be adequate to represent the aging tendencies and mostly performs similar to a linear property (section 6.3.1). The regression type is only applicable for feature representations Fr = 1 and Fr = 2.

- **Determination and evaluation of adequate age classes using time thresholds (Tt):** For a performance evaluation of the machine-learning based age estimation, suitable parameters need to be specified for the model training sub-step (Figure 43). The definition of adequate age classes Ac is one of these parameters. Only basic two-class problems are investigated here, specifying if a fingerprint is younger as or older than a certain time threshold Tt; 1 h ≤ Tt ≤ 24 h - Ti. Depending on such threshold Tt, features are separated into age classes Ac1 = [0,Tt] and Ac2 = [Tt,24h-Ti]. The here defined aging classes can also be extended in future work to more complex classes. In the scope of this thesis, the time threshold Tt is varied in steps of one hour {Tt = {1h,2h,...,24h - Ti), to allow for a comparison of the classification performance for varying pairs of age classes, such as ([0h,1h][1h,24h-Ti]), ([0h,2h][2h,24h-Ti]) up to ([0h,24h-Ti][24h-Ti,24h-Ti+1]).
- **Performance of different classifiers (Cl):** In the scope of model training (Figure 43), numerous classifiers are available for machine-learning based classification. Certain classifiers might be more suitable for the purpose of age estimation than others, leading to differences in their performance. In this work, four well known classifiers Cl = {NaiveBayes,J48,LMT,RotationForest} from the machine-learning toolbox WEKA [133] are exemplary chosen for evaluation.
- **Statistically significant amount of samples (Sas):** In the scope of model training (Figure 43), a certain amount of AEFs is assigned to each age class. It is important to show that for each investigated value of Tt, a statistically significant amount of training samples is present (Sas = true). This can be done empirically, by determining the learning curve of a classifier Cl. If it can be shown for each investigated Tt that the learning curve of the classifier has converged to a certain degree to its optimal value, the results can be considered as being of statistical significance. Otherwise, Sas = false indicates that the results are not statistically reliable.

In the scope of the conducted experiments, these six different parameters of interest are combined in the form of a configuration Conf. A suitable configuration needs to be determined by optimization and selection of the single parameters, which cannot be evaluated simultaneously:

$$Conf = (Ti, Fr, Re, Tt, Cl, Sas) \qquad (81)$$

For designing concepts for parameter optimization, statistical significance determination as well as age estimation performance evaluation, the model training and machine-learning based age estimation sub-steps are not a strictly sequential process (as might be concluded from Figure 43), but rather an iterative execution of several training and age estimation rounds. They are therefore visualized in more detail in Figure 44. At first, a captured test set is randomized. During such process, only complete time series are shuffled, while age estimation features AES within a time series remain in order. Such randomization is performed to equally distribute certain systematic influences during the capturing process (e.g. seasonal changes in temperature) over the complete test set.

After randomization, the first 80% of time series within the test set are cropped and used as model training set TRS, while the remaining 20% are used as performance evaluation set PES. The training set is split in a way to only include complete time series into TRS and PES, to assure that TRS and PES are independent of each other. For both sets TRS and PES, samples are divided into the two age classes Ac1 and Ac2, according to the specified time threshold Tt. The total number of AEF samples in the two age classes of TRS is equalized by determining the bigger class and randomly deleting samples until both age classes are of equal size. Such equal size is necessary to create a non-biased training model. The performance evaluation set PES remains unaltered, because the computed kappa quality measure accounts by design for different sizes of PES classes.

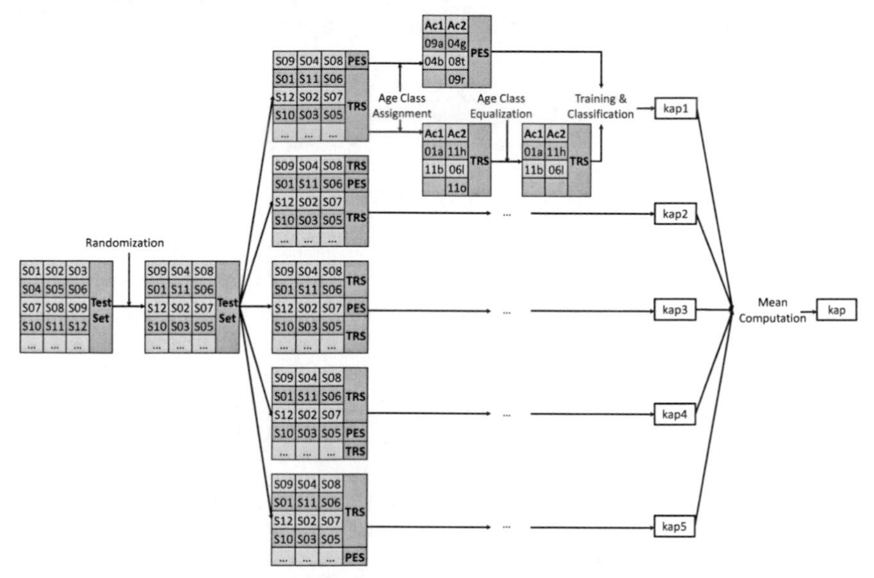

Figure 44: Model training and machine-learning based age estimation sub-steps of the proposed machine-learning based age estimation scheme. Both sub-steps are conducted in five rounds to compute the final classification performance kap (also known as five-fold cross validation). S01 - S012: exemplary fingerprint time series; TRS: training set; PES: performance evaluation set; 01a - 01z: exemplary consecutive age estimation features a - z of time series S01; Ac: age class; kap: classification performance kappa.

The classification model is then trained using TRS and its performance is evaluated using PES, by computing the kappa quality measure for the different age classes. The complete process of dividing the test set into TRS and PES is repeated five times, whereas for every round, the 20% of AEFs used for PES are disjunct from those of the other rounds. The kappa performance of these five rounds is averaged, resulting in the final kappa value kap, which is returned. Such test setup therefore realizes a five-fold cross-validation with the special constraint that only complete time series are mapped to TSR and PES.

In the scope of research objective O4, a concept is proposed for parameter optimization, statistical significance determination and age estimation performance evaluation using a machine-learning based classification approach. The concept is comprised of four different phases, where parameters Re and Fr are optimized in the first phase, followed by an optimization of parameters Ti and Cl in the second phase. In the third phase, the statistical significance Sas of a given test set is examined, while the fourth phase is designed to evaluate the performance of the machine-learning based age estimation scheme. The four phases are practically evaluated in the scope of test goal TG4 (section 5.1.4), according to Figure 44.

Optimization of parameters Re and Fr (Op1): In a first phase, the approximation type Re as well as the feature representation Fr are optimized. Both parameters are required in the scope of the age estimation feature AEF computation sub-step. For that purpose, the parameters Ti, Cl and Tt are set to fixed values, while Re and Fr are systematically varied (section 5.7.2). According to the age estimation performance of both devices and data types in respect to selected features Fk, optimal values O_Re and O_Fr are selected, which are then applied to all further optimizations.

Optimization of parameters Ti and Cl (Op2): After O_Re and O_Fr have been selected, the required amount of temporal images Ti (age estimation feature AEF computation sub-step) as well as the used classifier Cl (model training sub-step) are optimized in the second phase. Both parameters are systematically varied and need to be investigated for all considered time thresholds Tt, because they can vary for different age classes (section 5.7.2). In respected to the achieved age estimation performance of both capturing devices and data types and in respect to selected features Fk, optimal values O_Ti and O_Cl are selected, which are then applied to all further investigations.

Statistical significance determination (Ss): To show that the amount of AEFs computed for a certain test set is of statistical significance and therefore sufficient, the parameter Sas is investigated in the third phase. Only two different values Sas = {true,false} can be assumed, describing if the used test set is of statistical significance. All considered variations of time thresholds Tt need to be investigated, because different amounts of AEFs are available for different values of Tt (section 5.7.2). Furthermore, the age estimation performance needs to be evaluated for both capturing devices and data types in respect to all investigated features, because different features might require different amounts of samples for statistically significant results.

Age estimation performance evaluation (Ap): Having optimized the parameters Re, Fr, Ti and Cl as well as having shown the statistical significance Sas = true for a certain test set, the performance of the machine-learning based age estimation scheme can be evaluated in the fourth phase. Performance evaluation should consider different capturing devices, data types and aging features, as well as a combination of different features. (section 5.7.2).

4.4.3 Quality Measures for Digital Age Estimation Strategies

For evaluating the quality of age estimation strategies in the scope of research objective O4, different measures are proposed. For the formula-based age estimation, the estimated absolute age of a latent print t_est is computed. For evaluating the correctness of such age, the absolute time difference t_diff = |t_est - t_real| between the estimated age of a fingerprint t_est and its corresponding real age t_real (known from the experimental test setup) is regarded as a feasible measure. In the scope of machine-learning based age estimation, prints are classified into well-specified age classes Ac_est. Because the general classification accuracy (relative amount of correctly classified samples) might be subject to distortions caused by an unequal distribution of classes within the performance evaluation set PES, Cohens kappa [150] is proposed here as general quality measure. The kappa value kap considers the amount of AEFs in each age class of PES and therefore eliminates possible distortions of the classification accuracy. A kappa value of kap = 0 represents pure guessing of the trained model and kap = 1 represents perfect classification accuracy. For further reading about kappa please refer to [150]. The kappa quality measure is used during all four phases of the designed machine-learning based age estimation scheme, including parameter optimization, statistical significance determination as well as age estimation performance evaluation. Further details about the quality evaluation of age estimation strategies can be found in the experimental test setup (section 5.7.3).

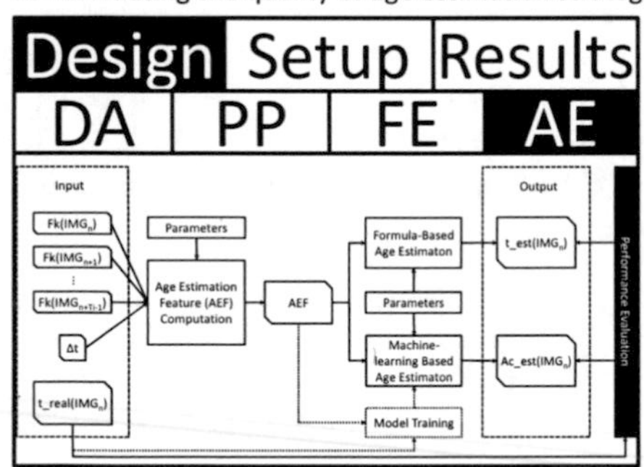

4.5 Modeling Influences on Fingerprint Aging (O5)

In section 2.4, different influences have been summarized from the literature, which have been reported to potentially influence the fingerprint aging process. However, their impact needs to be re-evaluated for the here investigated digital fingerprint time series of the non-invasive capturing devices, which is the aim of research objective O5. A re-evaluation of these influences seems to be necessary due to the following reasons:

- The reported influences on the aging process summarized in section 2.4 are mostly based on first qualitative studies rather than quantitative ones.
- The investigated time periods differ and often focus on long-term aging greater than 24 hours.
- The significantly higher resolution of the fingerprint images captured in the scope of this thesis provides the opportunity for a more detailed investigation of certain influences and the extraction of additional findings, only possible using high-resolution images.
- Statistical features are investigated in this thesis as a promising alternative to morphology based ones, which might be more robust towards certain influences than fingerprint morphology.
- Additional influences from the capturing devices seem to exist, potentially impacting the quality and accuracy of the captured time series, such as various kinds of distortions as well as possible internal (proprietary) preprocessing mechanisms not specified in the general sensor description.

For the given reasons, a re-evaluation of influences on the captured fingerprint aging series seems to be necessary. While this thesis provides only a starting point into such investigations by first qualitative studies on selected influences as well as the design of a first influence model, quantitative investigations using large scale test sets represent an important task for future research.

Influences are categorized in the scope of research objective O5 into the five groups of scan influences, environmental influences, sweat influences, surface influences and application influences. First qualitative evaluations of selected influences from these groups are designed in sections 4.5.1 to 4.5.4. While the investigations of section 4.5.1 have been published in the scope of this thesis and are therefore only briefly summarized and discussed here, additional evaluations are designed in sections 4.5.2 to 4.5.4, in respected to the presumably most relevant influences. Section 4.5.5 presents the quality measures used for examination and comparison.

4.5.1 Selected Influences from own Published CWL Studies

Several qualitative experiments have been conducted in the scope of this thesis to identify important as well as negligible influences on the aging behavior of non-invasively captured and digitally processed fingerprint time series. The investigations are based on the CWL device, which has been available for 3.5 years, while the CLSM device has only been available for 1.5 years. They are published in [95], [153], [154], [155], [156], [157], [158], [159] and are only briefly discussed here.

Scan influences are studied in [95], where the Binary Pixels feature (F1) is applied to time series of a common flat-bed scanner. Furthermore, the influence of different resolutions and measured area sizes on the Binary Pixels feature are investigated in [153], [156], [157] and [159]. The findings of these studies are summarized in section 6.5.1.3. **Environmental influences** of different ambient temperatures, relative humidity, UV radiation as well as wind on the aging behavior of latent prints are studied in [153] and [154]. The findings of these studies are summarized in section 6.5.1.2.

Studies published in [153] and [154] investigate different **sweat influences**, by evaluating the aging behavior of fingerprints after different types of sweating, e.g. prints of eccrine, sebaceous and sweat from daily work, prints after sweating under ambient high and low temperatures and after sports as well as prints after the consumption of alcohol. Furthermore, the aging of prints contaminated with skin-lotion and cooking oil is studied [155]. The aging behavior of a single droplet of water, skin-lotion and oil is compared in [153] and [154]. Further influences of the sweat composition of different donors on the degradation behavior of latent prints is studied in the scope of [153], where the aging behavior of prints from males and females is compared and the inter-/intra-person as well as inter-/intra-finger variation is investigated. The findings of these studies are summarized in section 6.5.1.3.

Surface influences of ten different substrates (car door, furniture, veneer, CD-case, brushed scissor blades, 5 Euro-Cent coin, socket cap, hard disk platter, mobile phone display and glass) on the aging of latent fingerprints are investigated in [153] and [158]. Furthermore, changes in size and amount of corrosion artifacts of fingerprints on copper coins fresh from the mint are studied in [157]. The findings of these studies are summarized in section 6.5.1.4 In [153] and [155], **application influences** of different contact times, contact pressures as well as smearing of a fingerprint are investigated. The findings of these studies are summarized in section 6.5.1.4.

For further information about the design of the various published studies, the reader may refer to [95], [153], [154], [155], [156], [157], [158] and [159]. The studies are based on CWL short-term aging periods, using the obsolete preprocessing techniques described in [154]. Although they cannot achieve the accuracy and reliability of the studies designed in sections 4.5.2 to 4.5.4 based on the preprocessing and feature extraction techniques of O2 and O3, they provide first qualitative results about potential influences on the short-term aging of latent prints.

In the scope of this thesis, the aim is not to re-evaluate such published experiments, but rather to conduct additional ones including both the CWL and the CLSM device. However, the findings of these earlier conducted studies are of major relevance towards the creation of a first qualitative influence model designed in the scope of test goal TG5 (section 6.5.5). They are therefore summarized in section 6.5.1 and are included into the created influence model.

4.5.2 Selected Scan Influences

In the scope of this thesis, several distortions from the used capturing devices are observed, which are addressed in the scope of the designed preprocessing methods (O2). However, it has not been investigated so far if the characteristic logarithmic aging property observed in section 6.3 for the best performing aging features is an exclusive consequence of fingerprint degradation, or if it is partially overlaid by certain characteristics of the capturing devices. For example, such characteristics can be caused by an increase in temperature after starting a capturing device until the operational temperature is reached. During such increase in temperature, changes in the optical systems of the devices might occur, leading to systematic changes in the brightness of the captured fingerprint images. Furthermore, the non-invasiveness of the devices is defined in section 3.2.1 as a basic assumption of this thesis. It is of interests to evaluate such non-invasiveness in the scope of an investigation, especially towards the influence of the radiation energy emitted by the light source.

To address these two issues, an evaluation is designed in the scope of O5 for both capturing devices. The study is limited to intensity images, because topography data seems to be of inadequate quality for reliable results in this scope. The main idea is to capture a latent print within three different time periods Tp1 - Tp3 (three different ages), while the capturing device is halted in between and started separately for each time period. If a significant decrease in the aging speed As can be observed between Tp1 and Tp2 as well as between Tp2 and Tp3, it can be assumed that the aging speed is indeed caused by fingerprint degradation. In case the measured speed is similar for Tp1, Tp2 and Tp3, the characteristic aging property is caused by an influence related to the starting of the capturing device and not a result of fingerprint degradation.

Furthermore, three different areas Ar1 - Ar3 of a single print (which are assumed to exhibit a comparatively similar aging speed) are exposed to different amounts of illumination by the light source. In case the observed aging speed is similar, the non-invasiveness of the capturing device is confirmed in respect to the emitted radiation. However, if a significant difference in the aging speeds of the three fingerprint areas is computed, correlating to the amount of illumination, an influence of the light source on the aging behavior of the prints can be assumed.

In the scope of test goal TG5 (section 5.1.5), the here designed investigation is transferred into a specific test setup and is experimentally evaluated. The two issues are combined in the form of six different time series Ts1 - Ts6, captured in three different time periods Tp1 - Tp3 and from three different areas Ar1 - Ar3 of a latent print. The specific test setup is introduced in section 5.8.2.

4.5.3 Selected Environmental Influences

In the study of [154] published in the scope of this thesis, the environmental influences of ambient temperature, relative humidity, UV radiation as well as wind are investigated. However, the experiments are based on a limited test set (three fingers of one donor for each investigated variation). Parts of the provided results are inconclusive, however indicating an influence of the examined climatic conditions. Furthermore, only CWL time series are evaluated, without studying the influence of ambient conditions on fingerprint time series captured with a CLSM device. Also, the studies are based on obsolete preprocessing methods, which seem to be less reliable than the methods designed in the scope of O2.

For these reasons, an investigation is designed here to re-evaluate the influence of ambient climatic conditions on the aging behavior of latent fingerprints in an extended scenario. While CWL as well as CLSM aging time series are included, an increased amount of fingerprints as well as the improved preprocessing methods designed in O2 are used. For the evaluation, the influences of UV light and wind are excluded, because they are regarded as less relevant for the here considered indoor scenario (section 3.2.1).

Evaluating the influence of ambient temperature and relative ambient humidity is the main aim of designing such investigation. While one influence is kept constant, the other one is systematically altered. Studying the fingerprint aging under such systematic variations, conclusions might be derived about the increase or decrease of the aging speed in respect to changing ambient conditions. However, temperature and relative humidity are interdependent. For example, when the ambient temperature increases, its water intake capacity does also increase. Therefore, the ambient relative humidity decreases. To achieve a reliable evaluation, an adaptation of the relative humidity needs to be performed in respect to the alterations of temperature and vice versa.

4.5.4 Selected Sweat Influences

In general, the studied CWL and CLSM devices are not able to capture substance-specific information of a trace. To still allow for the evaluation of the sweat influence on the aging of latent prints, different types of sweat as well as fingerprints after various sweating conditions are investigated in the study of [154], conducted in the scope of this thesis. However, the achieved results are inconclusive and no clear dependencies could be identified. Furthermore, the studies face similar restrictions than the environmental studies discussed in section 4.5.3 of comprising a small test set, being limited to obsolete preprocessing methods as well as not including CLSM studies. Therefore, an extended investigation is designed here to further evaluate the influence of different substances on the aging of fingerprints, based on an increased test set size, the advanced preprocessing techniques designed in O2 as well as both capturing devices.

The published study of [155] concludes that a significant difference of the aging speed seems to exist between substances containing fat and those containing water (fingerprints contaminated with skin-lotion and cooking oil are used in the study). Here, an experiment is designed to investigate the aging behavior of different water-based substances. The idea of printing artificial sweat with common ink-jet printers for achieving reproducible and comparable fingerprints was introduced by Schwartz in [161] and has been further investigated by Dittmann et al. in [162]. The approach is limited to non-fatty substances at this point of time, because fatty compounds seem to block the nozzles of such printer and therefore prevent a reliable printing process.

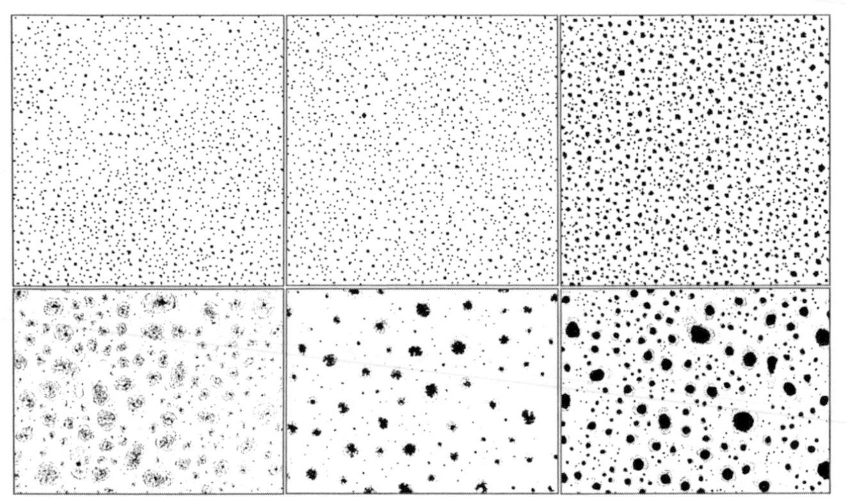

Figure 45: Exemplary images of the three investigated, printed substances (from left to right): distilled water DW, tap water TW, artificial sweat AS. CWL intensity images directly after printing are depicted in the upper row, CLSM intensity images directly after printing are depicted in the lower row.

For the here designed approach of investigating water-based substances, the usage of an inkjet printer is applicable. The three exemplary substances of distilled water DW, tap water TW and artificial sweat AS are printed onto a surface (Figure 45) and are evaluated towards their aging speed. Such investigation might provide findings about the difference in degradation behavior between these exemplary substances. However, such limited evaluation cannot replace a more detailed analysis of substance-specific aging behavior with more applicable devices in future work.

It can be observed from Figure 45 that the substance particles printed by the inkjet printer appear differently for the CWL images (upper row) than for the CLSM images (lower row). While for the CWL images, particles are represented mainly by small dark dots (which are significantly stronger for artificial sweat), substance particles captured by the CLSM device seem to form gray areas around a dark center point (with an increasing size and darkness of the center point from distilled water to artificial sweat). Therefore, the characteristic shape of the particles seems to be better represented by CLSM images. However, the binarization methods introduced in section 4.2.2 can successfully separate substance from background pixels, therefore allowing for the application of similar preprocessing steps than for real fingerprint images.

4.5.5 Quality Measures for Influences on Fingerprint Aging

For an objective evaluation and comparison of different influences on the fingerprint degradation, the aging speed As is used. Such speed is represented by the slope of an experimental aging curve and is equal to the parameter 'a' of the approximated regression function (section 4.3.3). According to the findings of section 6.3.1, a logarithmic aging property is dominant for most aging curves. Therefore, the parameter 'a' is computed in respect to a logarithmic function and is regarded as an adequate quality measure.

For the evaluation of influence studies published in the scope of this thesis, the earlier introduced Pearson correlation coefficient r is also used, describing how well an experimental aging curve correlates to its corresponding (logarithmic) aging function. Further details about the quality evaluation of different influences can be found in the experimental test setup (section 5.8.5).

5 Experimental Test Setup

Design Setup Results

In this chapter, the test setup of the experiments conducted in this thesis is specified. The earlier defined five research objectives are transferred into experimental test goals in section 5.1. Afterwards, a description of the software methods implementing the pipeline of chapter 4 is given in section 5.2. The general test setup is introduced in section 5.3, followed by a description of the test goal specific setups and their respective quality measures in sections 5.4 to 5.8.

To allow for a comprehensive discussion with other experts in the field and to evaluate the general acceptability of the proposed approach in the scientific community, parts of the here presented concepts and results have been previously published, which is a common practice for a dissertation project in the field of computer science. The corresponding publications are:

2013

- [151] presented at the 1st ACM Workshop on Information Hiding and Multimedia Security (IH&MMSEC), Montpellier, France, June 17th - 19th, 2013

2012

- [153] published in the journal Forensic Science International, Elsevier, Volume 222, Issues 1 - 3, pp. 52 - 70, October 10th, 2012
- [95] presented at the 14th ACM Workshop on Multimedia and Security (MMSEC), Coventry, UK, September 6th - 7th, 2012
- [154] presented at the SPIE conference on Optics, Photonics, and Digital Technologies for Multimedia Applications II, Brussels, Belgium, June 1st, 2012

2011

- [155] presented at the 2011 IEEE International Workshop on Information Forensics and Security (WIFS), Foz do Iguacu, Brazil, November 29th - December 2nd, 2011
- [156] presented at the 12th Joint IFIP TC6 and TC11 Conference on Communications and Multimedia Security (CMS), Ghent, Belgium, October 19th - 21st, 2011
- [157] presented at the SPIE conference on Optics and Photonics for Counterterrorism and Crime Fighting VII, Optical Materials in Defence Systems Technology VIII and Quantum-Physics-based Information Security, Prague, Czech Republic, October 13th, 2011
- [158] presented at the 13th ACM Workshop on Multimedia and Security (MMSEC), Niagara Falls, New York, USA, September 29th - 30th, 2011
- [159] presented at the 7th International Symposium on Image and Signal Processing and Analysis (ISPA 2011), Dubrovnik, Croatia, September, 4th - 6th, 2011

5.1 Test Goals

In this section, the five research objectives O1 - O5 are mapped to adequate test goals TG1 - TG5 for the experimental investigations. Because it is not possible in the limited scope of this thesis to evaluate each aspect of the designed processing pipeline in detail, selected issues are examined in the scope of TG1 -TG5. They are chosen in respect to their estimated importance for the overall age estimation performance and are introduced in the following sections.

5.1.1 Test Goal 1 (TG1): Selection of Basic Sensor Settings

Test goal TG1 aims at the selection of basic sensor settings, which is a part of the research objective O1. As discussed in section 3.1.6, the comprehensive investigation of the capturing devices and their properties is not the aim of this thesis. However, adequate settings of basic capturing parameters are required for the creation of reliable fingerprint time series. Especially the lower bounds of measured area size and capturing resolution are of great importance to an accurate observation of fingerprint degradation behavior, maintaining a certain image quality while maximizing the capturing speed and therefore the amount of investigated time series. Defining minimum requirements of these two parameters for both sensors is therefore the aim of test goal TG1.

5.1.2 Test Goal 2 (TG2): Selection of Digital Fingerprint Binarization Techniques

In test goal TG2, the aim is to experimentally investigate selected aspects of image preprocessing methods, designed in the scope of the research objective O2. The selection of optimal binarization methods seems to be most important for a successful preprocessing of fingerprints, because most proposed preprocessing techniques are based on the background area (background mask), computed from the binarized images. The main aim of TG2 is therefore to select the best performing binarization method for each capturing device and data type, to allow for an optimal time series preprocessing. Other aspects of the preprocessing step of the overall pipeline are also of interest, but remain subject to future research, such as investigating to which extent the error introduced by image binarization is propagated, increased or decreased throughout the other preprocessing substeps.

5.1.3 Test Goal 3 (TG3): Evaluation of Digital Aging Feature Sets

In the scope of the third research objective (O3), statistical as well as morphological feature sets have been proposed for measuring fingerprint degradation behavior of short and long aging periods. Consequently, the main aim of test goal TG3 is the empirical evaluation and comparison of the performance of such features, including a discussion of their usability for a potential age estimation scheme. Such evaluation has to be conducted in respect to each capturing device and data type.

5.1.4 Test Goal 4 (TG4): Evaluation of Digital Short-Term Age Estimation Strategies

Different age estimation strategies based on formulae as well as machine-learning are designed in the scope of research objective O4. The performance evaluation of these strategies is therefore the main aim of test goal TG4. However, the strategies are based on certain parameters, some of which require optimization prior to experimental age estimation. Such parameter optimization is therefore another aim of TG4. Age estimation performance is evaluated for both capturing devices, data types and in respect to the earlier defined features.

For statistically reliable results of the age estimation performance, a significant amount of test samples is required. Therefore, another aim of TG4 is to show the statistical significance of the used test sets and to evaluate the minimum amount of required samples for certain age classes. Because statistical significance cannot be provided for the limited amount of captured long-term aging series, only short-term aging periods are considered in the scope of TG4, leaving the experimental evaluation of long-term age estimation a subject to future research.

5.1.5 Test Goal 5 (TG5): Evaluation of Selected Short-Term Influences and Model Creation

Research objective O5 aims at understanding short-term influences on the fingerprint aging behavior. A comprehensive, quantitative investigation of the very complex set of potential influences is not possible in the limited scope of this thesis. Test goal TG5 therefore aims at the qualitative investigation of selected influences, which seem to have the greatest impact. The experimental findings of these investigations are included into a first qualitative influence model, which can be used as basis for future investigations. In the scope of test goal TG5, only the intensity images of both capturing devices are evaluated, because they are of higher quality and seem to be less sensitive to distortions. Furthermore, only short-term aging periods are investigated, because the systematic evaluation of long-term influences requires significantly higher capturing capacities as well as climatic chambers. However, long-term aging periods are also of interest in future work.

5.2 Software Implementation

Transforming the designed steps of the processing pipeline into a software implementation is an important prerequisite for the experimental evaluation of test goals TG1 - TG5. In Figure 46, an overview is given over the used digital modules. Certain modules have been implemented earlier and are re-used here (dashed boxes), while others are first implemented in this work (continuous boxes).

For the capture of fingerprint images in the data acquisition step, the proprietary software of the sensor manufacturers is used. For the CWL sensor, the provided software is named "Acquire [163]." Its CLSM complement is called "Observation Application [84]." Both software packages provide basic functionally for setting up the major capturing parameters, without support for the automated acquisition of time series (repeated scans). To allow for the consecutive automated capture of time series, basic tools have been implemented by Hildebrandt [164], [165] using Windows Script Host [166] and C# [167]. The tools extend the functionality of the proprietary manufacturer software to include time series related capturing tasks. Because the provided manufacturers software is proprietary, it is furthermore unknown to which extent internal, undocumented preprocessing of the captured data is performed (e.g. internal contrast enhancement), which is an important aspect in the scope of future research on the capturing devices.

After latent print digitization, images are processed according to the proposed preprocessing and feature extraction steps (sections 4.2 and 4.3). A general framework QT-FPX has been designed by Leich [168] in the scope of the Digi-Dak project [5]. It is based on the Qt framework [169], C++ [170]

and OpenCV [104]. The framework provides input and output as well as batch-processing routines and a basic framework for the visualization and combination of different image filters. In the scope of this thesis, the designed preprocessing and feature extraction steps are embedded into such framework, together with different performance evaluation routines.

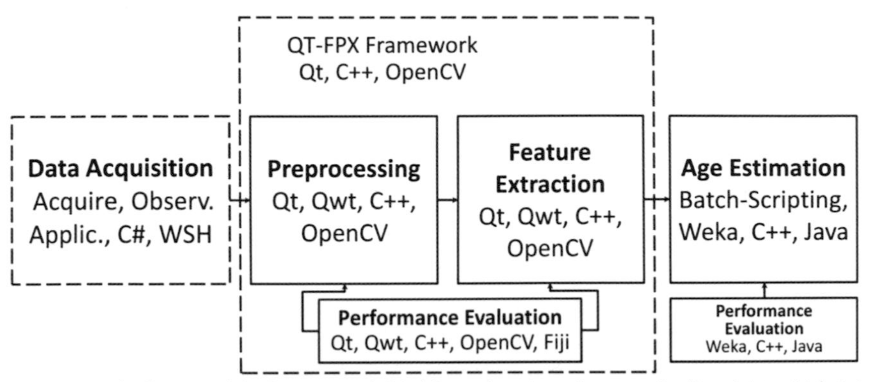

Figure 46: Used software modules for automated digital fingerprint aging series processing (from left to right): Data Acquisition based on the proprietary Acquire (CWL, [163]) and Observation Application (CLSM, [84]) software, extended for the capture of time series in [164] and [165] (based on Windows Script Host [166] and C# [167]); implemented preprocessing and feature extraction steps as described in sections 4.2 and 4.3, embedded into the QT-FPX framework [168] and respective evaluation using designed performance evaluation methods (based on Qt [169], Qwt [171], C++ [170], OpenCV [104] and Fiji [106], [107]); implemented age estimation and performance evaluation using the toolbox Weka [133] in combination with additional methods based on C++ [170] and Java [172].

The age estimation step is realized according to its design (section 4.4) for formula-based as well as machine-learning based strategies. Furthermore, performance evaluation routines are implemented. Both modules are realized outside the QT-FPX framework, by combining several designed methods with the well known machine-learning toolbox Weka [133]. Implemented routines include data preparation, age estimation feature AEF computation, formula-based age estimation as well as result evaluation. Used languages comprise Batch-Scripting, C++ and Java.

5.3 General Test Setup

In this section, the general test setup is introduced, where everything is regarded as general, which is similar for the conducted experiments of all test goals. Experiment-specific settings (potentially including minor deviations from the general setup) are introduced separately for each test goal in the following sections.

Surface: In the scope of test goals TG1 - TG5, hard disk platters are used as substrate for fingerprint deposition and capture. Hard disk platters are round discs commonly found in hard drives. They are metal-coated and therefore provide an ideal (well reflecting) surface. Such surface is chosen for a maximum separability of fingerprints from background. Exemplary other surfaces are studied in section 6.5.1.4.

Capturing devices: For the conducted experiments, three Chromatic White Light sensors of type MicroProf200 CWL600 [80] as well as one Confocal Laser Scanning Microscope Keyence VK-X110 [84] are available and are used in respect to their occupancy. Both types of devices are contactless and are considered non-invasive, due to the application of white light and visible (red) laser light (in contrast to contact-based methods or potentially destructive UV light). They are depicted in Figure 47.

Figure 47: Exemplary visualization of the used capturing devices and measurement table fixations: CWL sensor with substrate fixation using a metal plate and screws (left); CLMS device with substrate fixation using plasticine (right). Devices are provided in the context of the Digi-Dak [5] and DigiDak+ [145] projects.

The CWL device is based on the chromatic aberration of light at lenses [28] and can capture surfaces with a minimal lateral dot distance of 2 µm and a minimal longitudinal resolution of 6 nm. It can produce intensity images of a surface (12 bit quantization) by measuring the total amount of reflected light as well as topography images (15 bit quantization) by measuring the wavelength of the most intensely reflected light (from which height information can be derived based on the different focal points of the different wavelengths).

The light source is a white light LED, whose intensity is fixed during all experiments. The intensity and topography images are computed using a spectroscopic detector, which transforms the total amount as well as the peak intensity wavelength of reflected light into intensity as well as topography data. The measurement table moves a sample with a specific speed under the sensor. Such speed is defined by the measurement frequency, which is dependent on the surface material. Because hard disk platters are used for almost all experiments, a general measurement frequency of 2000 Hz is applied. Focusing is performed by vertical movement of the sensor. External vibrations are reduced by the hardware design of a massive marble panel. When operating the device, the standard settings of a longitudinal resolution of 2 nm and a total vertical scan distance of 660 µm are generally used. Lateral resolution and measured area sizes vary according to the specific experiments and are introduced separately for each test goal.

The CLSM device uses monochromatic laser light and the confocal measurement principle [83]. It produces intensity images (16 bit quantization) as well as topography images (32 bit quantization). For each lateral point, different vertical planes are systematically captured according to the longitudinal resolution and scan distance. The peak intensity of the focal plane is captured (intensity images) as well as the focal length of the peak intensity plane (topography images). The device can also capture color information using a CCD detector. However, the produced color images are rather blurred and therefore unsuitable for the investigations of this thesis. They are excluded in favor of the intensity and topography data.

Using deflection of the laser beam, tiles of up to 2048 x 1536 pixels can be captured without movement of the measurement table. For bigger images, the table can be moved in between the capture of tiles and adjacent tiles might be stitched together, forming the final image. The device can capture surfaces with four different available objective lenses (10x, 20x, 50x and 100x), resulting in a minimal lateral dot distance of 63 nm and a minimal longitudinal resolution of 10 nm.

The light source of the device is comprised of a red semiconductor laser diode (658 nm, 0.95 mW). Furthermore, samples are illuminated by a 100 Watt halogen lamp. The detector is comprised of a Photoelectron Multiplier Tube (PMT). In contrast to the CWL device, the measurement table is not moving during the measurement of a single tile. Focusing is performed by vertical movement of the sensor and the measurement table. The laser intensity is controlled by the brightness and ND-filter value. Together with the vertical scan distance, they are tuned separately for each measurement and can therefore slightly vary between different time series (see Appendix, Table XII). While the 10x objective lens and a longitudinal resolution of 200 nm is used for all conducted experiments, the lateral resolution and measured area size vary in respect to the specific experiments and are introduced separately for each test goal.

Measurement table fixation: In general, substrates are fixed onto the measurement tables of both devices using plasticine. However, for the long-term age investigations, substrates need to be removed from the table, stored and refitted at later points of time. Alignment therefore is a challenge. As introduced in section 4.2.1, long-term aging series are subject to a digital alignment in the designed preprocessing step. However, a rough spatial alignment, also assuring similar rotation angles, needs to be performed in the scope of fitting substrates onto the measurement tables.

For the CWL sensor, the measurement table provides the opportunity to fix substrates with screws. The hard disk platter substrates used for the long-term age investigation are therefore glued onto a metal plate, which can be fitted to the measurement table with the provided screws (Figure 47). The measurement table of the CLSM device does not provide such opportunity. Here, substrates are glued onto a board, which is manually adjusted at an edge of the measurement table and fixed with plasticine (Figure 47). Afterwards, a reference point on the substrate is used for manual alignment of the sensor coordinates with the substrate coordinates.

Capturing location and sample storage: In terms of environmental conditions, only indoor locations are used for the conducted experiments, limiting the environmental influences to a certain extent. For example, precipitation and strong winds are excluded, temperature and humidity fluctuations are limited. However, capturing locations comprise different laboratories and offices, which are accessible by employees and students and are therefore subject to a moderate amount of air circulation, dust and wind as well as temperature and humidity fluctuations due to opened windows or seasonal changes. Overall, they can therefore be regarded as common indoor scenarios, where some influences are limited, but others might vary to a certain extent. Table I gives the distributions of ambient temperature and relative humidity for exemplary locations and time periods, to provide an estimate of the degree of fluctuation of this variables during measurement and storage.

The table indicates that the standard deviation of ambient temperature is comparatively small during short-term aging periods of 24 hours (up to 0.6°C), which increases for longer time periods (up to 2.9°C). The relative ambient humidity seems to deviate up to 5.1% for short as well as long aging periods. The overall standard fluctuations of up to 2.9°C and 5.1% are considered as naturally occurring indoor fluctuations and are regarded as being acceptable for the here conducted experiments. Generally, short-term aging periods seem to be less affected than long-term periods. While short-term fluctuations can influence the temperature sensitive capturing devices, long-term fluctuations might rather influence the aging behavior of fingerprints.

Table I: Distributions of ambient temperature and relative humidity for exemplary locations and time periods, recorded in regular intervals. Time offsets vary between one and thirty minutes in respect to the used logging device.

Location	Time period	Temperature (°C)		Relative humidity (%)	
		Mean	StdDev	Mean	StdDev
Capturing device	24 hours (July)	28.1	0.6	44.3	3.3
Capturing device	24 hours (January)	23.5	0.3	26.2	0.6
Capturing device	1 week (July)	27.6	1.1	43.5	4.0
Capturing device	1 week (January)	23.0	1.2	29.5	1.8
Capturing device	6 months	24.2	2.2	39.9	5.1
Storage closet	6 months	23.2	2.9	43.7	4.3

For the long-term aging evaluation, investigated prints are stored in boxes or closets. These can limit the impact of environmental conditions, such as dust or air circulation. According to Table I, climatic storage conditions do not significantly vary from measurement conditions in the given examples. For the weekly capturing period, prints are fixed onto the measurement table of a capturing device for a period of one to three days, during which they are subject to the complete environmental influences.

Sweat composition, pre-treatment, application and fingerprint variability: The sweat composition of a person as well as the specific way in which a fingerprint is applied onto a surface can hardly be controlled. While it might be possible to a certain extent to control such factors in a laboratory, it seems unrealistic to control them at a practical crime scene. Therefore, the sweat composition of the donors as well as the way in which fingerprints are applied to a surface are not controlled throughout the experiments. Furthermore, no special treatment of fingers prior to a prints deposition is recommended (e.g. rubbing the neck or forehead). Using no such procedure, natural latent prints left under more realistic conditions are used for the experiments, avoiding artificially high amounts of fat in the prints (see also [61]). Also, a certain variety of donors is aimed at, to provide a realistic variation of investigated prints, which seems to be highest for different test subjects according to own studies conducted in [153]. By avoiding any pre-treatment of fingerprints, by not controlling the sweat composition and fingerprint application as well as by including different test subjects into the experiments, a natural variability of prints and their respective aging behavior is achieved. In the scope of TG5, selected aspects of fingerprint composition, -application and -variability are investigated using systematic variations of these factors.

5.4 Setup for TG1: Selection of Basic Sensor Settings

In the scope of test goal TG1, the lower bound of sensor settings for the measured area size and lateral dot distance are investigated in first studies. The optimization of these parameters is important to achieve an adequate image quality as well as a maximum capturing speed. For the

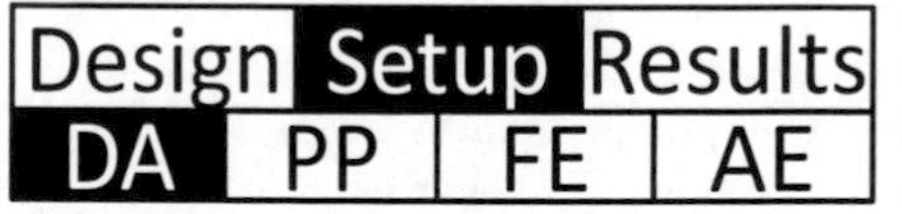

CWL sensor, which has been available for 3.5 years, the influence of different measured area sizes and resolutions is studied in several first experiments and has been published in [153], [156], [157] and [159]. Measured area sizes of 0.5 x 0.5 mm, 1 x 1 mm, 2 x 2 mm, 2.5 x 2.5 mm, 3 x 3 mm, 4 x 4 mm and 5 x 5 mm are investigated, as well as lateral dot distances of 2 µm, 3 µm, 5 µm, 10 µm, 20 µm, 50 µm and 100 µm. The specific test sets of these experiments are omitted here and the reader may refer to the publications for further reading. The experimental results of the studies are summarized in section 6.1.

For the CLSM device (which has been available for two years), the variability of measured area size and lateral dot distance is limited. Objective lenses of 10x, 20x, 50x and 100x are available, of which tiles with a resolution of 1024 x 768 pixels as well as 2048 x 1536 pixels can be captured. Furthermore, different tiles can be stitched together to bigger images. These limited capturing possibilities seem to not require an extensive experimental evaluation for determining the lower bound of measured area size and lateral dot distance. They are therefore selected by a discussion of the few possible hardware settings in respect to numerous preliminary observations in section 6.1.

Additional sensor-specific settings do exist, for which basic values need to be defined, such as measurement frequency, longitudinal resolution and scan distance as well as brightness of the light source. However, such settings are rather intuitive and their optimal value is either theoretically determined (usually to achieve a maximum capturing speed and precision) or set to standard values recommended by the sensor manufacturers. They are presented in the general test setup (section 5.3) and the Appendix (Table XI and Table XII).

5.4.1 Quality Measure Specifications for TG1

In the scope of test goal TG1, the Pearson correlation coefficient r is used to identify the lower bound of the measured area size as well as the lateral dot distance (section 4.1.3). Specifying a minimum required value of the correlation coefficient r, it can be ensured that the investigated capturing parameters can provide aging curves with a certain minimum correlation to a corresponding mathematical function. If the observed correlation coefficient drops below such minimum value for a certain amount of curves, the settings are considered as delivering images of insufficient quality. The experiments are conducted in respect to the well performing Binary Pixels feature designed in the scope of this thesis (section 4.3.1.1), based on a logarithmic aging function. Exemplary used minimum values of the correlation coefficient include $r \geq 0.8$ and $r \geq 0.9$. Because TG1 comprises first experiments (CWL sensor) as well as a selection based on theoretical discussions (CLSM device), such measure is only partially applied here. In some cases, a manual assessment of the quality of a computed aging curve is also performed.

5.5 Setup for TG2: Evaluation of Digital Fingerprint Binarization Techniques

Reliable binarization techniques seem to be most important for a successful preprocessing of fingerprint time series, because most preprocessing sub-steps are based on a successful separation of fingerprint from background pixels. Therefore, the 17 introduced intensity-based thresholding methods (section 4.2.2.1) as well as the three topography-based techniques designed for each device (section 4.2.2.2) are experimentally evaluated and compared in the scope of test goal TG2. For that purpose, test sets TSA_CWL (for the CWL sensor) and TSA_CLSM (for the CLSM device) are created. Each test set includes intensity as well as topography images of its respective capturing

device (time series are not required). An amount of 100 fingerprint images per device is considered to be of adequate size.

Three main types of fingerprint images are included, according to their importance: images captured under comparatively similar conditions Sim (50 images, used in TG3 and TG4), images of printed fingerprints Pri using distilled water, tap water and artificial sweat (10 images, used in TG5) and images captured under significantly varying influences Inf, such as different temperature, humidity, contact time, contact pressure, smearing, capturing resolution, size of the measured area as well as fresh and aged prints (40 images, used in TG5). Images vary in age between 0 and 24 hours (except for group Inf, which also contains prints aged up to two years). The test set is exclusively focused on well reflecting hard disk platter surfaces, which are considered an ideal substrate for the used capturing devices. Test set TSA is summarized in Table II. Additional capturing parameters can be found in the Appendix (Table XI and Table XII).

Table II: Experimental test sets TSA_CWL and TSA_CLSM for the investigation of optimal image binarization techniques. Each test set consists of 100 single fingerprint images captured from hard disk platters with the given variations.

TSA		CWL	
Group	**Samples**	**Parameters**	**Variation**
Sim	50	Measured size: 4 x 4 mm Lateral dot distance: 20 µm	Test subjects Fingers
Pri	10	Measured size: 4 x 4 mm Lateral dot distance: 20 µm	Printed substance (artificial sweat, distilled water, tap water)
Inf	40	Measured size: 15 x 20 mm - 2 x 2 mm Lateral dot distance: 20 µm - 2 µm	Environmental conditions Capturing parameters Type of application
TSA		**CLSM**	
Group	**Samples**	**Parameters**	
Sim	50	Measured size: 1.3 x 1 mm Lateral dot distance: 1.3 µm	Test subjects Fingers
Pri	10	Measured size: 1.3 x 1 mm Lateral dot distance: 1.3 µm	Printed substance (artificial sweat, distilled water, tap water)
Inf	40	Measured size: 1.3 x 1 mm - 0.13 x 0.1 mm Lateral dot distance: 1.3 µm - 0.13 µm	Environmental conditions Capturing parameters Type of application

Based on test set TSA, the experimental investigations of test goal TG2 are conducted, with the aim of selecting an optimal image binarization method for each capturing device and data type. For the experimental evaluation, the open source image toolbox Fiji [106], [107], is used. In such toolbox, 16 of the 17 intensity-based binarization methods are implemented and used for evaluation (the 17th method RelRange is implemented and evaluated accordingly). However, some of the provided threshold determination methods can only be applied to 8 bit images. Because the captured CWL and CLSM images are of much higher quality, providing bit-depths between 12 bit and 32 bit (section 5.3), the experiments are conducted in two phases. In a first preselection phase (Ps), the 17 intensity-based methods are evaluated based on 8 bit images using the Fiji toolbox, of which the four best performing methods are preselected. In a second selection phase (S), the four preselected methods as well as the three topography-based methods designed for each device are implemented in the experimental software modules of this thesis (section 5.2) to work with the full provided bit-depth. Based on such implementation, the best performing method for each capturing device and data type is chosen in the second phase. The test setup of both phases is given in Table III.

Table III: Image binarization methods investigated in the scope of TG2.

Device	Phase	Intensity image based methods																	Topography image based methods					
		Default	Huang	Intermodes	IsoData	Li	MaxEntropy	Mean	MinError(I)	Minimum	Moments	Otsu	Percentile	RenyiEntropy	Shanbhag	Triangle	Yen	RelRange	CWL_Topo_DoG	CWL_Topo_Line	CWL_Topo_Var	CLSM_Topo_Thresh	CLSM_Topo_Blur	CLSM_Topo_Var
CWL	Ps	x	x	x	x	x	x	x	x	x	x	x	x	x	x	x	x	x	-	-	-	-	-	-
	S	-	x	-	-	-	-	x	-	-	-	-	-	-	-	x	-	x	x	x	x	-	-	-
CLSM	Ps	x	x	x	x	x	x	x	x	x	x	x	x	x	x	x	x	x	-	-	-	-	-	-
	S	-	x	-	-	-	-	x	-	-	-	-	-	-	-	x	-	x	-	-	-	x	x	x

Preselection of binarization methods (Ps): In the preselection phase, the four best performing binarization methods of the introduced intensity-based techniques (section 4.2.2.1) are selected based on 8 bit images and the open source image toolbox Fiji [106], [107]. The 200 test images of test set TSA (100 for TSA_CWL and 100 for TSA_CLSM) are converted into 8 bit images, of which the optimal threshold (regarded as ground truth) is manually determined by a human scientist, based on the intensity data. This is done by moving a threshold slider over the image histogram until an optimal separation of fingerprint and background pixels is reached in the opinion of the human scientist (see section 4.2.7 for a discussion on the feasibility of this method for the determination of ground truth). The corresponding automatically binarized images are then computed by the Fiji toolbox and compared to the ground truth images. The four best performing methods are selected in respect to CWL as well as CLSM intensity data. According to the experimental results of section 6.2, the following four methods perform best for both devices and are preselected for further investigations: Huang, Mean, Triangle, RelRange.

Selection of binarization methods (S): After a preselection of the four best performing intensity image based binarization methods, a more precise evaluation is conducted in the selection phase, where all conducted examinations are based on the full bit-depth images of TSA_CWL and TSA_CLSM, for the manual creation of ground truth as well as the automated binarization. For that purpose, the four best performing intensity-based techniques as well as the three designed topography-based binarization methods for each device are implemented in the experimental software modules of this thesis (section 5.2). From TSA_CWL and TSA_CLSM, the optimal threshold is again manually determined by a human scientist using a threshold slider and the ground truth images are created accordingly. The original images are then automatically binarized by the implemented methods and the results are compared to their corresponding ground truth images.

From the investigated binarization methods of Table III, the best-performing technique is selected for each capturing device and data type. The chosen methods are then applied to all remaining experiments of this thesis in the scope of time series preprocessing.

5.5.1 Quality Measure Specifications for TG2

For the comparison between automatically binarized images and their corresponding ground truth images, the specification of an adequate error rate is required. As introduced in section 4.2.7, the binarization error (b_err) is defined as a quality measure for the evaluation and comparison. It computes the difference between two images by determining the relative amount of corresponding pixels being different between the two binary images. The smaller b_err becomes, the smaller is the difference between both images and the more accurate a binarization method performs. The values of b_err are compared separately for each capturing device and data type.

Manual determination of the ground truth images by a human scientist is regarded as a feasible technique in the scope of this thesis, because well reflecting hard disk platter surfaces are used. As discussed in section 4.2.7, the judgment of a dactyloscopic expert is regarded as the most reliable measure for evaluating fingerprints in forensic police work and most identification and verification results are only accepted after manual confirmation of a dactyloscopic expert. The manual binarization of a captured fingerprint is therefore regarded as an adequate method for ground truth creation in the scope of this thesis.

5.6 Setup for TG3: Evaluation of Digital Aging Feature Sets

In the scope of test goal TG3, a set of seven statistical aging features (section 4.3.1) is evaluated separately for the short- (section 6.3.1) and long-term aging (section 6.3.2). Furthermore, a set of eight morphological aging features (section 4.3.2) is investigated only for the long-term aging (section 6.3.3). Both feature sets are computed from fingerprint time series acquired and preprocessed according to the designed procedures of objectives O1 and O2, using the best performing parameters and methods selected in test goals TG1 and TG2. The experimental evaluation of these aging feature sets is based on the general test setup introduced in section 5.3. Two different test sets are designed for the short- and long-term aging in respect to each capturing device. The overall aim is to analyze the change of a computed feature value over time to extract possible correlations to a linear, logarithmic or exponential mathematical aging function.

5.6.1 Setup for Short-Term Aging Features

For the short-term age estimation, the statistical features F1 - F7 are proposed in section 4.3.1. Two test sets TSB_CWL and TSB_CLSM are created for a systematic feasibility evaluation of these features. Each test set is comprised of a total of 770 fingerprint time series, capturing the 10 different fingers of 77 different test subjects from hard disk platter surfaces every hour over a total period of t_{max} = 24 hours. Values of Δt and t_{max} vary slightly, depending on the specific settings of each time series. Based on the general test setup described in section 5.3, fingers are not subject to any pretreatment (such as washing or touching the forehead) and the application process is not controlled. Therefore, natural sweat from daily activity is present, applied in a natural way, which seems to be the most realistic scenario for such investigation. The test sets are given in Table IV. Additional capturing parameters can be found in the Appendix (Table XI and Table XII).

Table IV: Test sets TSB_CWL and TSB_CLSM for the short-term evaluation of the statistical aging features F1 - F7, each consisting of all 10 fingers from 77 different test subjects (in total 770 time series). The time offset between the scans is $\Delta t = 1$ h and the total capturing time $t_{max} = 24$ h. Capturing conditions and parameters are chosen according to the general test setup described in section 5.3.

Test set	Gender (m/f)	Age	Lateral dot distance	Measured area size	Parallel scanned series	Total captured images
TSB_CWL	50/27	21-73	20 µm	4 x 4 mm	20	21933
TSB_CLSM	57/20	18-83	1.3 µm	1.3 x 1 mm	30	18480

It can be seen from Table IV that males are dominant in the test set in respect to women with a ratio of approximately 2:1, depending on the distribution of available donors. However, the difference in aging behavior between males and females seems to be minor according to first studies in the scope of this thesis [153]. A wide variety of different ages is present, adding to the representativeness of the test sets. The different lateral dot distances and measured area sizes are chosen in respect to the results of test goal TG1 (section 6.1). The parallel scanned series describe the amount of fingerprints which can be captured using one measurement run. The number is limited by the size of the measurement table as well as the capturing time, which cannot exceed Δt. When summing up all scans of the 770 time series for each device, the total amount of captured images is received for the respective test set. For the CWL device, some time series exceed $t_{max} = 24$ h, leading to a slightly increased total amount of images for this sensor.

5.6.2 Setup for Long-Term Aging Features

For the long-term age estimation, the statistical features F1 - F7 are extended by different morphological features F8 - F15. These morphological features are limited to intensity images of selected capturing devices (section 4.3.2). Two long-term aging test sets TSC_CWL and TSC_CLSM are specified here for the evaluation of F1 - F15. Fingerprints are captured in intervals of $\Delta t = 1$ week over a total time periods of $t_{max} = 2.3$ years (CWL) and $t_{max} = 30$ weeks (CLSM). Because the CLSM device was obtained 1.5 years later than the CWL sensor, its total investigation period inevitably is significantly shorter. Values of Δt and t_{max} vary slightly, depending on the specific settings of each time series. Prints are scanned for the first time after 24 hours (to exclude changes caused by the short-term aging). Test set TSC_CLSM has furthermore been captured in the scope of a student thesis [160]. Because comparatively long time periods are studied, only a small number of 20 different fingerprints can be investigated per device, left by 20 different test subjects. Based on the general test setup described in section 5.3, fingers are not pre-treated (such as washing or touching the forehead) and the application process is not controlled. The test sets are given in Table V. Additional capturing parameters can be found in the Appendix (Table XI and Table XII).

Table V: Test sets TSC_CWL and TSC_CLSM for the long-term evaluation of the aging features F1 - F15, each consisting of one fingerprint of 20 different test subjects. The time offset between the scans is $\Delta t = 1$ w and the total capturing time is $t_{max} = 2.3$ y for the CWL sensor and $t_{max} = 30$ w for the CLSM device. Capturing conditions and parameters are chosen according to the general test setup described in section 5.3.

Test set	Gender (m/f)	Age	Lateral dot distance	Measured area size	Parallel scanned series	t_{max}	Total captured images
TSC_CWL	17/3	24-42	10 µm	8 x 8 mm	10	2.3 years	1998
TSC_CLSM	15/5	21-47	1.3 µm	1.3 x 1 mm	20	30 weeks	546

From Table V it can be observed that in comparison to TSB_CWL (short-term aging), the lateral dot distance of TSC_CWL is decreased and the measured area size is increased. The reason for this can be seen in the nature of morphological features, which require a higher resolution than statistical ones. For TSC_CLSM, the lateral dot distance and measured area size are equal to TSB_CLSM, because they provide sufficient precision for extracting statistical as well as morphological features. Ten prints are captured in parallel by the CWL sensor, whereas 20 prints are scanned using the CLSM device.

5.6.3 Quality Measure Specifications for TG3

In the scope of TG3, mathematical aging functions are approximated from the experimental aging curves using regression and the sum of squared errors (section 4.3.3). Three different aging properties Pr = {lin,log,exp} are investigated, approximating linear, logarithmic and exponential aging functions for each feature in respect to each capturing device and data type. For the exponential regression of the Binary Pixels feature F1, the feature values are inverted, to allow for a better approximation: F1' = 1 - F1. This is only done for F1, because its general aging property is well known from own studies, which have been conducted in the scope of this thesis [151], [153], [156], [158], [159].

In some cases, the experimental aging functions exhibit negative values, mostly caused by the best-fit plane subtraction of topography images. Furthermore, some feature values are very small, e.g. when already small feature values are used for further variance computation. Such feature values might be infeasible for further processing, because negative values cannot be used for exponential regression and very small numeric values lead to a loss in precision when being stored. Negative and very small values are therefore rescaled by multiplication or addition of all investigated values of a certain feature with an adequate scaling factor.

For the statistical aging features F1 - F7, the Pearson correlation coefficient is computed, indicating the correlation between the experimental aging curve $Fk_E(t)$ and its corresponding mathematical function $Fk_M(t)$ in respect to Pr = {lin,log,exp}. For an objective evaluation, the average correlation coefficients r_avrg as well as the relative frequency of curves exhibiting a strong correlation to that property $h(r \geq 0.8)$ or $h(r \leq -0.8)$ are computed. Based on such objective measures, the features can be compared in respect to each capturing device, data type and aging period. The morphological long-term aging features F8 - F15 are evaluated based on a visual inspection of their aging curves, to discuss certain aspects in more detail. This is possible because the used test set TSC consists of only 20 time series for each device. However, the limited amount of this test set also prevents a sufficient size for experimental age estimation. Age estimation in the scope of TG4 is therefore only evaluated for short-term periods.

5.7 Setup for TG4: Evaluation of Digital Short-Term Age Estimation Strategies

Design	Setup	Results	
DA	PP	FE	AE

In the scope of test goal TG4, age estimation strategies are evaluated in respect to the short-term aging. Long-term age estimation is also of interest, however, has to be evaluated separately in future research using test sets of statistically significant size (the 20 time series of the long-term test set TSC introduced in section 5.6.2 are considered too small for a reliable age estimation).

The evaluated short-term age estimation strategies are based on test sets TSB_CWL and TSB_CLSM, each consisting of 770 time series (section 5.6.1). They are captured and preprocessed in respected to the procedures designed in O1 and O2 and based on the optimal parameters and methods selected in TG1 and TG2. Features F1 - F7 designed in O3 are computed and the resulting series of feature values are evaluated in respect to different formula-based (section 5.7.1) and machine-learning based (section 5.7.2) age estimation strategies.

5.7.1 Setup for Formula-Based Age Estimation

Four different formula-based age estimation strategies are evaluated for the short-term aging in two experimental phases Pk and Npk. For both designed phases, the corresponding age estimation features AEF are computed according to section 4.4.1. In case parameters 'a' or 'b' of the logarithmic function are required as input (phase Pk), they are averaged from all samples of the respective test set. Because formula-based age estimation requires reliable features, the best performing feature Fb of each capturing device and data type (section 6.3.1) is exemplary evaluated, which is the CWL_Int_F1, CWL_Topo_F1, CLSM_Int_F2 as well as CLSM_Topo_F1 feature. A total amount of 15400 AEFs is computed for the evaluation of each phase Pk and Npk.

5.7.2 Setup for Machine-Learning Based Age Estimation

In the scope of machine-learning based age estimation, four different experimental phases Op1, Op2, Ss and Ap are designed (section 4.4.2). Parameters are described according the configuration Conf introduced in formula (81).

Optimization of parameters Re and Fr (Op1): In the first phase of the experiments, optimal values of two basic parameters for the age estimation feature AEF computation sub-step are determined. The approximation type Re = {lin,log} as well as the optimal feature representation Fr = {0,1,2} are investigated. For that purpose, the parameters Ti, Tt and Cl are fixed during the evaluation. Values of Ti = 5, Tt = 5 h and Cl = J48 are exemplary chosen, which have been shown in first tests to deliver acceptable results. The parameter Sas describes if sufficient samples are present for statistically reliable age estimation and is undefined at this point of time. It is investigated in phase Ss. The following configuration is obtained, where variations of a parameter are marked by the delimiter '|':

$$Conf(Op1) = (5{,}0|1|2, lin|log, 5, J48, unknown) \qquad (82)$$

Five variations of Fe and Re result from such configuration: (Fe,Re) = (0,-);(1,lin);(1,log);(2,lin);(2,log). The feature representation Fe = 0 represents a single feature value and cannot be used for regression. For each capturing device and data type, the best performing feature Fb of the feature evaluation experiments (section 6.3.1) is used, which is CWL_Int_F1, CWL_Topo_F1, CLSM_Int_F2 as well as CLSM_Topo_F1, leading to 20 different investigated combinations. The resulting performances are compared and the best combination of Fr and Re is selected in the scope of phase Op1. According to the results of section 6.4.2, the variables lead to optimal performances for values of O_Fr = 1 and O_Re = lin, which are fixed for the following experimental phases.

Optimization of parameters Ti and Cl (Op2): The second experimental phase is designed to optimize the required amount of temporal images Ti (age estimation feature AEF computation sub-step) as well as the classifier Cl (model training sub-step). In the experiments, variations of Ti = {2,3,4,7,9,10} and Cl = {NaiveBayes,J48,LMT,RotationForest} are exemplary investigated. The parameters Fr and Re are fixed by their optimal values O_Fr and O_Re, determined in Op1. While Sas is still unknown and examined in phase Ss, Ti and Cl need to be investigated for all variations of Tt = {1,2,...,24-Ti}, because different amounts of consecutive temporal images can be optimal for different ages of the sample. The resulting configuration is formalized as follows (NaiveBayes: NaBa, RotationForest: RotFor):

$$Conf(Op2) = (2|3|5|7|9|10, O\_Fr, O\_Re, 1|2|\ldots|24 - Ti, NaBa|J48|LMT|RotFor, unknown) \quad (83)$$

Because the CWL and the CLSM device differ in their acquisition technique and precision, different values of O_Ti can be optimal for the different devices. In total, 96 variations are investigated, based on the AEFs of the four best performing aging features Fb of each capturing device and data type, which are CWL_Int_F1, CWL_Topo_F1, CLSM_Int_F2 and CLSM_Topo_F1 (section 6.3.1). From the experimental results, the best performing combination of Ti and Cl is selected (section 6.4.2), leading to the optimized value O_Cl = LMT as well as O_Ti = 5 (CWL sensor) and O_Ti = 3 (CLSM device). Such optimized values are fixed for the remaining experimental phases Ss and Ap.

Statistical significance determination (Ss): To show that the amount of AEFs computed from test sets TSB_CWL and TSB_CLSM is of statistical significance, the parameter Sas = {true,false} is investigated in the experimental phase Ss. In particular, the learning curves of the classifier O_Cl are evaluated. Sas is studied for the optimized parameters O_Fr, O_Re and O_Ti. All possible variations of Tt = {1,2,...,24 O_Ti} need to be investigated, because different amounts of AEFs are available for different values of Tt. The investigated configuration is formalized as follows:

$$Conf(Ss) = (O\_Ti, O\_Fr, O\_Re, 1|2|\ldots|24 - Ti, O\_Cl, true|false) \quad (84)$$

All features F1 - F7 of CWL_Int, CWL_Topo, CLSM_Int and CLSM_Topo data need to be examined, leading to a total of 560 investigated combinations. Evaluating the learning curves of O_Cl in respect to each of these configurations, the statistical significance of the used test set TSB can be shown. Furthermore, also the minimum required amount of AEFs can be determined. For such purpose, the computed training set TRS (section 4.4.2) is systematically reduced to smaller sub-sets according to a split value spl = {0.01,0.05,0.1,0.15,0.2...0.95,1}, where for each sub-set, TRS contains between 1% and 100% of AEF training samples, depending on the split value. During the split, it is ensured that the amount of training samples per age class is equal. The age estimation performance is then computed according to section 4.4.2 for each value of spl, describing the learning curve of the classifier O_Cl in respect to a specific configuration. From the learning curves of O_Cl, the statistical significance as well as the minimum amount of required AEFs is computed using the range ra. The value of ra is defined as the range of the five consecutive age estimation performances kap (section 4.4.3) of a sliding window moved along the learning curve of the classifier, where i represents the number of the last element of this window according to spl:

$$min\_kap_i = min\,(kap_{i-4}, kap_{i-3}, kap_{i-2}, kap_{i-1}, kap_i); i = 5,6,\ldots,21 \quad (85)$$

$$max\_kap_i = max\,(kap_{i-4}, kap_{i-3}, kap_{i-2}, kap_{i-1}, kap_i); i = 5,6,\ldots,21 \quad (86)$$

$$ra(i) = max\_kap_i - min\_kap_i; i = 5,6,\ldots,21 \quad (87)$$

The systematic increase of the amount of training samples TRS is expected to lead to an increased classification performance, which eventually converges towards a certain maximum value, considered the desired optimal performance. The state of convergence of such learning curve is of interest to determine the statistical significance as well as the required amount of training samples. It can be measured using the range ra of the last five increases of the training set. If ra is smaller than a certain maximum value ra_max, the performance of the classifier has not changed more than ra_max within the last five steps and can therefore be considered as having converged (Sas = true) in respect to ra_max. By traversing the different split positions spl in increasing order, it can be determined when ra ≤ ra_max for the first time, leading to this split position being considered as representing the minimum amount of samples required for a statistical significance in respect to ra_max. If ra > ra_max for all split positions, the learning curve is considered to not have been converged yet (Sas = false) and therefore the results are unreliable. Different values of ra_max = {0.01,0.02,...,0.06} are investigated. The experimental results of section 6.4.2 show that statistical significance is achieved for all 560 evaluated combinations in respect to ra_max = 0.06 and the results are therefore considered as reliable.

Age estimation performance evaluation (Ap): After optimization of the parameters Fr, Re, Ti and Cl as well as showing the statistical significance Sas of the used test set, the final classification performance of the proposed machine-learning based age estimation approach is evaluated and compared in phase Ap:

$$Conf(Ap) = (O\_Ti, O\_Fr, O\_Re, 1|2|\ldots|24 - Ti, O\_Cl, true) \quad (88)$$

Age estimation performances are evaluated in respect to the different capturing devices, data types, age class pairs according to Tt and all seven statistical features F1 - F7. Furthermore, an additional classification approach combining all seven features is examined. In total, 600 different variations are evaluated in section 6.4.2.

5.7.3 Quality Measure Specifications for TG4

In the scope of test goal TG4, two different quality measures area applied in respect to the investigated age estimation strategy (section 4.4.3). For formula-based age estimation, the absolute time difference t_diff = |t_est - t_real| between the estimated age t_est of a fingerprint and its corresponding real age t_real is used. Such time difference is not normally distributed and varies widely between the time series of a test set. Therefore, the mean value and standard deviation are not regarded as being adequate quality measures. Instead, the median value of t_diff is used. However, the range of the computed values needs to be considered when assessing the performance of formula-based age estimation strategies.

For machine-learning based age estimation, the kappa value kap is applied as objective quality measure. It has the benefit of excluding the influence of unequal amounts of samples within the different age classes of the performance evaluation set PES. However, at selected points of this thesis, the classification accuracy equivalent to such kappa value for equally distributed classes of PES is given for comparison. In the scope of experimental phase Ss, the quality measure kap is extended to the range ra of five consecutive kappa values of the learning curve of a classifier. This parameter allows for the investigation of kappa value changes in respect to systematically increased test set sizes. If these changes are smaller than a certain maximum range ra_max, the learning curve is considered to have been converged to its maximum performance. In the scope of test goal TG4, values of ra_max = {0.01,0.02,...,0.06} are investigated.

5.8 Setup for TG5: Evaluation of Selected Short-Term Influences and Model Creation

In test goal TG5, several experiments are performed for evaluating different influences on the short-term aging of latent prints. As introduced in section 2.4, influences are divided into the five main categories of sweat influence, environmental influence, surface influence, application influence and scan influence. Several qualitative experiments are conducted in the scope of this thesis to identify major influence factors and rule out negligible ones, which have been published in [95], [153], [154], [155], [156], [157], [158] and [159]. They are based on the CWL device, which has been available for 3.5 years, while the CLSM device has only recently become available (two years ago). Here, these experiments are not re-visited in detail, in favor of conducting additional evaluations, including both the CWL as well as the CLSM device. However, their findings are of major relevance for the design of a first qualitative influence model in the scope of test goal TG5. They are therefore briefly summarized (section 6.5.1) and included into the created influence model (section 6.5.5).

Three additional test setups are designed in the scope of O5 (sections 4.5.2 to 4.5.4) and are evaluated in the scope of TG5. They are each conducted for both capturing devices, but limited to intensity images. As shown in sections 4.2.2.2 and 6.2 for the binarization of captured prints, intensity images are of much higher quality than their topography counterparts. Furthermore, the significant temperature and humidity changes occurring during the experiments with varying ambient climatic conditions (section 5.8.3) seem to have a particularly strong influence on the topography images of both devices. Also, the proposed topography segmentation methods seem to fail for the differently shaped particles of printed substances (section 5.8.4) to correctly segment particle from background pixels. Therefore, topography data seems to be much more unreliable than intensity data and is therefore not considered in the experimental evaluations of TG5. For examining influences on the intensity images of each capturing device, the best performing feature Fb as determined in section 6.3.1 is used, which is the CWL_Int_F1 as well as the CLSM_Int_F2 feature.

In section 5.8.1, important aspects concerning the setups of the various experiments published in the scope of this thesis are briefly summarized. The experimental setups of the three additionally investigated, presumably very important influences on the aging of latent fingerprints, are given in sections 5.8.2 to 5.8.4. Apart from the general settings described in section 5.3, similar measured area sizes and lateral dot distances (CWL: 4 x 4 mm, 20 µm; CLSM: 1.3 x1 mm, 1.3 µm) are used for all three experiments, capturing fingerprint images every hour over a total time period of t_{max} = 24 hours.

5.8.1 Setup for Selected Influences from own Published CWL Studies

Various experiments on influences of the capturing devices, environmental conditions, sweat composition, used substrate as well as fingerprint application are conducted in the scope of this thesis, based on CWL short-term aging series. They have been published in [95], [153], [154], [155], [156], [157], [158] and [159]. Their varying test setups are omitted here in favor of the three additional experiments described in sections 5.8.2 to 5.8.4. For information about the different test setups of published studies, the reader may refer to the specific publications. The results of these studies are briefly summarized in section 6.5.1. The experiments are based on obsolete preprocessing techniques and do not provide experimental results for the CLSM device, which has only recently become available (two years ago). In some cases, their test set size is comparatively small and therefore of preliminary nature. However, numerous observations throughout the experiments have shown that CWL intensity images achieve a certain accuracy of their experimental results, even when using obsolete preprocessing techniques (in contrast to CLSM images). Therefore, the published findings are valuable for the here conducted additional investigations as well as the creation of a first qualitative influence model in section 6.5.5.

5.8.2 Setup for Selected Scan Influences

Several short-term influences of the capturing devices on the acquired fingerprint time series exist, some of which are introduced in section 4.1.2 (PC2, PC4 and PC5). These influences have been addressed and reduced by different preprocessing techniques in the scope of objective O2 (section 4.2). Here, two additional, potential influences of the capturing devices on the observed fingerprint aging behavior are investigated, according to the design of section 4.5.2. Because all short-term aging investigations of this thesis are based on the consecutive capture of prints right after application until a total time of 24 hours has passed, it is of importance to experimentally investigate to which extent the starting of a capturing device influences the observed aging behavior. Furthermore, showing the non-invasiveness of the capturing devices in respect to their radiation energy is another aim of this experiment.

For this purpose, a latent print is applied to the hard disk platter surface, from which three different areas Ar1 - Ar3 are selected. According to the findings introduced in section 6.5.1, the variation of aging speed between these three areas is expected to be comparatively small. Furthermore, the overall short-term aging period of t_{max} = 24 hours is divided into three equal time periods Tp1 - Tp3. Six different time series Ts1 - Ts6 are then captured for such fingerprint according to the specifications of Table VI.

Table VI: Experimental variations of fingerprint areas (Ar1 - Ar3) and time periods (Tp1 - Tp3) of a single fingerprint for evaluating the starting influence as well as the non-invasiveness of a capturing device using six different time series Ts1 - Ts6. Prints are captured in regular time intervals of Δt = 1 hour.

	Tp1 (0-8h)	**Tp2 (8-16h)**	**Tp3 (16-24h)**
Ar1	Ts1	Ts2	Ts3
Ar2	-	Ts4	Ts5
Ar3	-	-	Ts6

Using the time periods Tp1 - Tp3 from Table VI, it can be determined to what extent a measured aging property does indeed represent the aging behavior of the print and not certain influences caused by starting the capturing device. Because the device is started three times, three different capturing processes are present, each capturing a time series from Ar1 (Ts1 - Ts3) of a different age (Tp1 - Tp3). It is expected that the degradation behavior is fastest for Tp1, leading to the highest aging speed for Ts1, followed by Ts2 and Ts3 in descending order. If the aging speeds observed for time periods Tp1 - Tp3 are not in this particular order (e.g. if they are similar), then the aging behavior of a feature Fk is overlaid by influences of the capturing device. However, the experiment does only produce reliable results, if well performing aging features Fk are used. The best performing intensity-based features CWL_Int_F1 and CLSM_Int_F2 (section 6.3.1) are therefore used for the experiment.

Apart from investigating the influence of starting a capturing device, this test setup also allows to evaluate the non-invasiveness of the device in respect to the radiation energy of its light source. After time period Tp3, three time series (Ts1 - Ts3) have been captured from area Ar1, two series (Ts4 - Ts5) from area Ar2 and one series (Ts6) from area Ar1. Therefore, area Ar1 has been subject to the triple amount of radiation from the light source in respect to Ar3. As a consequence Ts3, Ts5 and Ts6 as well as Ts2 and Ts4 can be compared to each other regarding their aging speed. In case the light source of a device is invasive, an increased degradation (aging speed) would be observed for the print area being exposed to the most radiation. In such case, area Ar1 would exhibit a higher aging speed than area Ar2 and area Ar3 (in descending order). In case Ts3, Ts5 and Ts6 as well as Ts2 and Ts4 can be shown to exhibit a similar aging speed of their aging curves, the non-invasiveness of the capturing device has been confirmed to be comparatively small.

All time series are captured and processed according to the optimal settings and methods selected in TG1 and TG2, with one exception for the preprocessing of CWL images, where the original background mask B is used instead of its dilated version B', because an insufficient amount of pixels remains for some of this particular time series when using B'. Such influence of the amount of pixels selected for background mask creation on the performance of the preprocessing methods should be investigated in more detail in future work. The experiments are conducted for 7 prints captured with the CWL sensor (male / female: 6/1, age: 21 - 47) and ten prints captured with the CLSM device (male / female: 9/1, age: 20 - 47). More subjects could be used with the CLSM device, because its capturing time is slightly faster in this test setup.

From all donors, a random thumb fingerprint is applied to a hard disk platter. Areas Ar1 - Ar3 are randomly chosen and the six time series Ts1 - Ts6 are captured for each fingerprint over the three time periods Tp1 - Tp3, acquiring a fingerprint image every hour ($\Delta t = 1$ hour). All fingerprints are evaluated in parallel using one measurement run (to avoid potential distortions caused by other influence factors, which might occur during different measurement runs, such as different ambient climatic conditions). With such test setup, a total of 7 x 6 time series is obtained for the CWL sensor and a total of 10 x 6 series for the CLSM device. Each time series consists of 8 consecutive scans of a sample for both devices. The test sets are referred to as test set TSD_CWL and TSD_CLSM. Additional capturing parameters can be found in the Appendix (Table XI and Table XII).

5.8.3 Setup for Selected Environmental Influences

In this section, the test setup of qualitative studies on the influence of different ambient temperature and humidity conditions is introduced (section 4.5.3). Ambient temperature and relative humidity are investigated, because they seem to represent the most important environmental influences occurring in an indoor scenario. Their impact does not only manifest itself in changes to the fingerprint aging behavior, but also in the form of sensor distortions (see also [154], published in the scope of this thesis). However, measures have been taken in the preprocessing step (section 4.2) to reduce such impact. For investigating the influence of temperature Te and relative humidity Hu on the aging behavior of latent prints, different variations of each variable are evaluated. While one variable is systematically varied, the other one is kept as constant as possible. This is especially challenging, because Te and Hu are related to each other: if the temperature rises, the ambient air is able to absorb more water, leading to a decrease in humidity and vice versa.

In the scope of this thesis, no climatic chamber is available. Therefore, the means of controlling climatic conditions are limited. However, such scenario can be considered as close to a practical indoor crime scene. Temperature and humidity are controlled using desk lamps, spotlights, a common air humidifier (beurer LB 50 [173]) as well as the controlled opening and closing of windows and doors. All these devices and actions are common for practical indoor scenarios, rendering the aging conditions realistic. Depending on the ambient conditions of each capturing device, temperature and humidity are each set to three different values in the range of common indoor scenarios. A total of six climatic variations is therefore evaluated for each device, using an automated capture and preprocessing according to the results of TG1 and TG2. The distributions of temperature Te and humidity Hu during all experiments are depicted in Table VII. The aging speeds of the resulting curves are used to investigate if certain ambient conditions accelerate or decelerate the print degradation process.

Table VII: Distribution of temperature Te and humidity Hu during the investigation of selected environmental influences. Systematically varied conditions are marked in bold.

	Temperature (Te) in °C		Relative Humidity (Hu) in %	
Device	**Mean**	**StdDev**	**Mean**	**StdDev**
CWL - Te1	**16.7**	0.5	42.8	2.6
CWL - Te2	**24.5**	0.5	38.7	0.8
CWL - Te3	**42.4**	0.4	40.0	4.4
CWL - Hu1	24.3	0.3	**33.0**	0.3
CWL - Hu2	24.5	0.4	**37.9**	0.4
CWL - Hu3	26.3	0.7	**51.2**	4.6
CLSM - Te1	**23.4**	0.6	11.4	0.5
CLSM - Te2	**30.0**	0.3	10.1	0.5
CLSM - Te3	**40.9**	0.2	9.3	0.8
CLSM - Hu1	31.3	0.3	**6.5**	1.2
CLSM - Hu2	33.0	0.3	**20.9**	0.9
CLSM - Hu3	35.8	0.3	**47.6**	6.6

It can be seen from Table VII that the standard deviation of the temperature takes a maximum value of 0.7°C for all experiments. Furthermore, the relative humidity has a maximum standard deviation of 4.4% in all but two cases (which are discussed later). This is the maximum extent to which the ambient climatic conditions can be controlled without using a dedicated climatic chamber. They seem to be accurate enough for the experimental evaluations in the scope of this thesis and represent realistic indoor conditions. However, for the precise determination of climatic ambient influences, additional experiments using a climatic chamber are required in future work.

The temperature is systematically varied for both devices (CWL - Te1, CWL - Te2, CWL - Te3; CLSM - Te1, CLSM - Te2, CLSM - Te3). Different Temperature offsets are chosen in respect to the sensors location, where different temperature equilibriums can be achieved in dependence on the surroundings of the capturing device. Similar principles apply to the systematic humidity variations (CWL - Hu1, CWL - Hu2; CLSM - Hu1, CLSM - Hu2), where certain humidity values are produced according to the predominant ambient conditions and kept as constant as possible.

When investigating high humidity values (CWL - Hu3; CLSM - Hu3), an air humidifier is used. In this case, it is observed that the humidity is strongly fluctuating depending on the location and time, varying between a medium humidity and a total saturation of 100%. The humidity and temperature logger is placed in a distance of approximately 20 cm to the measurement table. Therefore, the humidity of CWL - Hu3 and CLSM - Hu3 might be significantly higher than the logged values for certain times and locations of the measurement table. The mean humidity values specified in Table VII for CWL - Hu3 and CLSM - Hu3 are therefore regarded as minimum values, which are strongly fluctuating and can be significantly higher at times.

From a pool of 12 test subjects (males / females: 10/2; ages: 20 - 47), five different donors are randomly selected for each variation (at least one female). No pre-treatment of fingers is conducted (e.g. rubbing the forehead) and one random thumb of each donor is applied to the hard disk platter surface. From each fingerprint, four (CWL) and six (CLSM) random areas are selected and captured every hour (Δt = 1 hour) for a total time period of t_{max} = 24 hours. Because the CLSM device has a higher capturing speed in this experimental setup, more latent prints can be captured in parallel for this device. A total of 60 (CWL) and 90 (CLSM) temperature varied time series as well as 60 (CWL) and 90 (CLSM) humidity varied time series is captured in total. All 20 (CWL) and 30 (CLSM) time series captured under a single climatic condition are scanned in parallel, to assure similar climatic conditions. The test sets are referred to as test sets TSE_CWL and TSE_CLSM. Additional capturing parameters can be found in the Appendix (Table XI and Table XII).

5.8.4 Setup for Selected Sweat Influences

The chemical composition of fingerprint residue has been widely studied in the area of chemistry and also degradation studies exist (chapter 2). The influence of different sweat compositions therefore seems to be of great importance. It is of major interest to investigate if such influences also manifest themselves over short-term aging periods of only 24 hours. Furthermore, it is unclear which changes in chemical composition actively contribute to changes of the overall fingerprint appearance. Because the CWL as well as the CLSM device are not substance-specific, only the overall aging behavior of the complete residue can be studied. The here conducted experiments are based on the design of section 4.5.4.

Three different substances of distilled water DW, tap water TW and artificial sweat AS are printed with the Canon PIXMA iP4500 inkjet photo printer [174] onto a hard disk platter. The complete platter is evenly coated with material instead of printing fingerprint patterns, to apply a sufficient amount of each substance. A fast evaporation of some of the material within the first 60 seconds after printing can be observed by visible inspection. From the remaining material, 20 (CWL) and 30 (CLSM) random locations are captured in parallel every hour (Δt = 1 hour) over a total time period of t_{max} = 24 hours. The procedure is repeated until each substance has been evaluated. Time series are captured and preprocessed according to the optimal settings and methods selected in TG1 and TG2. In total, 60 (CWL) and 90 (CLSM) time series are captured, each consisting of 24 temporal images. The composition of the artificial sweat AS can be found in [161]. It is to note here that no fatty substances are contained in AS. The test sets for investigating selected sweat composition influences are referred to as TSF_CWL and TSF_CLSM. Additional capturing parameters can be found in the Appendix (Table XI and Table XII).

5.8.5 Quality Measure Specifications for TG5

The fingerprint aging speed As is used in the scope of test goal TG5 as an objective quality measure (section 4.5.5). It is equal to the parameter 'a' of an approximated mathematical function. Regression and the sum of squared errors are used for approximation in respect to a logarithmic function (which represents the best performing aging property according to section 6.3.1). Because the aging speed is normally distributed, its mean value and standard deviation from all curves of a certain investigated influence represent adequate means of comparison.

6 Experimental Results: Discussion of Findings

In this chapter, the experimental results of this thesis are presented, according to the defined test goals TG1 - TG5. In test goal TG1 (section 6.1), basic sensor settings are chosen, based on which the time series of all remaining test goals are captured. Adequate binarization methods are selected in test goal TG2 for each capturing device and data type (section 6.2). Such methods are applied in the scope of image preprocessing to the time series of all remaining test goals. Age estimation features for the short- and long-term aging are evaluated in TG3 (section 6.3), followed by a performance evaluation of formula-based and machine-learning based short-term age estimation strategies (section 6.4). Influences on the short-term aging of latent prints are evaluated in first qualitative studies in test goal TG5 (section 6.5), including the design of a first influence model. The results of all test goals are evaluated and compared in accordance to their respective, earlier defined quality measures.

Design | Setup | Results

To allow for a comprehensive discussion with other experts in the field and to evaluate the general acceptability of the proposed approach in the scientific community, parts of the here presented concepts and results have been previously published, which is a common practice for a dissertation project in the field of computer science. The corresponding publications are:

2013

- [151] presented at the 1st ACM Workshop on Information Hiding and Multimedia Security (IH&MMSEC), Montpellier, France, June 17th - 19th, 2013

2012

- [153] published in the journal Forensic Science International, Elsevier, Volume 222, Issues 1 - 3, pp. 52 - 70, October 10th, 2012
- [21] presented at the 7th DPM International Workshop on Data Privacy Management, Pisa, Italy, September 13th - 14th, 2012
- [95] presented at the 14th ACM Workshop on Multimedia and Security (MMSEC), Coventry, UK, September 6th - 7th, 2012
- [154] presented at the SPIE conference on Optics, Photonics, and Digital Technologies for Multimedia Applications II, Brussels, Belgium, June 1st, 2012
- [79] presented at the SPIE conference on Three-Dimensional Image Processing (3DIP) and Applications II, Burlingame, California, USA, February 9th, 2012

2011

- [155] presented at the 2011 IEEE International Workshop on Information Forensics and Security (WIFS), Foz do Iguacu, Brazil, November 29th - December 2nd, 2011
- [156] presented at the 12th Joint IFIP TC6 and TC11 Conference on Communications and Multimedia Security (CMS), Ghent, Belgium, October 19th - 21st, 2011
- [157] presented at the SPIE conference on Optics and Photonics for Counterterrorism and Crime Fighting VII, Optical Materials in Defence Systems Technology VIII and Quantum-Physics-based Information Security, Prague, Czech Republic, October 13th, 2011
- [158] presented at the 13th ACM Workshop on Multimedia and Security (MMSEC), Niagara Falls, New York, USA, September 29th - 30th, 2011
- [159] presented at the 7th International Symposium on Image and Signal Processing and Analysis (ISPA 2011), Dubrovnik, Croatia, September, 4th - 6th, 2011

6.1 Results for TG1: Selection of Basic Sensor Settings

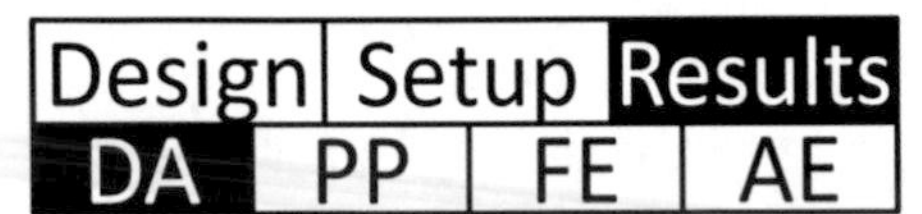

In the scope of test goal TG1, the two main capturing parameters of measured area size and lateral dot distance are investigated in first studies. For the CWL sensor (available for 3.5 years), several experimental findings have been published in [153], [156], [157] and [159]. These studies derive a lateral dot distance of 20 µm in combination with a measured area size of 4 x 4 mm as the minimum reliable capturing setting, providing images of adequate size and quality. Using such lower bound of sensor settings, correlation coefficients of $r \geq 0.8$ can be achieved for most aging curves of the exemplary investigated, well performing Binary Pixels feature. Higher resolutions or sizes are equally possible and might even be able to provide slightly more accurate results. However, a significant decrease in the quality of aging curves is observed for lower resolutions or measured area sizes.

For the CLSM device (available for two years), a lower bound of the size of measured area and lateral resolution is predefined by the fixed size of a captured tile. Single tiles can be stitched to larger images. However, the usage of images larger than one tile seems to not be recommendable, because the proprietary stitching process of the device introduces significant distortions at the edges of tiles. Therefore, only one tile is used for the age estimation of fingerprints. When investigating the available magnifications of the different objective lenses (10x, 20x, 50x and 100x), it can be observed that a maximum of three horizontal ridges is usually contained within one tile when using the 10x objective lens. Using higher magnifications therefore leads to a reduction of the investigated ridges to two or less, which might not be desired for investigating an adequate amount of ridges in the scope of aging studies. Although it seems feasible in future work to investigate the aging behavior of a latent print based on two or less ridges, the 10x objective lens is selected in the scope of this thesis to assure that at least three ridges are included into each experimental time series.

For capturing a single tile with a size of 1.3 x 1 mm, two resolutions of 1024 x 768 pixels and 2048 x 1536 pixels are available. However, the capturing time as well as the storage requirements quadruple for the higher resolution, preventing the efficient capture, storage and processing of an adequate amount of time series. For example, the amount of time series captured in parallel during one measurement run based on the general parameters specified in section 5.3 decreases from 30 to

8, while the storage capacity of a single short-term aging series increases from approximately 230 MB to 900 MB. However, using the lower resolution of 1024 x 768 pixels is found in numerous observations to exhibit an acceptable quality of the resulting aging curves. Therefore, a single tile captured with a 10x objective lens and a resolution of 1024 x 768 pixels (measured area size: 1.3 x 1 mm, dot distance: 1.3 µm) is chosen as the lower bound of CLSM capturing parameters for the studies of this thesis. Exemplary aging curves of both devices captured with different lateral dot distances are depicted in Figure 48 for intensity time series.

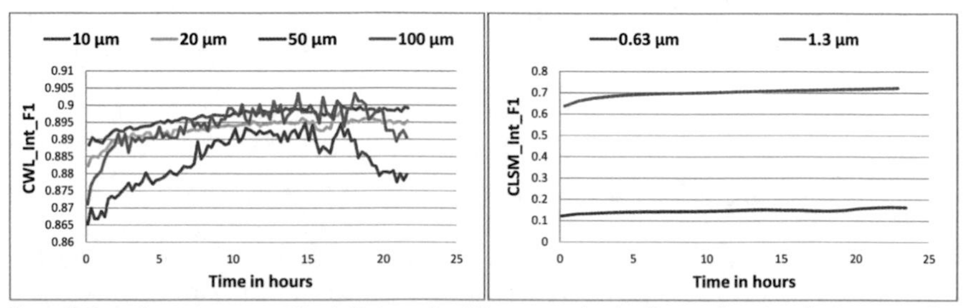

Figure 48: Exemplary aging curves of both devices using different capturing dot distances based on the CWL_Int_F1 feature (left image) as well as the CLSM_Int_F1 feature (right image).

Figure 48 depicts exemplary aging curves of four different lateral dot distances for the CWL device (left image). It can be observed that the curves captured with 50 µm (red) and 100 µm (blue) are subject to severe distortions and therefore of insufficient quality for a reliable age estimation. Curves computed from 10 µm (purple) and 20 µm (green) images exhibit a logarithmic aging property with only minor distortions. They are therefore considered of adequate quality for age estimation.

For the CLSM device (right image), only two different resolutions of 1024 x 768 pixels (1.3 µm, blue) and 2048 x 1536 pixels (0.63 µm, red) are available. It can be observed that the curve computed from images of higher dot distances (0.63 µm, red) is not necessarily of better quality than its lower resolution (1.3 µm, blue) counterpart. It is therefore assumed in the scope of this thesis that a lateral dot distance of 1.3 µm is sufficient for reliable age estimation.

In general, it is not possible to compare the CWL curves of Figure 48 with their CLSM counterparts, because different devices and dot distances are used and the specific properties of the randomly selected fingerprint samples can vary significantly between both devices. Furthermore, the y-axes are not equally scaled, to achieve better visibility. However, it can be observed that the CLSM curves seem to be much smoother than their CWL counterparts. This difference is attributed to the significantly different dot distances of both devices and therefore the different precision of the captured images. It does not necessarily indicate that the CLSM device can produce fingerprint time series of a higher reliability than the CWL device, which remains to be investigated.

The finding that for both devices only very small measured area sizes are required for an adequate evaluation of changes occurring during fingerprint aging has significant implications to forensic investigations and daily police work. Because such small fingerprint areas do not seem to be sufficient for identification at this point of time [22], age estimation can be performed without capturing person-related data, potentially improving the data privacy protection of crime scene investigations as well as in preventive applications. These implications are further discussed in section 7.3 and have been published in the scope of this thesis [21].

6.2 Results for TG2: Selection of Digital Fingerprint Binarization Techniques

The main aim of test goal TG2 is to select suitable binarization methods for each capturing device and data type. For that purpose, 17 different intensity-based methods as well as 3 topography-based methods for each device are evaluated and compared in respect to the binarization error b_err (section 4.2.7). The experiments are conducted in two phases, where a preselection of intensity-based images is performed in the preselection phase (Ps) based on 8 bit images, followed by the selection phase (S) of adequate methods for both devices and data types using the full pixel-depths of the captured time series.

Preselection of binarization methods (Ps): In the preselection phase of evaluating the quality of different binarization methods, 8 bit fingerprint intensity images of test sets TSA_CWL and TSA_CLSM (section 5.5) are compared between a manual, optimal segmentation and 17 different automated segmentation methods. The evaluation is conducted in respect to three different fingerprint classes, which are defined by fingerprints aged under approximately similar conditions (Sim), printed fingerprints (Pri) and fingerprints aged under significantly different influences (Inf). The results are depicted in Figure 49 and Table VIII.

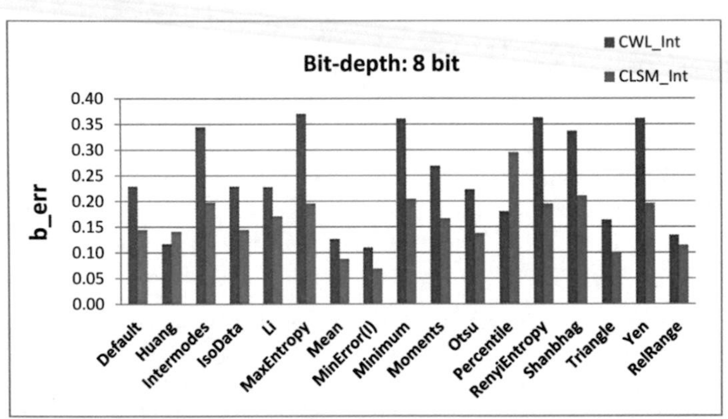

Figure 49: Binarization error (b_err) of the 17 investigated intensity-based methods. Errors are computed from test sets TSA_CWL and TSA_CLSM using images of a reduced pixel bit-depth of 8 bit and are weighted and averaged over all three classes Sim, Pri and Inf.

Comparing the b_err rates of Figure 49 and Table VIII, it can been seen that the b_err of CLSM images is in most cases lower than its CWL counterpart, showing that the automated threshold segmentation methods perform significantly better on such data. Merely the Huang and Percentile methods perform better for CWL images. When comparing b_err between the different binarization methods, it can be seen that the MinError(I), Mean, Huang, Triangle and RelRange methods perform best for both devices, showing b_err rates smaller than 15% (with the exception of the Triangle method for CWL images, where b_err = 16%). Amongst these five methods, the MinError(I) technique exhibits the best results for the CWL as well as the CLSM device.

Table VIII: Binarization error (b_err) of the 17 investigated intensity-based methods. Errors are computed from test sets TSA_CWL and TSA_CLSM using images of a reduced pixel bit-depth of 8 bit and are weighted and averaged over all three classes Sim, Pri and Inf. The best performing method for each class is marked in bold.

		Default	Huang	Intermodes	IsoData	Li	MaxEntropy	Mean	MinError(I)	Minimum	Moments	Otsu	Percentile	RenyiEntropy	Shanbhag	Triangle	Yen	RelRange
CWL	**Sim**	0.23	0.10	0.38	0.23	0.23	0.44	0.13	0.11	0.39	0.29	0.23	0.17	0.44	0.35	0.20	0.44	**0.10**
	Pri	0.16	0.14	0.27	0.16	0.16	0.16	0.14	0.14	0.28	0.16	0.16	0.29	0.15	0.21	0.15	0.15	**0.13**
	Inf	0.24	0.13	0.32	0.24	0.24	0.34	0.12	**0.11**	0.34	0.27	0.23	0.16	0.33	0.36	0.12	0.32	0.18
	All	0.23	0.12	0.34	0.23	0.23	0.37	0.13	**0.11**	0.36	0.27	0.22	0.18	0.36	0.34	0.16	0.36	0.13
CLSM	**Sim**	0.17	0.11	0.23	0.17	0.16	0.23	0.07	**0.06**	0.23	0.20	0.16	0.27	0.22	0.23	0.12	0.23	0.10
	Pri	0.05	0.39	0.08	0.04	0.30	0.07	0.22	0.10	0.09	0.05	0.04	0.42	0.07	0.15	**0.01**	0.07	0.20
	Inf	0.13	0.11	0.19	0.13	0.15	0.19	0.08	**0.07**	0.20	0.16	0.13	0.29	0.19	0.20	0.10	0.19	0.11
	All	0.14	0.14	0.20	0.14	0.17	0.20	0.09	**0.07**	0.20	0.17	0.14	0.30	0.20	0.21	0.10	0.20	0.11

Because such first quality evaluation is performed using 8 bit data only, the accuracy of the best performing methods needs to be verified in the selection phase based on the complete information of each pixel. Therefore, these methods are re-implemented for the application to full pixel-depth images. However, the MinError(I) method exhibits some major flaw, making it unfeasible for further investigation. As described in [118], the algorithm of this method does not always converge. In such case, the result of the Mean algorithm is returned instead. When investigating the phenomenon in more detail, it is observed that for the used test set TSA, the MinError(I) algorithm does not converge in many cases, using 8 bit as well as full bit-depth images. Such performance is inacceptable for an automated preprocessing. Furthermore, the Mean algorithm returned in case of non-convergence performs almost similarly well and is also amongst the best performing algorithms. For these two reasons, the MinError(I) algorithm is excluded from further investigation and the Mean algorithm is considered as its substitute (which is furthermore much less computationally expensive).

Considering such exclusion, the Mean, Huang, Triangle and RelRange methods are selected as performing best for both devices, with error rates of 0.13, 0.12, 0.16 and 0.13 for the CWL sensor and 0.09, 0.14, 0.10 and 0.11 for the CLSM device. For the CLSM device, the Default, IsoData and Otsu methods perform equally well than the Huang method. However, they perform poorly for the CWL device and are therefore discarded.

Selection of binarization methods (S): In the selection phase of the experiment, the four selected intensity-based thresholding methods are evaluated on the images of TSA_CWL and TSA_CLSM. In addition, the proposed topography-based binarization methods of section 4.2.2.2 are evaluated. In this phase, the full bit-depths of the images are used for a more accurate evaluation. The ground truth for all images of TSA_CWL and TSA_CLSM is manually determined on the full bit-depth intensity images. The results are depicted in Figure 50 and Table IX. They show the best performance for CWL intensity images when using the Huang method (b_err = 11%). However, the Mean and RelRange methods perform only one percentage point worse than their predecessor, showing that the three methods perform almost equally well. Also, the Triangle method (b_err = 15%) is in comparatively close proximity.

Comparing the different types of fingerprints, it can be seen from Table IX that most methods perform for most fingerprint classes with b_err rates between 11% and 13% and that except for the Mean method, each algorithm performs significantly worse than the others for one specific fingerprint type. While the Huang method exhibits a b_err of 15% for printed fingerprints, the RelRange method exhibits a b_err of 17% for variations of different influences. The Triangle method shows a b_err of 18% for prints captured under similar conditions. Summarizing these findings, the Huang method exhibits the lowest overall error rate and is selected as the best performing CWL_Int binarization method. It is therefore used in the remaining evaluations of this thesis (TG3 - TG5) for the binarization of fingerprint images in the scope of the preprocessing step.

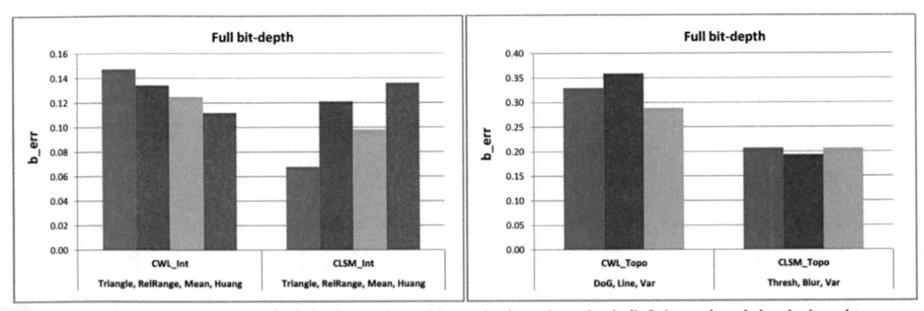

Figure 50: Binarization error (b_err) of the four selected intensity-based methods (left image) and the designed topography-based methods (right image). Errors are computed from test sets TSA_CWL and TSA_CLSM using the full pixel bit-depths provided by the data types and are weighted and averaged over all three classes Sim, Pri and Inf.

When examining the test results for CLSM intensity images (Figure 50, Table IX), it can be seen that the Triangle method significantly outperforms the RelRange, Mean and Huang methods with an overall b_err of 7% (in comparison to 12%, 10% and 14%). While the latter three methods do not perform much better than the methods for CWL intensity images, the Triangle method performs 4% better than the previously selected, best performing CWL_Int method (Huang). The method furthermore performs best in all but one fingerprint categories (b_err = 0% for printed fingerprints, b_err = 5% for prints under different influences) and is only outperformed by the Mean algorithm for fingerprints aged under similar conditions. In summary, the Triangle method is selected for the preprocessing of CLSM_Int images as the best performing technique, being significantly better than its competing algorithms and the best performing method selected for CWL_Int images.

Table IX: Binarization error (b_err) of the four selected intensity-based methods and the designed topography-based methods. Errors are computed from test sets TSA_CWL and TSA_CLSM using the full pixel bit-depths provided by the data types and are weighted and averaged over all three classes Sim, Pri and Inf. The best performing method for each class is marked in bold.

		Triangle	RelRange	Mean	Huang	CWL_Topo_DoG	CWL_Topo_Line	CWL_Topo_Var
CWL	Inf	**0.11**	0.17	0.11	0.11	0.3	0.35	**0.26**
	Pri	0.11	**0.11**	0.13	0.15	**0.32**	0.39	0.42
	Sim	0.18	0.11	0.13	**0.1**	0.35	0.36	**0.29**
	All	0.15	0.13	0.12	**0.11**	0.33	0.36	**0.29**
		Triangle	RelRange	Mean	Huang	CLSM_Topo_Thresh	CLSM_Topo_Blur	CLSM_Topo_Var
CLSM	Inf	**0.05**	0.1	0.09	0.11	0.19	**0.18**	0.22
	Pri	**0**	0.17	0.23	0.39	0.08	**0.06**	0.08
	Sim	0.1	0.13	**0.08**	0.11	0.25	0.23	**0.22**
	All	**0.07**	0.12	0.1	0.14	0.21	**0.19**	0.21

The experimental results of Figure 50 and Table IX for CWL topography images show much higher error rates. This result is not surprising, because the much smaller differences in the topography image gray values are subject to a much higher distortion (section 4.2.2.2) and consequently are much more difficult to use for fingerprint binarization. The CWL_Topo_Var feature exhibits the best overall performance with at total b_err of 29% (more than twice the error rate of the Huang method for CWL_Int images). The CWL_Topo_DoG (33%) and CWL_Topo_Line (36%) features perform significantly worse. Only for printed fingerprints Pri, the CWL_Topo_Var feature is outperformed by the other features. Summarizing such results, the CWL_Topo_Var feature is selected for the preprocessing of CWL_Topo images in the remainder of this thesis.

Comparing the investigated binarization methods based on CSLM_Topo images (Figure 50, Table IX), it can be observed that the performance is significantly better than that of the corresponding algorithms for CWL_Topo images. However, also here the error rate for topography-based methods is significantly higher than that of the CLSM_Int based algorithms. The CLSM_Topo_Blur method performs best with a b_err rate of 19% and is only slightly outperformed by the CLSM_Topo_Var method for fingerprints captured under similar conditions. In total, the other methods perform only two percentage points worse and are therefore considered as being almost similarly accurate. Summarizing the test results, the CLSM_Topo_Blur method is selected as the best performing binarization method for CLSM_Topo images and is therefore used for all remaining experiments in the scope of this thesis.

6.3 Results for TG3: Evaluation of Digital Aging Feature Sets

In this section, the experimental results of test goal TG3 are presented and discussed. The results are divided into short- and long-term aging evaluation, including the features F1 - F7 for the short-term aging (section 6.3.1) and F1 - F15 for the long-term aging (sections 6.3.2 and 6.3.3).

6.3.1 Evaluation of Statistical Short-Term Aging Features

In this section, the experimental results of the feasibility evaluation of seven statistical short-term aging features are presented in respect to the intensity as well as the topography images of the CWL and the CLSM sensor and three different aging properties Pr = {lin,log,exp}. As introduced in sections 4.3.3 and 5.6.3, the Pearson Correlation Coefficient r is used as a quality measure to decide on the feasibility of a feature towards short-term age estimation. The 770 different fingerprint time series of each test set TSB_CWL and TSB_CLSM (section 5.6.1) are evaluated and the correlation coefficient for each series is determined for a total time period of t_{max} = 24 hours. Results are compared in the form of the average correlation coefficient r_avrg of the complete test set as well as the relative amount of test set curves exhibiting a strong correlation h(r≥0.8) or h(r≤-0.8) in respect to the investigated aging property. A general overview over r_avrg as well as h(r≥0.8) and h(r≤-0.8) is given in Figure 51 and Figure 52. It can be observed from Figure 51 that the average correlation coefficient r_avrg can assume positive as well as negative values for the different features, but is similarly signed for the different aging properties of a single feature. One exception can be seen for feature F1 (Binary Pixels), where the general trend of the feature is known beforehand and is therefore inverted for the exponential property, leading to better results (section 5.6.3).

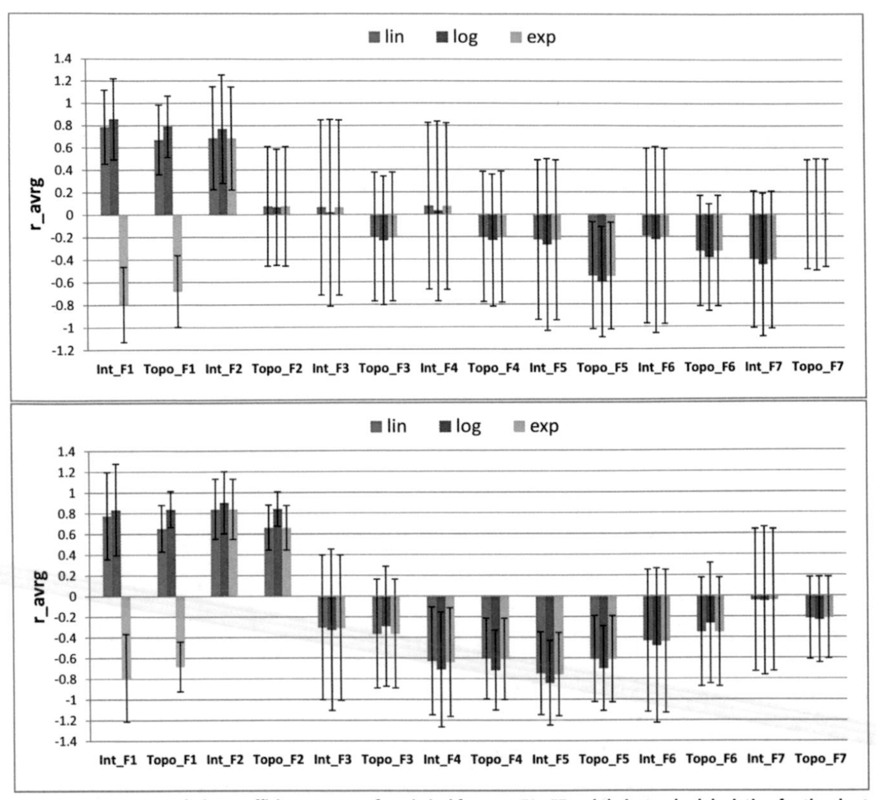

Figure 51: Average correlation coefficients r_avrg of statistical features F1 - F7 and their standard deviation for the short-term aging captured with a CWL sensor (TSB_CWL, upper image) as well as a CLSM device (TSB_CLSM, lower image) in respect to a linear, logarithmic and exponential aging property.

The closer r_avrg is to the value of one, the more characteristic is the monotonic increasing aging property of the feature. The closer r_avrg is to the value of minus one, the more characteristic is the monotonic decrease of the features aging property. In case r_avrg is close to the value of zero, no clear monotony of the aging property can be derived. When determining the relative amount of curves showing a strong characteristic aging property for a certain feature Fk, the results of Figure 51 are used for deciding if an increasing $h(r \geq 0.8)$ or decreasing $h(r \leq -0.8)$ aging property is used, depending on the value of r_avrg being negative or positive. In case r_avrg is close to zero, no clear monotony seems to exist and the feature is therefore not suitable for the age estimation. Furthermore, the standard deviation should be considered. The higher the standard deviation (marked by the error bars in Figure 51), the higher is the fluctuation of the aging properties of the 770 different curves in respect to a certain feature. Small standard deviations are desired to achieve a maximum reproducibility of an aging property.

When comparing the results of Figure 51, it can be seen that logarithmic aging properties are usually stronger than their linear and exponential counterparts for most features of both capturing devices and data types. The average linear and exponential properties seem to be rather similar in their absolute values. Standard deviations range between 0.27 and 0.83 for the CWL sensor and between 0.17 and 0.78 for the CLSM device, being very high for certain features. In such cases, the achieved results cannot be considered reliable, because the standard deviation is too high for the accurate determination of an aging property. However, such high standard deviations mostly exist for features with an r_avrg comparatively close to zero, therefore not being feasible features after all.

The best performance for the CWL sensor (upper image of Figure 51) is achieved by the features Int_F1 (Binary Pixels; r_avrg = 0.86), Topo_F1 (Binary Pixels; r_avrg = 0.79) and Int_F2 (mean gray value; r_avrg = 0.77). For the CLSM device, the feature Int_F2 (mean gray value) performs best with r_avrg = 0.90, followed by Int_F1 (Binary Pixels), Topo_F1 (Binary Pixels), Topo_F2 (mean gray value) and Int_F5 (mean gradients), all exhibiting values of r_avrg = 0.84 or r_avrg = -0.84 respectively. Overall, the CLSM device exhibits higher average correlation coefficients than the CWL sensor. The relative frequency of time series showing a strong increasing h(r≥0.8) or decreasing h(r≤-0.8) correlation to a certain aging property are depicted in Figure 52 for the test sets TSB_CWL and TSB_CLSM.

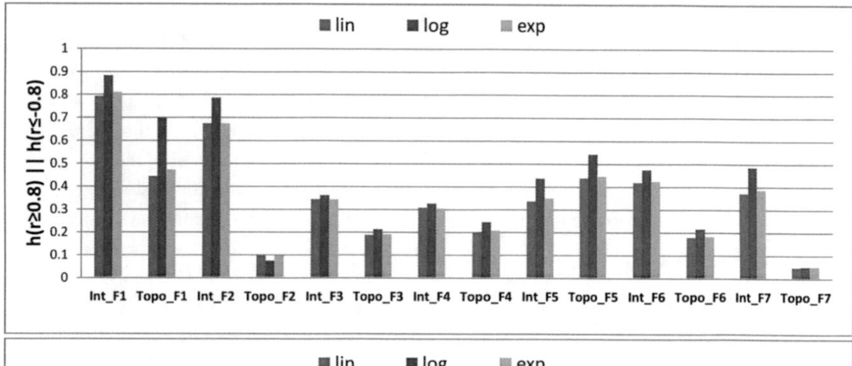

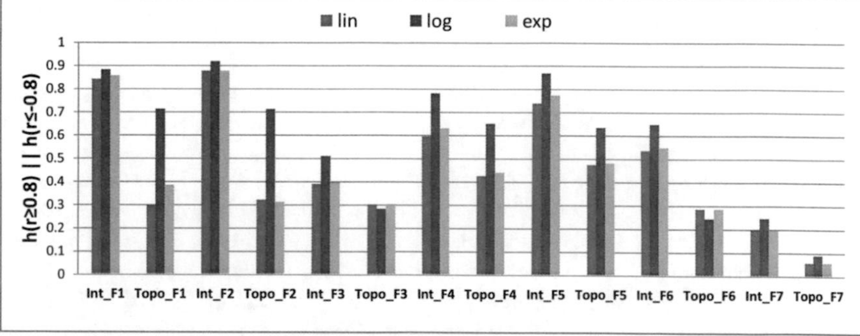

Figure 52: Relative frequency of experimental short-term aging curves (F1 - F7) exhibiting a strong increasing h(r≥0.8) or decreasing h(r≤-0.8) correlation to a certain aging property Pr (linear, logarithmic exponential). Results are depicted for the CWL sensor (TSB_CWL, upper image) as well as the CLSM device (TSB_CLSM, lower image).

From Figure 52 it can be confirmed that indeed a logarithmic aging property seems to be most characteristic for the majority of aging features of both devices, while linear and exponential properties exhibit approximately similar results. When comparing the performance of the single features between each other, it can be observed from Figure 52 that Int_F1 (Binary Pixels) and Int_F2 (mean gray value) perform best for the CWL sensor, while Topo_F1 (Binary Pixels) performs almost as good, however only for a logarithmic aging property Pr. All other features perform significantly worse. For these three best performing features, a strong correlation to a logarithmic aging property is present in 88% (Int_F1), 70% (Topo_F1) and 78% (Int_F2) of all aging curves of TSB_CWL.

For the CLSM device, Int_F1 (Binary Pixels), Int_F2 (mean gray value) and Int_F5 (mean gradients) perform best. They exhibit a strong correlation to a logarithmic function in 88% (Int_F1), 92% (Int_F2) and 87% (Int_F5) of TSB_CLSM time series. Overall, it can be concluded that the CLSM device performs significantly better than the CWL sensor for most features. A more detailed evaluation of the different features is presented in the following sections. The CWL results correlate largely to those published in [153] in the scope of this thesis based on obsolete preprocessing methods. This highlights that the features computed from CWL images are comparatively robust and work adequately even when using obsolete preprocessing methods (which cannot be confirmed for the CLSM device from numerous observations throughout the experiments).

To examine the specific causes of aging curves showing an aging property different from the general trend (i.e. exhibiting a low correlation coefficient) is a complex task and not possible at this point of time. A wide variety of different influences impacts the aging behavior of fingerprints and potentially distorts a certain aging property. Different sweat compositions might be a source of significant differences, which cannot be examined with the here investigated devices. Many other influences might also be of importance, such as environmental conditions or influences from the capturing devices. It is therefore not feasible in the scope of this thesis to investigate and identify these causes in detail, independently from the examined aging feature. It can only be evaluated to which extent the aging property of a certain feature is reproducible for a significant amount of different samples in the scope of r_avrg, h(r≥0.8) and h(r≤0.8). A first attempt to investigate the complex set of influences on the aging behavior of fingerprints in more detail is performed in the scope of test goal TG5. The following sections examine the features F1 - F7 in more detail. Correlation coefficients r are given for each time series of test sets TSB_CWL and TSB_CLSM in respect to the three aging properties.

6.3.1.1 Evaluation of Binary Pixels (F1)

The Binary Pixels feature F1 is specifically designed to capture the characteristic changes occurring during fingerprint aging. Its most characteristic aging property is shown in Figure 51 to be a logarithmic one, confirming the findings of earlier studies published in the scope of this thesis, based on obsolete preprocessing techniques [153], [156], [159]. The correlation coefficients of all 770 time series of test set TSB are presented in Figure 53. It is to note that as argued in section 4.3.3, the exponential aging property is computed for the inverse Binary Pixels tendency (F1'=1-F1), because F1 measures the amount of increasing background pixels and an exponential decay should naturally be applied to the decreasing print tendency F1'=1-F1. However, this procedure is only performed for F1, because it is the only feature whose aging property is well known in advance, due to extensive prior tests.

It can be observed that for CWL as well as CLSM intensity data (left images of Figure 53), F1 exhibits a comparatively strong correlation to all three aging properties, while the correlation coefficients of a logarithmic aging property seem to be closest to one. A logarithmic tendency therefore seems to be the dominant aging property. However, such property is found to exhibit a relatively small slope, still allowing for a comparatively well linear approximation. The correlation to an exponential aging property seems to be similar to that of a linear property (yet inverted due to the use of F1'=1-F1). As is demonstrated in more detail in the next section (evaluation of feature F2), the observed exponential aging properties are extremely close to their linear counterparts, rendering a separate investigation of such exponential property unnecessary. A few outliers exist for all three aging properties, scattered throughout the complete diagram.

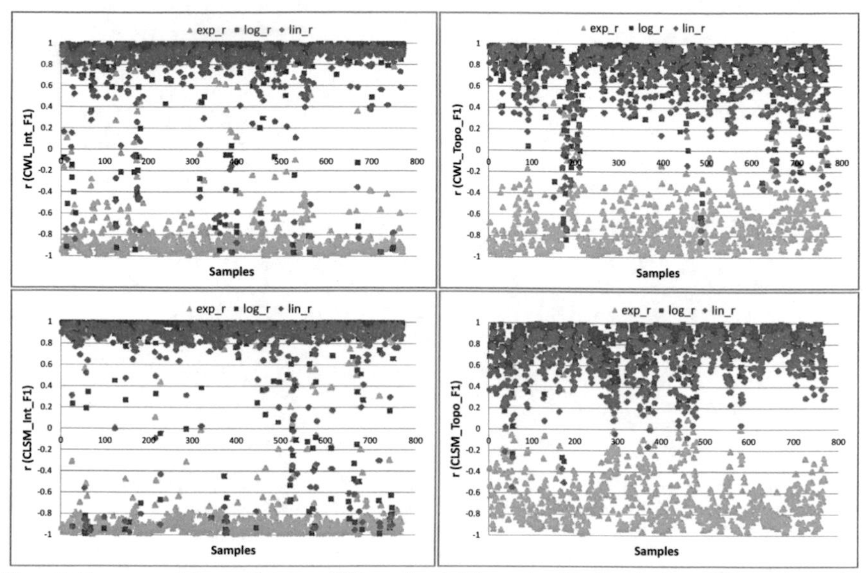

Figure 53: Pearson correlation coefficients r of 770 experimental time series of the intensity data of TSB_CWL (upper left image), topography data of TSB_CWL (upper right image), intensity data of TSB_CLSM (left lower image) and topography data of TSB_CLSM (right lower image) in respect to feature F1 for a linear, logarithmic and exponential aging property.

For the topography data of both devices (right images of Figure 53), feature F1 exhibits a similar trend than for the intensity images, however with a much higher deviation of the correlation coefficients. Such observation demonstrates the significantly less quality of the topography data in comparison to intensity images, probably resulting in a significantly decreased performance of the later investigated age estimation approaches. Possible reasons for the worse performance of topography data might be found in the worse quality of the separation of fingerprint pixels from noise and other background distortions (section 6.2).

6.3.1.2 Evaluation of Mean Image Gray Value (F2)

The mean image gray value feature F2 is another promising aging feature. In case fingerprint pixels become brighter during aging, the average fingerprint pixel value might increase. To investigate if such increase follows a characteristic aging property is the aim of this section. The correlation coefficients of all 770 time series are depicted in Figure 54 for CWL as well as CLSM intensity and topography images. It can be seen that for CWL images, a significant difference exists between intensity (upper left image) and topography (upper right image) data. While intensity images show the highest correlation coefficients for a logarithmic aging property, followed by an almost similar correlation for linear and exponential aging properties, the correlation coefficients of the CWL topography time series are widely scattered throughout the complete diagram for all three aging properties. It can therefore be concluded that CWL topography samples to not exhibit any characteristic aging property according to feature F2.

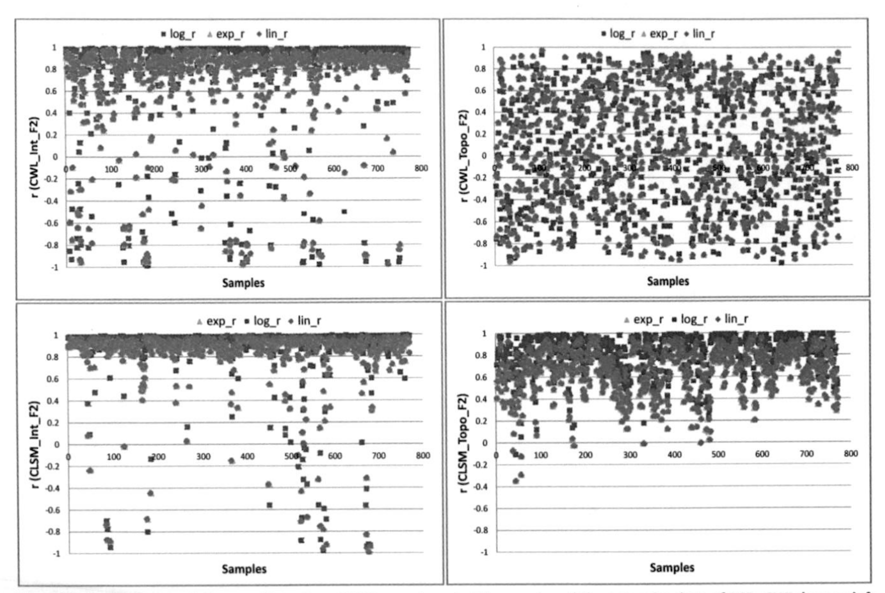

Figure 54: Pearson correlation coefficients r of 770 experimental time series of the intensity data of TSB_CWL (upper left image), topography data of TSB_CWL (upper right image), intensity data of TSB_CLSM (left lower image) and topography data of TSB_CLSM (right lower image) in respect to feature F2 for a linear, logarithmic and exponential aging property.

For CLSM images (lower row of Figure 54) a high correlation can be observed for the intensity series (left image) to aging functions of all three properties. CLSM_Int_F2 therefore seems to be a feasible feature for the age estimation of latent prints and is the overall best performing feature in the scope of TG3 short-term age investigation. For CLSM topography data, the correlation coefficients are much more scattered, yet still comparatively close to a value of one.

It is to note that for intensity as well as topography images of both devices, the correlation coefficients of an exponential aging property are almost similar to their linear counterparts (they can barely be seen in Figure 54, because they lie almost exactly behind the linear ones). In such case, an approximated exponential curve is almost linear. Examples for such behavior are depicted in Figure 55. It can be concluded that the approximation of an exponential aging property is in general not useful for the here investigated monotonic increasing aging tendencies. However, also in the case of a monotonic decreasing tendency, the exponential approximation is close to a linear one, as is exemplary shown in Figure 55. A similar behavior is observed for topography images (Figure 55). It can therefore be concluded that an exponential aging property does not seem to provide any added value in comparison to a linear property. It is therefore not of interest to further investigate such aging property, because all important aspects can be modeled using a linear property. This conclusion is supported by most of the following investigated aging features F3 - F7, exhibiting similar results.

From Figure 55 it can be seen that both intensity-based curves (left images) exhibit a characteristic logarithmic progression (CWL_Int_F2: r = 0.991; CLSM_Int_F2: r = 0.997), whereas the exemplary topography-based curves (right images) show a progression less well correlated to a logarithmic function (CWL_Topo_F2: r = -0.754; CLSM_Topo_F2: r = 0.823). It can furthermore be seen that the topography-based time series seem to be much more vulnerable to distortions and noise, confirming the earlier findings of a worse correlation to certain aging properties.

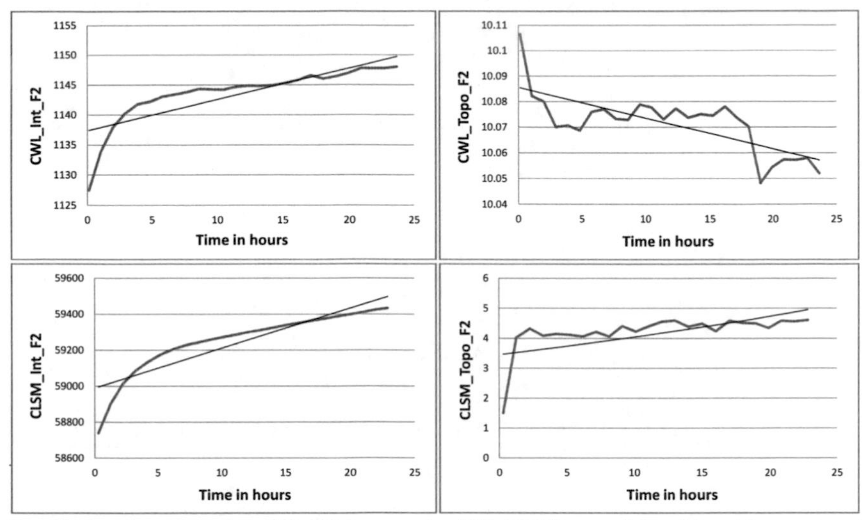

Figure 55: Exemplary aging curves of TSB_CWL (upper row) and TSB_CLSM (lower row). From left to right: curves based on feature Int_F2; curves based on feature Topo_F2. The regression line of an exponential approximation is given for each curve (black lines).

When studying the exponential regression curves depicted in Figure 55 (black lines), it can be seen that they all have the shape of a straight line, which is close to a linear approximation, not exhibiting any exponential property at all. It can therefore be concluded from such results that an approximated exponential aging property is similar to its corresponding linear approximation. Because such behavior occurs for most investigated aging features, an exponential aging property does not need to be further studied and is discarded in the remainder of this thesis in favor of a linear aging property.

6.3.1.3 Evaluation of Global Standard Deviation (F3)

The global standard deviation of fingerprint image pixels is investigated in the scope of feature F3. Such standard deviation might change during fingerprint aging, in case the gray values of image pixels become more similar over time. The results are depicted in Figure 56. It can be concluded that although the intensity images of both devices show a certain correlation to a linear or logarithmic aging property, these correlations are not similar for all time series. Some curves seem to exhibit a rather strong increasing correlation, while others exhibit a rather strong decreasing one. The different monotony of occurring strong aging properties indicates that such properties are not the product of a systematic aging behavior, but rather some unidentified influence. The feature therefore does not seem to be feasible for age estimation. The topography images of both devices exhibit widely scattered correlation coefficients and are therefore even less suitable. The investigation of potential causes for such high variation of aging tendencies remains a subject of more detailed studies in the future. Similar to the observations for feature F2, most correlation coefficients using exponential approximation are similar to their linear counterparts (lying exactly behind them in Figure 56) and are therefore discarded.

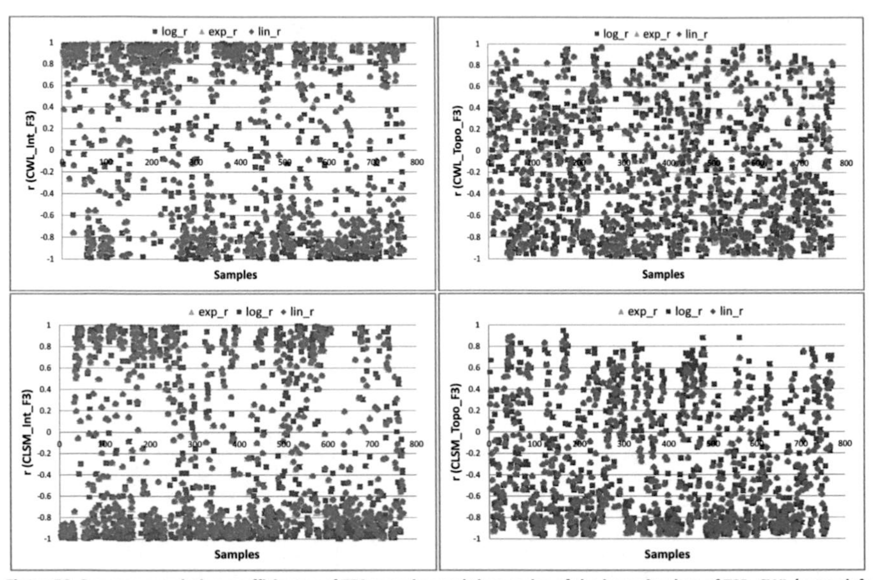

Figure 56: Pearson correlation coefficients r of 770 experimental time series of the intensity data of TSB_CWL (upper left image), topography data of TSB_CWL (upper right image), intensity data of TSB_CLSM (left lower image) and topography data of TSB_CLSM (right lower image) in respect to feature F3 for a linear, logarithmic and exponential aging property.

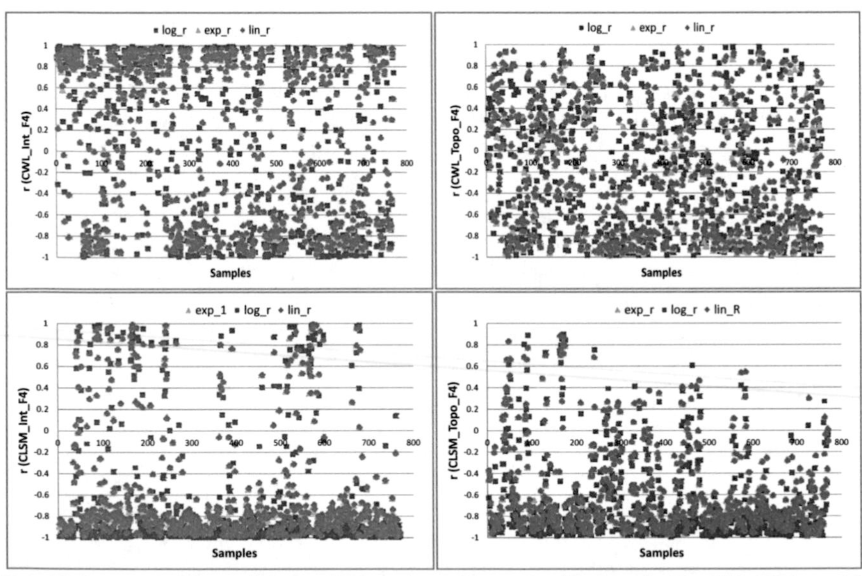

Figure 57: Pearson correlation coefficients r of 770 experimental time series of the intensity data of TSB_CWL (upper left image), topography data of TSB_CWL (upper right image), intensity data of TSB_CLSM (left lower image) and topography data of TSB_CLSM (right lower image) in respect to feature F4 for a linear, logarithmic and exponential aging property.

6.3.1.4 *Evaluation of Mean Local Variance (F4)*

The mean local variance (F4) is a feature based on the local differences between image pixels. Areas of high variance are likely to contain fingerprint residue, which consists of many small particles, accounting for such variance. During aging, such variance might either become smaller (when residue is slowly disappearing) or higher (in cases particles are collapsing into smaller particles). The experimental results for this feature are depicted in Figure 57. It shows a poor performance concerning the CWL device (upper row) for intensity as well as topography images. While intensity-based curves seem to exhibit strong increasing as well as decreasing tendencies at the same time, topography-based correlation coefficients are widely scattered. Time series captured with the CLSM device (lower row) show a common trend towards a monotonic decreasing aging behavior for both data types. However, a significant amount of outliers exist, decreasing the overall feasibility of the feature. To understand the specific causes of such outliers is an important research topic for future studies.

6.3.1.5 *Evaluation of Mean Gradient (F5)*

Feature F5 is characterized by the gradients of a fingerprint image. Because degradation processes potentially lead to fingerprint pixels turning into background pixels, the gradients of a latent print image might gradually decrease. However, the decomposition of fingerprint compounds can also lead to the creation of novel edges and therefore an increase of the gradients. The results of this feature are depicted in Figure 58.

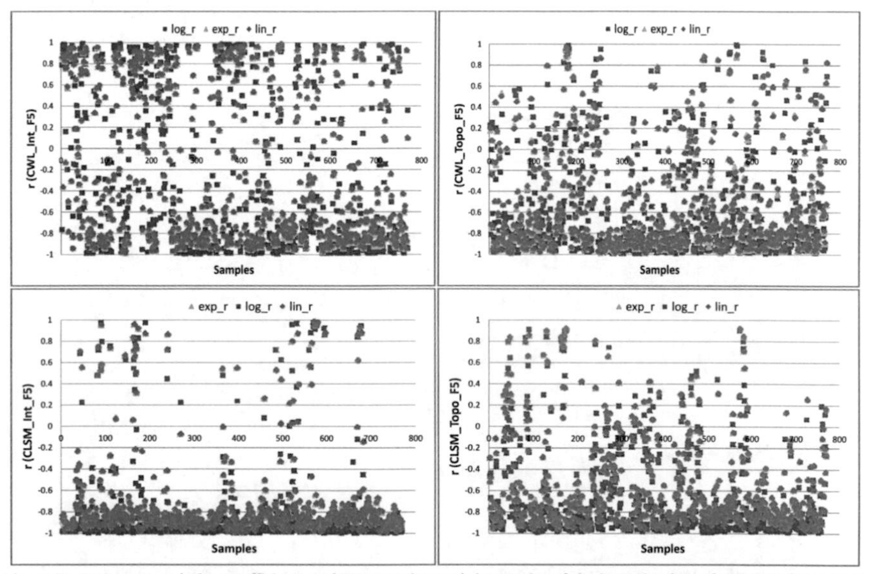

Figure 58: Pearson correlation coefficients r of 770 experimental time series of the intensity data of TSB_CWL (upper left image), topography data of TSB_CWL (upper right image), intensity data of TSB_CLSM (left lower image) and topography data of TSB_CLSM (right lower image) in respect to feature F5 for a linear, logarithmic and exponential aging property.

The results presented in Figure 58 show a monotonic decreasing trend for all for data types. However, a comparatively high amount of outliers is present for the CWL_Int, CWL_Topo and CLSM_Topo series, rendering feature F4 rather infeasible for these data types. Merely the CLSM_Int time series exhibit only a few outliers, making CLSM_Int_F4 a potential feature for age estimation.

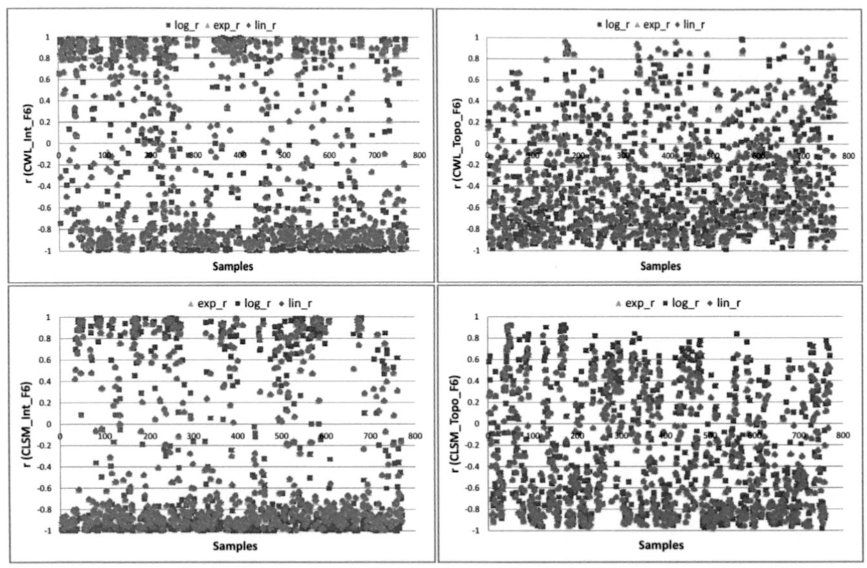

Figure 59: Pearson correlation coefficients r of 770 experimental time series of the intensity data of TSB_CWL (upper left image), topography data of TSB_CWL (upper right image), intensity data of TSB_CLSM (left lower image) and topography data of TSB_CLSM (right lower image) in respect to feature F6 for a linear, logarithmic and exponential aging property.

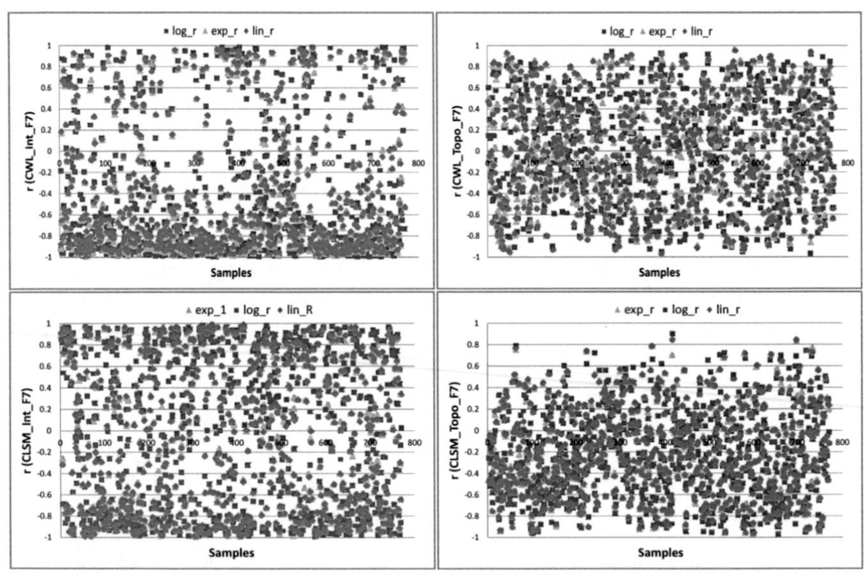

Figure 60: Pearson correlation coefficients r of 770 experimental time series of the intensity data of TSB_CWL (upper left image), topography data of TSB_CWL (upper right image), intensity data of TSB_CLSM (left lower image) and topography data of TSB_CLSM (right lower image) in respect to feature F7 for a linear, logarithmic and exponential aging property.

6.3.1.6 *Evaluation of Mean Surface Roughness (F6)*

The surface roughness represented by feature F6 might be a potential feature for the age estimation of latent prints. Because fingerprint pixels seem to disappear over time, the surface roughness could decrease. However, also an opposite trend is possible, where big particles are decomposed into smaller ones, leading to an increase of the surface roughness. The results for feature F6 are depicted in Figure 59. Examining such results, feature F6 does not seem to be a feasible representation of latent fingerprints aging behavior. For both intensity data types, a monotonic decreasing trend seems to be present. However, it is biased by a significant amount of time series with a high positive correlation coefficient and therefore a positive trend, preventing the identification of a clear overall aging property. The topography-based series of both devices show widely scattered correlation coefficients and are therefore also infeasible for practical age estimation.

6.3.1.7 *Evaluation of Coherence (F7)*

The statistical feature F7 is based on the coherence of fingerprint images. Whether the coherence exhibits characteristic changes during the short-term aging of a latent print is unknown and therefore here investigated. The experimental results are depicted in Figure 60. Concluding from them, coherence seems to be infeasible for the age estimation of latent fingerprints. While CWL_Int time series exhibit a certain negative trend with a significant amount of outliers, CWL_Topo, CLSM_Int and CLSM_Topo series produce correlation coefficients widely scattered. The cause of such variation is hard to determine at this point of time and requires more detailed studies in future work.

6.3.2 Evaluation of Statistical Long-Term Aging Features

In this section, the statistical features F1 - F7 are evaluated towards their performance for the long-term fingerprint aging. Test sets TSC_CWL and TSC_CLSM (section 5.6.2) are used, each comprising 20 time series of different fingerprints. The prints are examined once every week ($\Delta t = 1$ week) over time periods of $t_{max} = 2.3$ years (TSC_CWL) and $t_{max} = 30$ weeks (TSC_CLSM). Prints are first examined after one day, to exclude the strong changes during the drying process of the prints (short-term aging). Similar to the quality evaluation of short-term aging features, the correlation coefficient r is used in the form of r_avrg as well as $h(r \geq 0.8)$ and $h(r \leq -0.8)$. Because the approximation of an exponential function has been shown in section 6.3.1 to produce similar results than a linear approximation, only the aging properties Pr = {lin,log} are investigated here. A general overview over r_avrg as well as $h(r \geq 0.8)$ and $h(r \leq -0.8)$ is given in Figure 61 and Figure 62.

The long-term aging results depicted in Figure 61 of the statistical features F1 - F7 show a significantly worse performance for the CWL device (upper image) in comparison to the earlier reported short-term aging series (Figure 51) for some of the features (such as Int_F1, Topo_F1 and Int_F2). Reasons for such behavior can be seen in the significantly longer time period of $t_{max} = 2.3$ years. During such long period of time, fluctuating environmental conditions, such as (among others) seasonal temperature and humidity changes, could become more evident. Furthermore, dust can contribute significantly to certain distortions of the experimental aging curves, despite the elimination of large amounts of dust particles in the preprocessing step (section 4.2.5.6). Other influences, such as changes in the sensor itself (e.g. slow degeneration of the light source or the mechanics of the measurement table) might contribute as well. The repeated storage, transportation as well as fixation and removal of the prints from the measurement table can add certain distortions (e.g. increase air movement during transportation, slight differences in the repeated fixation of the platter onto the measurement table). Also, the fingerprints have undergone a 24 hour period of drying before their first capture, to exclude the short-term aging effects. Therefore, long-term aging changes are of smaller magnitude, being more vulnerable to distortions than for the investigation of short-term aging behavior.

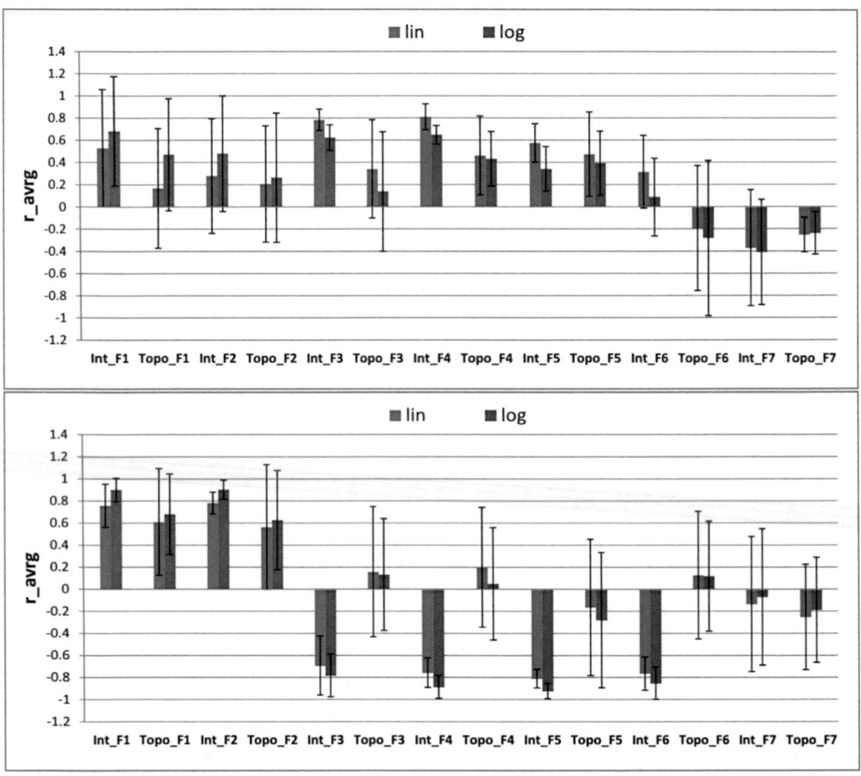

Figure 61: Average correlation coefficients r_avrg of statistical features F1 - F7 and their standard deviation for the long-term aging captured with a CWL sensor (TSC_CWL, upper image) as well as a CLSM device (TSC_CLSM, lower image) in respect to a linear and logarithmic aging property.

Apart from CWL time series showing a worse performance for certain features, other features exhibit a significantly increased performance, such as Int_F3, Int_F4, Topo_F4 and Int_F5. Possible reasons for such behavior might be seen in certain features targeting degradation processes which manifest themselves only over long time periods. Such features are therefore more suitable for the long-term aging. However, due to the significant amount of distortions possibly influencing the observed long-term aging behavior, characteristic changes could also be caused by certain distorting influences, such as an increased amount of dust.

Some features of the CWL time series change the monotony of their aging curves (e.g. Topo_F3, Topo_F4, Int_F5, Topo_F5 and Int_F6) in respect to the earlier investigated short-term aging (Figure 51). Most of this features have exhibited average correlation coefficients r_avrg close to zero for the short-term aging (such as Topo_F3, Topo_F4, Int_F5 and Int_F6, Figure 51) and are therefore considered as not being characteristic for representing the short-term aging behavior of fingerprints. However, feature Topo_F5 can be regarded as an exception, because its average correlation coefficient changed from r_avrg = -0.6 (short-term aging) to r_avrg = 0.39 (long-term aging) for a logarithmic aging property. Possible reasons for such change might be seen in certain differences in the degradation process between short and long time periods. However, a detailed analysis of possible reasons for changes in monotony between short- and long-term aging series remains subject to future research and possibly requires the inclusion of chemical analyses of the fingerprint degradation, e.g. using FTIR spectroscopy.

Examining the long-term aging results of Figure 61 in respect to the CLSM time series, intensity- and topography-based series need to be distinguished. For intensity-based series, the average correlation r_avrg seems to be similar or better than for the short-term aging (Figure 51). Such observation might be explained by the respective aging features being more suitable to capture fingerprint changes over long aging periods. However, it seems more likely that systematic influences of the capturing device during the aging period lead to these systematic changes in feature values, such as challenge PC2, identified in section 4.1.2. Although this challenge has been addressed in the scope of O2 (section 4.2), it cannot be guaranteed that the occurring distortions are completely excluded, which are significantly stronger for the long-term aging. For example, the temporal normalization might overcompensate for such distortions over long time periods, leading to a systematic error. Therefore, the CLSM long-term aging series are regarded as less reliable than their CWL counterparts. For the topography-based CLSM series, the features Topo_F1 and Topo_F2 show only small differences of r_avrg in comparison to the short-term aging (Figure 51), while all other topography-based features perform particularly bad. A possible reason for such behavior could be the increased sensitivity of the topography-based features F3 - F7 to certain distortions. However, a more detailed investigation of such phenomenon remains an important challenge for future research.

Comparing the aging properties of Figure 61, it can be seen than a significantly higher amount of aging features now exhibit their best performance using a linear property. This seems reasonable taking into consideration that the first 24 hours of aging (exhibiting the most changes due to the drying process) have been excluded from the long-term aging time series. The remaining changes in aging speed are much smaller, leading to an optimal linear approximation in many cases, especially for the CWL time series, which are captured over a much longer time period.

Because significant changes have been observed for r_avrg between the earlier investigated short-term aging and the here investigated long-term aging, significant changes are also expected for the relative frequencies of curves $h(r \geq 0.8)$ and $h(r \leq -0.8)$ showing a strong increasing or decreasing correlation. The experimental results of such frequencies are depicted in Figure 62. They show significant changes of $h(r \geq 0.8)$ and $h(r \leq -0.8)$ in comparison to the short-term aging evaluation (Figure 52). For the CWL-based time series, the amount of curves showing a strong logarithmic property has significantly decreased, while a linear property becomes more dominant. Furthermore, well performing features in the scope of short-term aging (e.g.Int_F1, Topo_F1 and Int_F2) perform significantly worse. At the same time, other features perform significantly better, such as Int_F3 and Int_F4. Over the investigated time period of 2.3 years, many aging curves are subject to severe distortions, especially for the topography-based time series.

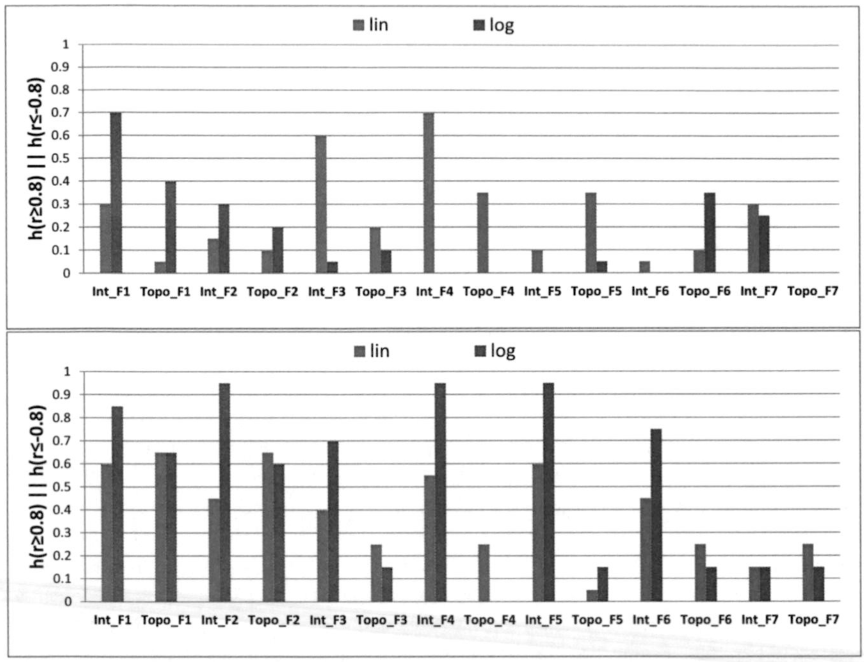

Figure 62: Relative frequency of experimental long-term aging curves (F1 - F7) exhibiting a strong increasing h(r≥0.8) or decreasing h(r≤-0.8) correlation to a certain aging property Pr (linear, logarithmic). Results are depicted for the CWL sensor (TSC_CWL, upper image) as well as the CLSM device (TSC_CLSM, lower image).

For CLSM-based results, some features remain comparatively constant in the amount of series exhibiting a strong correlation to a logarithmic property (e.g. Int_F1, Int_F2), while others show a significantly increased performance (e.g. Int_F3, Int_F4, Int_F5, Int_F6). Topography-based images perform particularly bad for features F3 - F7, confirming what has been observed for r_avrg (Figure 61). It can furthermore be noted that a logarithmic aging property performs significantly better for CLSM long-term series than a linear one. Possible reasons for such behavior are difficult to determine and range from a potentially higher precision of the CLSM device to a (far more likely) systematic influence of the device on the captured fingerprints. A systematic influence of the capturing device would also explain why such comparatively high amount of aging curves exhibit a strong correlation for several features, which has not been observed for the short-term aging. The issue is discussed in more detail in the scope of test goal TG5 (section 6.5.2) and remains an important topic for future research.

The following sections examine the features F1 - F7 in more detail. Because the size of TSC_CWL and TSC_CLSM is significantly smaller than the size of its short-term aging counterparts TSB_CWL and TSB_CLSM, time series are investigated by plotting 10 exemplary aging curves for each feature instead of depicting the correlation coefficients.

6.3.2.1 Evaluation of Binary Pixels (F1)

From the 20 investigated time series, 10 exemplary experimental aging curves are depicted in Figure 63 for the features CWL_Int_F1, CWL_Topo_F1, CLSM_Int_F1 and CLSM_Topo_F1.

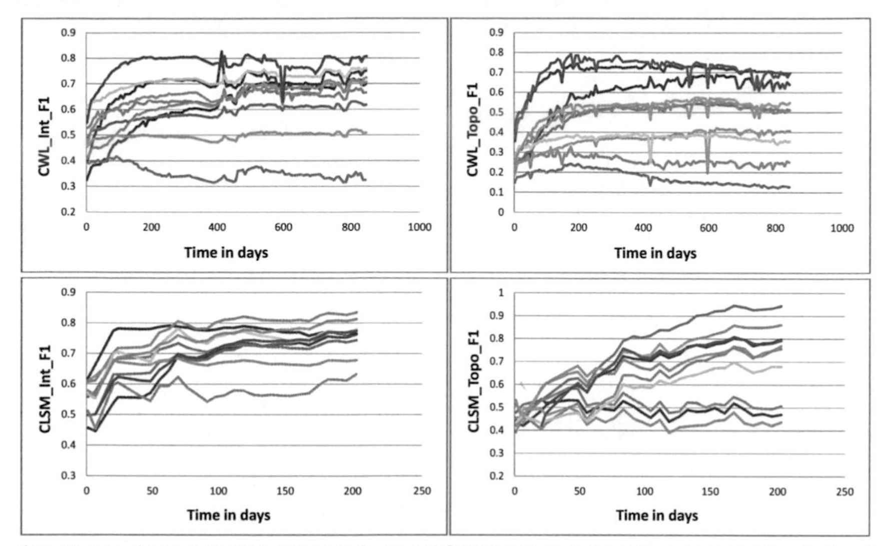

Figure 63: Exemplary experimental aging curves from 10 of the 20 investigated time series based on the intensity data of TSC_CWL (upper left image), topography data of TSC_CWL (upper right image), intensity data of TSC_CLSM (left lower image) and topography data of TSC_CLSM (right lower image) in respect to feature F1.

When analyzing the experimental results of Figure 63, it can be seen that an increasing monotony seems to be the dominant aging trend, which confirms the expected behavior of the Binary Pixels feature (F1). When comparing the aging properties, the logarithmic nature of the aging behavior using feature F1 can be clearly recognized for CWL_Int_F1 and CWL_Topo_F1 (upper row), confirming the results published in [151] using obsolete preprocessing methods. For CLSM_Int_F1, a logarithmic progression also seems to be present, however significantly distorted. For CLSM_Topo_F1, the aging property seems to be rather linear and also shows a significant variation in its slope. The reason for such behavior could be an inadequate quality of CLSM topography time series when investigating the long-term aging, where different influences (e.g. from the capturing device) might significantly overlay the observed aging tendency.

For most experimental aging curves of both devices and data types, systematic influences can be observed. For example, spikes in the aging curves can be seen for all curves of a certain device and data type (e.g. for CWL_Int_F1: t = 790 days; CLSM_Int_F1: t = 25 days). Because these spikes occur similarly throughout all series of a certain data type, external influences (e.g. sensor or environmental influences) seem to be the reason for such distortions. In the scope of test goal TG5, some of such influences are exemplary studied in first qualitative tests.

6.3.2.2 Evaluation of Mean Image Gray Value (F2)

For the long-term aging based on feature F2, 10 exemplary aging curves are depicted in Figure 64. According to the figure, feature F2 seems to be characteristic for CWL as well as CLSM intensity-based time series (CWL_Int_F2 and CLSM_Int_F2). A clear logarithmic progression can be observed for most curves of both devices. Systematic influences on all intensity-based curves exist (e.g. for CWL_Int_F2: t = 790 days; CLSM_Int_F2: t = 25 days). Overall, the intensity-based curves of both devices seem to exhibit a characteristic trend, which confirms the high performance of F2 in the scope of the short-term aging (section 6.3.1).

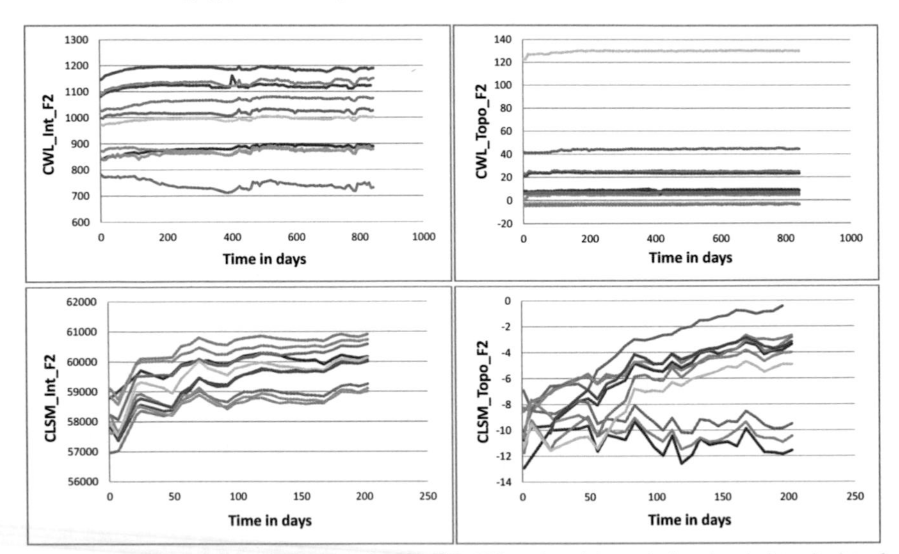

Figure 64: Exemplary experimental aging curves from 10 of the 20 investigated time series based on the intensity data of TSC_CWL (upper left image), topography data of TSC_CWL (upper right image), intensity data of TSC_CLSM (left lower image) and topography data of TSC_CLSM (right lower image) in respect to feature F2.

When studying the topography-based results of Figure 64 (right row), a mixed performance can be observed. The CWL topography-based series (CWL_Topo_F2) show, apart from a monotonic increase in the beginning, almost no changes in their feature values, which is representative for a logarithmic progression. However, these logarithmic curves seem to converge much faster to their maximum value than the corresponding intensity-based curves, potentially capturing different properties (e.g. the decrease of droplet size in contrast to the light intensity reflected at droplets). The causes of such different behavior needs to be investigated in more detail in future studies, including also substance-specific analyses. The CLSM topography-based series (CLSM_Topo_F2) show a high variation in the slopes of the curves and also severe distortions of many curves. A logarithmic progression cannot be concluded for most curves. However, it needs to be considered that the aging period of t = 30 weeks is considerably smaller than the period used for CWL images (t = 2.3 years).

Summarizing the findings for feature F2, a characteristic logarithmic progression can be observed for CWL_Int_F2, CLSM_Int_F2 and CWL_Topo_F2, whereas CLSM_Topo_F2 exhibits a significant variation. Furthermore, it can be seen from Figure 64 that after a certain time of logarithmic increase (t = 150 days), most curves have converged, therefore not providing characteristic information about the age of a print after that time.

6.3.2.3 Evaluation of Global Standard Deviation (F3)

The experimental results for feature F3 are provided in Figure 65 for 10 exemplary experimental aging curves. Examining such results, an opposing trend can be observed between CWL and CLSM intensity-based series (CWL_Int_F3 and CLSM_Int_F3). While the CWL intensity-based curves exhibit a heavily distorted increase of feature values (which seems to be subject to severe systematic distortions), the CLSM intensity-based curves exhibit a similar, yet decreasing trend.

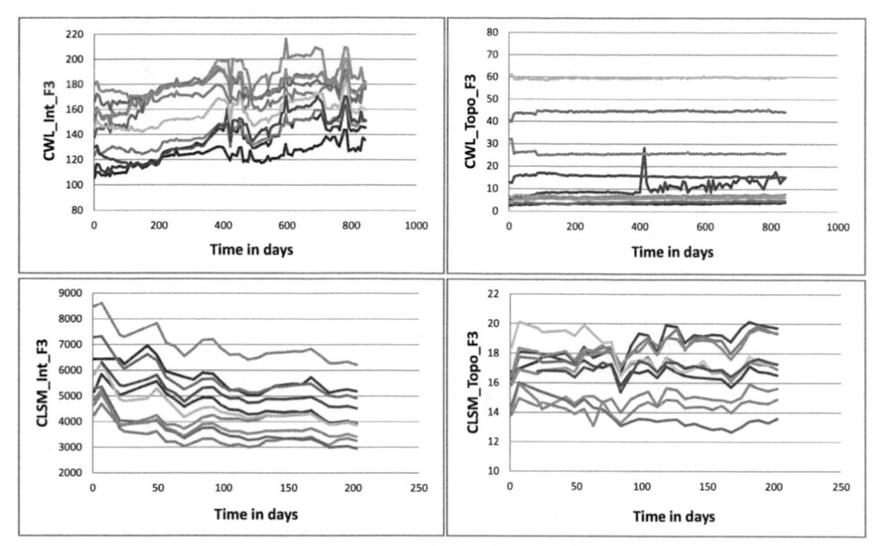

Figure 65: Exemplary experimental aging curves from 10 of the 20 investigated time series based on the intensity data of TSC_CWL (upper left image), topography data of TSC_CWL (upper right image), intensity data of TSC_CLSM (left lower image) and topography data of TSC_CLSM (right lower image) in respect to feature F3.

Topography-based series show indifferent trends for the CWL device (CWL_Topo_F3), where logarithmically increasing as well as logarithmically decreasing aging tendencies do occur. Such indifferent trends make the feature infeasible for age estimation, confirming the corresponding findings for the short-term aging. Topography-based CLSM series (CLSM_Topo_F3) show irregular and heavily distorted progressions and therefore seem to also be infeasible for the long-term age estimation.

6.3.2.4 Evaluation of Mean Local Variance (F4)

The experimental results for the long-term aging evaluation of feature F4 are depicted in Figure 66 using 10 experimental aging curves. The intensity-based results seem to be similar to those of feature F3. While Figure 66 shows a monotonic increasing, yet heavily distorted trend for feature CWL_Int_F4, a characteristic logarithmic decrease can be observed for all curves of CLSM_Int_F4, exhibiting a certain potential for fingerprint age estimation. It is of interest to note that the systematic distortions of both CWL_Int_F4 and CLSM_Int_F4 can also be found for feature F3 (left column of Figure 65) as well as in later investigated features. They might therefore be attributed to certain systematic sensor or external influences, impacting all fingerprints in a similar way.

When examining the topography-based series of feature F4, the experimental curves of the CWL device (CWL_Topo_F4) exhibit comparatively linear progressions of the curves, varying significantly in their slope. Some of them are heavily distorted. The topography-based series of the CLSM device (CLSM_Topo_F4) do not show any clear trend and also seem to be subject to heavy distortions. The topography-based series of both devices therefore seem to be unsuitable to represent latent print long-term aging.

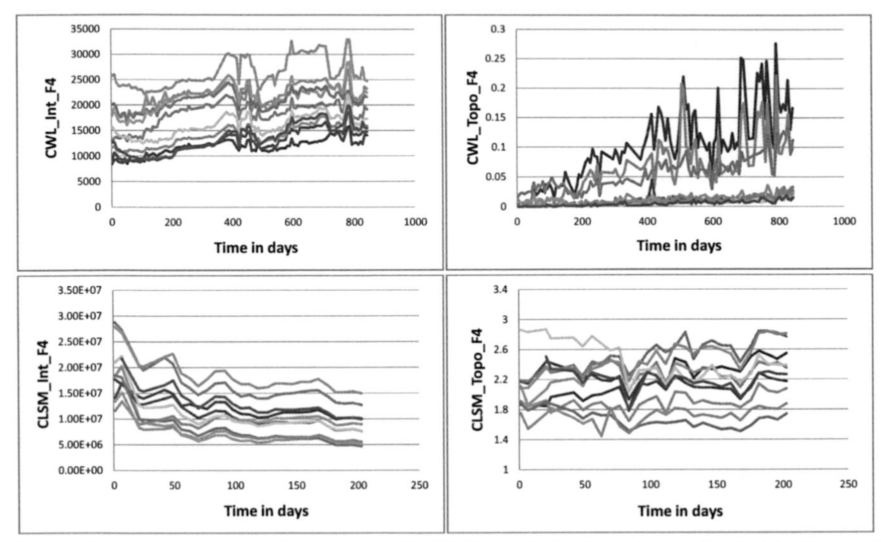

Figure 66: Exemplary experimental aging curves from 10 of the 20 investigated time series based on the intensity data of TSC_CWL (upper left image), topography data of TSC_CWL (upper right image), intensity data of TSC_CLSM (left lower image) and topography data of TSC_CLSM (right lower image) in respect to feature F4.

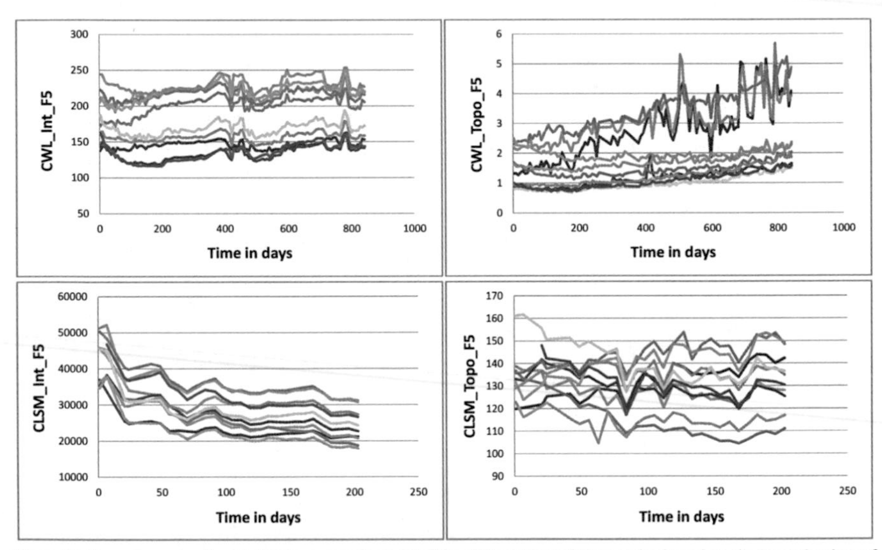

Figure 67: Exemplary experimental aging curves from 10 of the 20 investigated time series based on the intensity data of TSC_CWL (upper left image), topography data of TSC_CWL (upper right image), intensity data of TSC_CLSM (left lower image) and topography data of TSC_CLSM (right lower image) in respect to feature F5.

6.3.2.5 Evaluation of Mean Gradient (F5)

In the scope of the long-term aging evaluation based on feature F5, the experimental results are exemplary depicted in Figure 67 using 10 different measured aging curves. The results seem to be almost identical to those of feature F4 in their general trends. This finding confirms the earlier observation that systematic influences of external factors seem to be dominant for these features in the scope of long-term aging. However, because both features measure changes occurring at the edges of particles, the strong similarity of the curves might also be attributed to a certain similarity in the design of the features.

6.3.2.6 Evaluation of Mean Surface Roughness (F6)

The experimental results for feature F6 are visualized in Figure 68 using 10 exemplary experimental aging curves.

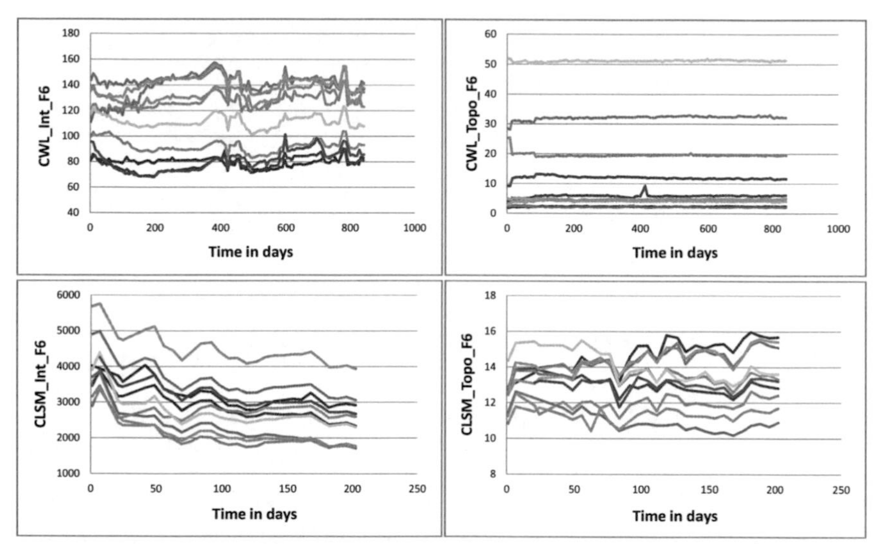

Figure 68: Exemplary experimental aging curves from 10 of the 20 investigated time series based on the intensity data of TSC_CWL (upper left image), topography data of TSC_CWL (upper right image), intensity data of TSC_CLSM (left lower image) and topography data of TSC_CLSM (right lower image) in respect to feature F6.

Studying the experimental results of Figure 68, a strong similarity to those of feature F3 (Figure 65) can be observed. As has been previously discussed, strong similarities between the results of certain features might be attributed to strong external influences on the captured prints as well as similarities in the feature design. Similar to feature F3, intensity-based time series of the CLSM device seem to be feasible for potential age estimation, whereas CWL-based intensity images as well as topography-based series of both devices seem to be rather infeasible.

6.3.2.7 Evaluation of Coherence (F7)

The experimental results of feature F7 investigated in the scope of fingerprint long-term aging are visualized in Figure 69.

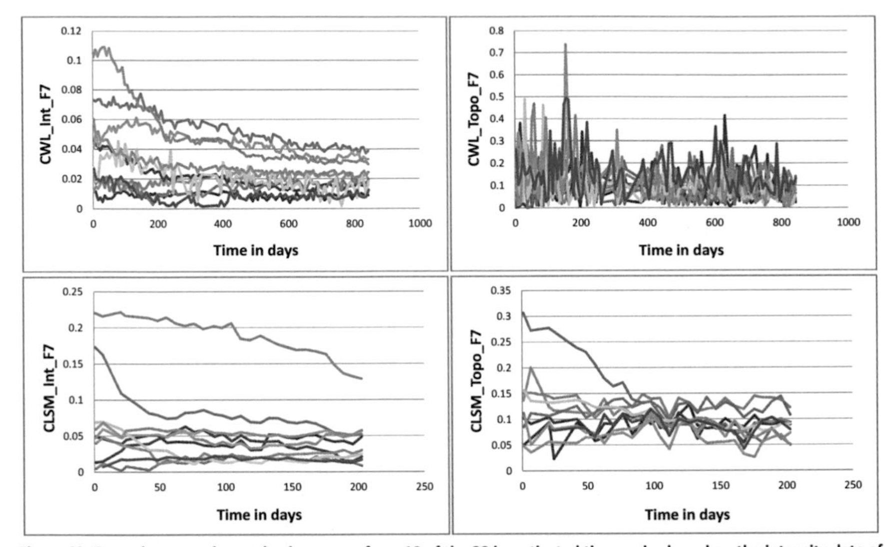

Figure 69: Exemplary experimental aging curves from 10 of the 20 investigated time series based on the intensity data of TSC_CWL (upper left image), topography data of TSC_CWL (upper right image), intensity data of TSC_CLSM (left lower image) and topography data of TSC_CLSM (right lower image) in respect to feature F7.

While Figure 69 shows opposing trends for the experimental curves of all four data types, heavy distortions render most of the curves unsuitable for representing a certain aging trend. Especially the CWL topography-based time series (CWL_Topo_F7) produce heavily distorted aging curves, clearly demonstrating the infeasibility of the features to represent a certain aging property.

6.3.3 Evaluation of Morphological Long-Term Aging Features

In this section, the morphological aging features F8 - F15 are investigated in addition to the earlier examined statistical features F1 - F7. Morphological features are only evaluated for the long-term aging, because significant changes of certain fingerprint structures or particles seem to occur only over long time periods. The features F8 - F15 are introduced in section 4.3.2 and are only investigated for intensity-based time series of selected capturing devices. The test sets TSC_CWL and TSC_CLSM are used, similar to the earlier evaluation of long-term aging using features F1 - F7. If not stated otherwise, all 20 experimental aging curves are depicted for result examination, because only the intensity images of a single capturing device are used for each feature.

6.3.3.1 Evaluation of Mean Ridge Thickness (F8)

In the scope of feature F8, the mean ridge thickness of 20 fingerprint time series of intensity CWL images is investigated over a total time period of t = 2.3 years. The resulting experimental aging curves are depicted in Figure 70. It can be concluded from the results that most fingerprints exhibit a decrease in ridge thickness of less than one pixel over the investigated 2.3 years of aging, which equals a decrease of less than 10 µm within this period. Only few examples exhibit a bigger decrease, which is in most cases accompanied by considerable fluctuations, suggesting that these curves are highly distorted by certain influences, most likely not excluded dust.

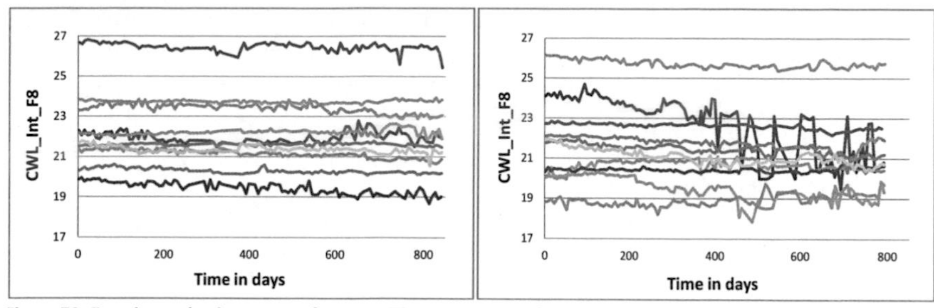

Figure 70: Experimental aging curves of test set TSC_CWL, captured in time intervals Δt = 1 w over a total time period t_{max} = 2.3 y in respect to feature F8 (mean ridge thickness). Left image: Ts1 - Ts10; right image: Ts11 - Ts20.

Popa et al. [75] have previously investigated changes in the ridge thickness of latent fingerprints on glass substrates over a time period of t_{max} = 0.5 years (section 2.1.3). Although their results are comparatively vague (only the minimum and maximum value of the observed ridge thicknesses are given; it is not stated at which point of the print and how the ridge thickness was measured), a comparison is attempted here. The minimum and maximum value of ridge thickness given by Popa et al. are averaged and used for comparison, explicitly noting that the average of such range does not necessarily represent the real mean value of the (normally distributed) ridge thickness. However, it is the only possibility for a comparison of Popa et al. with the here investigated feature F8.

When approximating the mean ridge thickness observed by Popa et al., a decrease of such thickness of 60 µm (20% of total fingerprint ridge width) is observed over the investigated time period of t_{max} = 0.5 years. In contrast, the experimental results of Figure 70 show a decrease in ridge thickness of less than 10 µm for a time period of t_{max} = 2.3 years. Therefore, the findings in the scope of this thesis are in sharp contrast to those of Popa et al. Differences might be explained by the different surfaces used (glass vs. hard disk platter). However, first studies in the scope of this thesis have shown similar results for the visualization of fingerprints on glass and hard disk platter substrates.

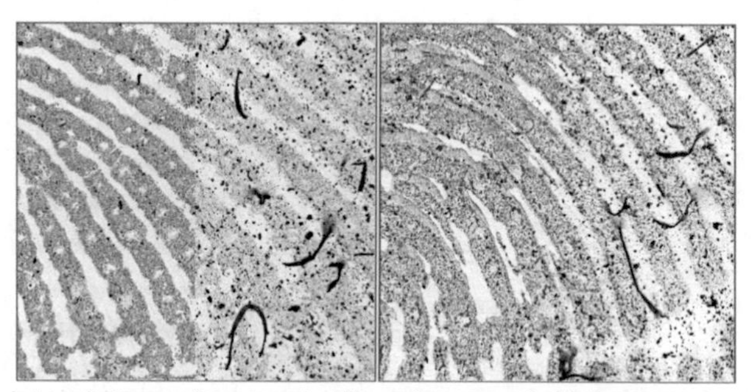

Figure 71: Grayscale intensity images of two exemplary fingerprints, captured with the CWL sensor at times t = 1 d (left half of each image) and t = 2.3 y (right half of each image). The transition between the two temporal versions of each fingerprint (vertical line in the middle of each image) can be used for a visual comparison of the change in ridge thickness, occurring during the investigated aging period.

Figure 71 provides a visual comparison of two exemplary fingerprints, captured one day after deposition and again after 2.3 years. It confirms that indeed the ridge thickness does not exhibit a significant decrease during this time period, but rather fingerprint pixels disappear at random over the complete image, leading to a decreased image contrast. Therefore, the ridge thickness does not seem to be a feasible measure for estimating a fingerprints age, in contrast to statistical features, which can detect disappearing fingerprint pixels everywhere in the complete image and are therefore much more promising.

6.3.3.2 Evaluation of Pores (F9-F11)

The features F9 - F12 are based on pores and their surrounding regions and are investigated for CWL intensity images only. Because the investigations within this thesis are focused exclusively on latent prints, pores are not visible in every case, because not all latent prints exhibit pores (in comparison to exemplary prints, where the finger itself is present and pores can be imaged in most cases). The experimental results for features F9 - F11 are depicted in Figure 72, Figure 73 and Figure 74 for all experimental aging curves where pores have been identified.

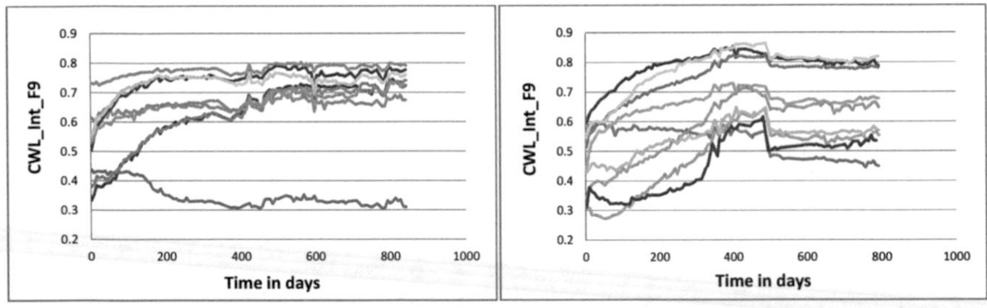

Figure 72: Experimental aging curves of test set TSC_CWL in which pores have been identified, captured in time intervals $\Delta t = 1$ week over a total time period $t_{max} = 2.3$ y in respect to feature F9 (relative frequency of white pixels in pore areas). Left image: Ts1 - Ts3, Ts5, Ts7, Ts8 - Ts10; right image: Ts11, Ts13 - Ts20.

Feature F9 represents the amount of white pixels in the pore areas of a binarized fingerprint image. It therefore represents the feature F1 (Binary Pixels) applied to the pore regions. Because pores are gaps in the fingerprint ridge line, they are represented by white pixels, similar to background (surface) pixels, where no fingerprint material is present. The feature therefore effectively measures the aging behavior of the fingerprint material in close proximity to pores.

As can be seen from Figure 72, most curves show a monotonic increasing tendency, which appears to be close to a logarithmic property. Such aging behavior of fingerprint material close to pores seems to exhibit a similar trend than earlier investigated for feature F1 (Figure 63, upper left image) applied to the complete fingerprint image. Such correlation shows that the overall aging tendency of features F1 and F9 seems to be similar, regardless if applied to the complete image or only the pore regions.

Several distortions can be observed in Figure 72, some of which seem to be of systematic nature (e.g. for t = 480 days, right image). Also, a certain variation of aging properties is evident. Especially for one curve of Figure 72 (left image), a decreasing monotony can be observed. The reason for this phenomenon cannot be clearly identified at this point of time. Possible reasons might be the presence of external substances within the fingerprint residue or strong environmental influences.

In Figure 73, the amount of pores is depicted in respect to the time of aging. It can be seen that no significant change in the amount of pores does occur, which confirms prior findings of this thesis, in which the loss of overall image contrast is identified to be the main source of changes occurring during aging, rather than pores disappearing.

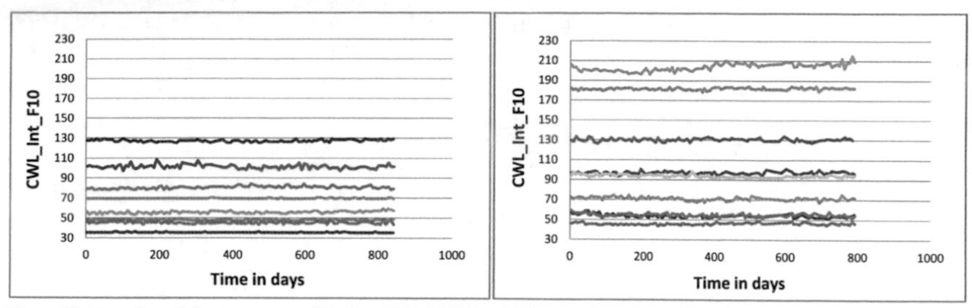

Figure 73: Experimental aging curves of test set TSC_CWL in which pores have been identified, captured in time intervals $\Delta t = 1$ w over a total time period $t_{max} = 2.3$ y in respect to feature F10 (amount of pores). Left image: Ts1 - Ts3, Ts5, Ts7, Ts8 - Ts10; right image: Ts11, Ts13 - Ts20.

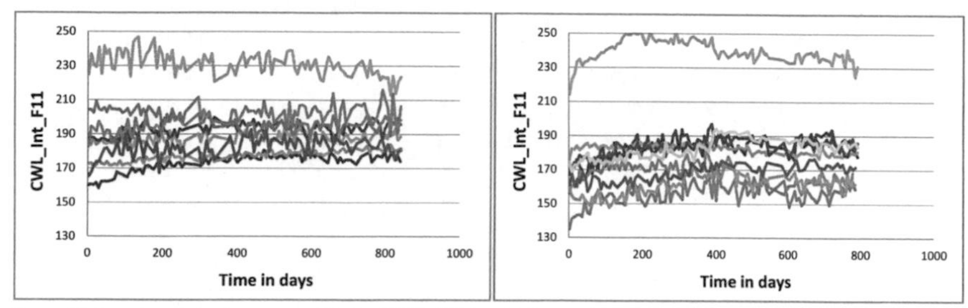

Figure 74: Experimental aging curves of test set TSC_CWL in which pores have been identified, captured in time intervals $\Delta t = 1$ w over a total time period $t_{max} = 2.3$ y in respect to feature F11 (mean pore size). Left image: Ts1 - Ts3, Ts5, Ts7, Ts8 - Ts10; right image: Ts11, Ts13 - Ts20.

Figure 74 depicts the mean pore size in respect to the time (feature F11). During the investigated time period of $t_{max} = 2.3$ years, no common trend seems to exist. Some curves depict and increase in pore size, while others remain constant or even exhibit a decrease. Furthermore, a significant amount of distortions is present, because only slight changes of a few pixels might lead to the pore extraction algorithm spilling over the corrugated boundaries of a pore. Because no systematic increase of pores can be observed, it is assumed that pores do not change in their size during aging. Observed changes in pore size are rather regarded as a product of pixels disappearing in a pseudo-random manner all over the fingerprint image, steadily corrugating the edges of pores, resulting in the extraction algorithm spilling over their boundaries in some cases. For a visual confirmation of such phenomenon, Figure 75 depicts the detected pores for feature F11 using an enlarged exemplary time series. It can be seen that the rather circular-shaped pores for time $t_1 = 1$ day (left image) often turn into irregular shapes over time, caused by the algorithm spilling over the pore boundaries.

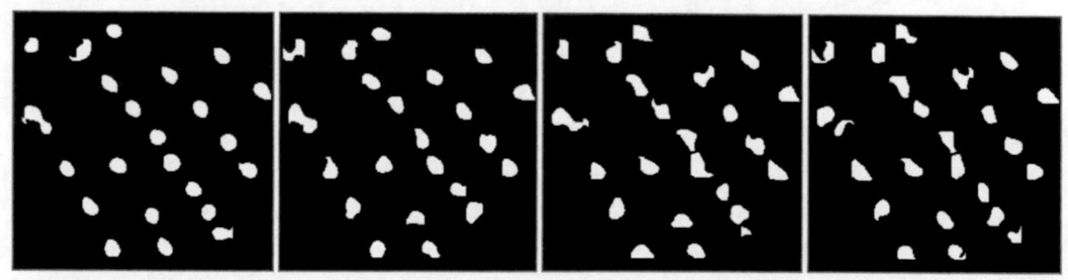

Figure 75: Exemplary visualization of a CWL intensity long-term aging time series, preprocessed according to feature F11 and enlarged for visual comparison of changes in mean pore size ($t_1 = 1$ d, $t_2 = 35$ w, $t_3 = 79$ w, $t_4 = 120$ w).

Summarizing the experimental findings for features F8 - F11, the morphological structures of ridges and pores do not seem to age in a significantly different manner than the overall fingerprint material. Therefore, neither the ridge thickness (F8) nor the amount (F10) or mean size (F11) of pores could be shown to exhibit a characteristic change during aging. The area around pores (F9) ages in a similar way than the complete image in respect to the statistical feature F1, demonstrating that a statistical approach seems to be more promising than the here investigated morphological traits. These results confirm the findings of [151] published in the scope of this thesis, using obsolete preprocessing methods.

6.3.3.3 Evaluation of Dust, Particles and Puddles (F12-F15)

The features F12 - 15 investigated in this section are exclusively based on CLSM intensity time series, because a comparatively high resolution is required. They represent the aging behavior of certain internal and external structures, such as dust (F12), particles (F13 and F14) as well as puddles (F15). For all four features, the targeted structures have been segmented and binarized according to sections 4.2.4.4 and 4.3.2.3. Their relative frequency is examined here, representing the relative amount of segmented pixels belonging to a certain particles class. The experimental results are depicted in Figure 76 (F12), Figure 77 (F13), Figure 78 (F14) and Figure 79 (F15).

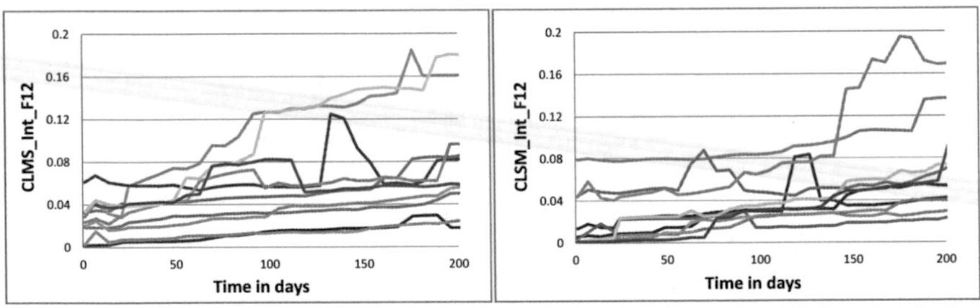

Figure 76: Experimental relative frequency curves of dust (F12) for all time series of test set TSC_CLSM, captured in time intervals Δt = 1 w over a total time period t_{max} = 30 w. Left image: Ts1 - Ts10; right image: Ts11 - Ts20.

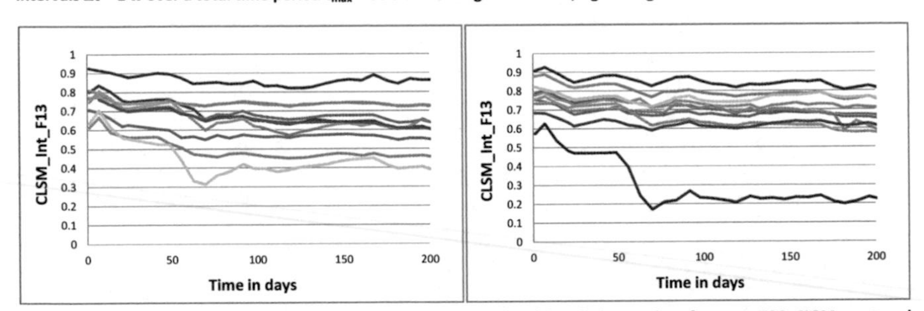

Figure 77: Experimental relative frequency curves of big particles (F13) for all time series of test set TSC_CLSM, captured in time intervals Δt = 1 w over a total time period t_{max} = 30 w. Left image: Ts1 - Ts10; right image: Ts11 - Ts20.

A monotonic increase in the frequency of dust particles can be observed for almost all experimental curves of Figure 76. Naturally, the total amount of dust settling over time on a surface does increase. The main limitation of using dust as an aging feature can be seen in the unreliability of dust settlement. Some curves of Figure 76 show a sudden sharp increase in their feature value, where bigger dust particles, fluffs or hairs might have settled on the surface. Furthermore, dust can also suddenly disappear, e.g. caused by air circulation. Such drastic changes pose a major limitation to the feasibility of dust as an aging feature, which could otherwise be regarded as a feasible approach for fingerprint age estimation.

Figure 77 shows a characteristic decrease of the big particles frequency over time, which seems to be similar for most time series. The big particles therefore seem to contribute to the characteristic changes observed for the statistical long-term aging of the CSLM intensity time series (section 6.3.2). A few outliers and systematic distortions seem to exist, which need to be investigated in more detail. In the scope of test goal TG5, a few possible influences are examined. Others remain subject to future research.

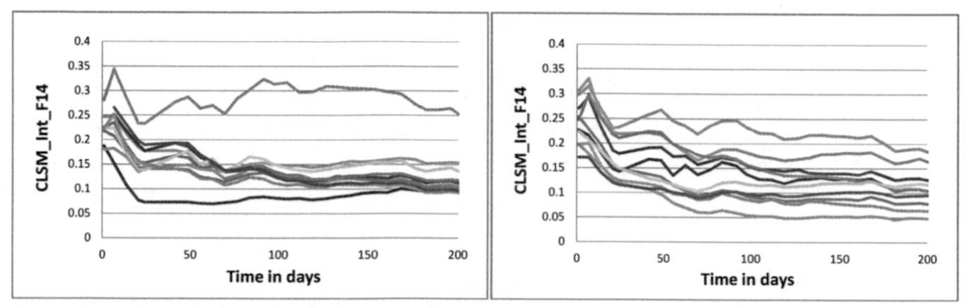

Figure 78: Experimental relative frequency curves of small particles (F14) for all time series of test set TSC_CLSM, captured in time intervals $\Delta t = 1$ w over a total time period $t_{max} = 30$ w. Left image: Ts1 - Ts10; right image: Ts11 - Ts20.

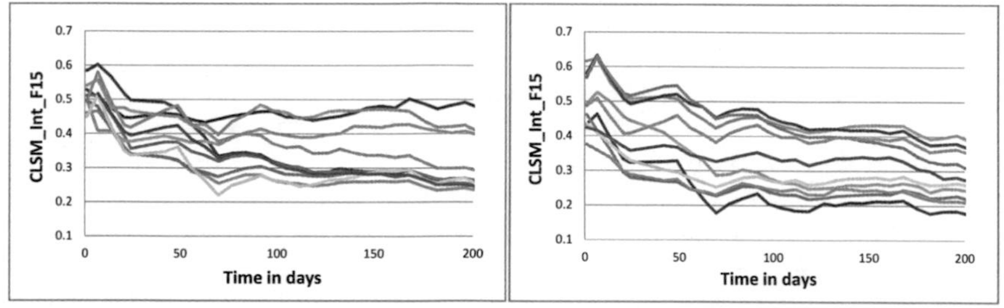

Figure 79: Experimental relative frequency curves of puddles (F15) for all time series of test set TSC_CLSM, captured in time intervals $\Delta t = 1$ w over a total time period $t_{max} = 30$ w. Left image: Ts1 - Ts10; right image: Ts11 - Ts20.

The relative frequency of small particles depicted in Figure 78 shows a characteristic, logarithmic decrease over time for almost all time series. It therefore also seems to significantly contribute to the characteristic long-term aging behavior observed for the statistical features (section 6.3.2). Systematic distortions can be seen for certain points of time. The relative frequency of puddles is shown in Figure 79 to also decrease in a characteristic way. Similar to big and small particles, pixels of puddles seem to contribute significantly to the overall aging property observed for the CLSM intensity-based long-term aging. Systematic distortions can also be observed.

Summarizing the findings of this section, dust might be considered as a potential feature for age estimation in future work, if only small dust particles are considered. However, additional factors such as air circulation can significantly influence such feature. When comparing big and small particles as well as puddles, all three are shown to exhibit a characteristic decrease in their relative frequency over time. Therefore, no single particle type can be identified as being the dominant source of the characteristic changes during aging, which have been observed for the statistical features F1 and F2. Consequently, morphological features based on certain particle types could not been shown to exhibit a more characteristic aging trend than the statistical features (e.g. F1 and F2). While additional experiments on morphological features should be performed in future work, so far they have not been shown to outperform their statistical counterparts.

6.4 Results for TG4: Evaluation of Digital Short-Term Age Estimation Strategies

In this section, the experimental results of the proposed digital age estimation strategies (TG4) are presented in respect to the short-term aging. The results provide for the first time an objective measure of age estimation performance, which can be used as a reference point for further studies on the age estimation of latent fingerprints. The formula-based age estimation results are presented in section 6.4.1, whereas the machine-learning based results are given in section 6.4.2.

6.4.1 Evaluation of Formula-Based Age Estimation

Based on the short-term aging test sets TSB_CWL and TSB_CLSM, four different formula-based age estimation approaches are examined in two phases Pk and Npk. All evaluations are based on the best performing feature Fb of each data type, which are CWL_Int_F1, CWL_Topo_F1, CLMS_Int_F2 and CLSM_Topo_F1.

Formula-based age estimation based on prior knowledge of aging characteristics (Pk): In the first experimental phase Pk, a general logarithmic aging function is used for a formula-based age estimation in respect to three different feature representations FR = {0,1,2}. The feature representations describe the derivatives of the logarithmic aging function (formulae (69) - (71)). The parameters 'a' and 'b' of the function are averaged over the complete test sets TSB_CWL and TSB_CLSM. The following values are computed and used for the evaluation: CWL_Int_F1: a = 0.019, b = 0.476; CWL_Topo_F1: a = 0.017, b = 0.452; CLSM_Int_F2: a = 177.677, b = 58582.258; CLSM_Topo_F1: a = 0.039, b = 0.540.

For evaluating the experimental formula-based age estimation results, the absolute difference t_diff between the estimated age t_est and the corresponding ground truth age t_real is used (section 5.7.3). Such measure t_diff is not normally distributed, which can be seen in Figure 80, exemplary visualizing the histogram of all t_diff values from the 15400 samples of the CWL_Int_F1 feature using Fr = 0. Many t_diff values form a normal-like distribution between 0 h ≤ t_diff ≤ 20 h. However, also a significant amount of values is arbitrarily distributed over a wide range of the histogram, with a maximum of t_diff = 1024062298 h. All values of t_diff ≥ 50 h are included in the last column of the histogram (5613 samples in total). The resulting t_diff values of all other investigated feature representations Fr show similar trends in their histograms.

The reason for such high amount of AEFs exhibiting a difference to their ground truth age of 50 hours and more seems to be the logarithmic property of the used mathematical function. Because the nature of a logarithmic curve is to assume extreme y-values for x-values close to zero, also extreme differences in t_est are obtained, if the computed aging features AEF exhibit only slight deviations from their expected values. Because t_diff is not normally distributed, the mean value, variance and standard deviation do not seem to be adequate measures for comparing t_diff. Therefore, the median value is selected for comparing the resulting t_diff values of the investigated features, devices and data types.

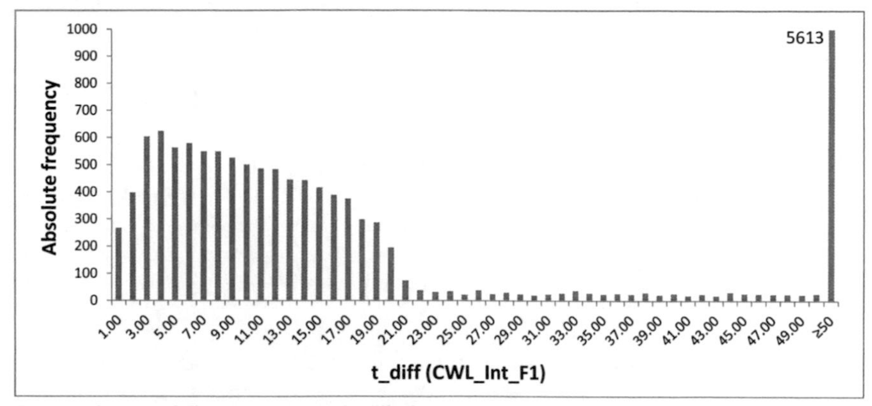

Figure 80: Histogram of all 15400 experimental t_diff values computed from TSB_CWL with the proposed formula-based age estimation approach using Fr = 0 in respect to the CWL_Int_F1 feature.

For Fr = 2, a significant amount of t_diff values is undefined, because slight deviations of the measurement results from their expected value produce invalid mathematical operations (e.g. the computation of the square root of negative values; please also refer to the formulae of section 4.4.1). Undefined values of t_diff are computed in 41.3%, 45.6%, 41.8%, and 29.9% of AEFs for the features CWL_Int_F1, CWL_Topo_F1, CLMS_Int_F2 and CLSM_Topo_F1 respectively. However, such invalid values also seem to represent an adequate filter mechanism for extreme outliers. From the remaining valid values of t_diff for the four investigated features in respect to Fr = 2, only between 0.6% and 2.9% of t_diff values are greater than 20 hours. The median results for t_diff in respect to the three feature representations Fr = {0,1,2} and the four investigated features CWL_Int_F1, CWL_Topo_F1, CLMS_Int_F2 and CLSM_Topo_F1 are presented and compared in Figure 81.

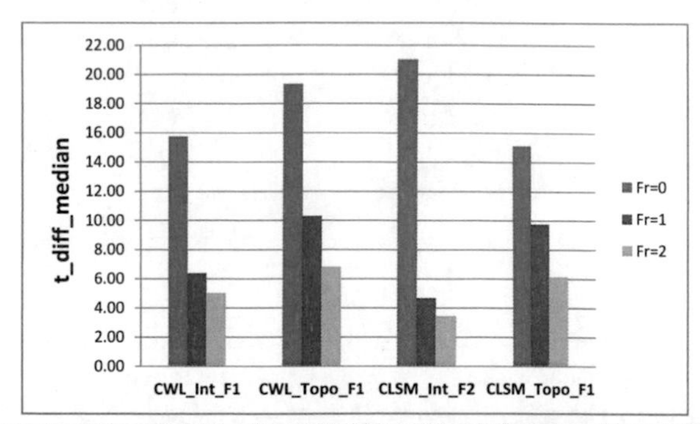

Figure 81: Median experimental results for t_diff computed from TSB with the proposed formula-based age estimation approaches of phase Pk using Fr = {0,1,2} in respect to the four aging features CWL_Int_F1, CWL_Topo_F1, CLMS_Int_F2 and CLSM_Topo_F1.

It can be seen from Figure 81 that for all four investigated features, t_diff is comparatively high for Fr = 0, rendering this feature representation unfeasible. When comparing Fr = 1 and Fr = 2, the best performance is achieved for Fr = 2 in respect to all four features, with a median deviation of the estimated ages of 3.4 h to 6.8 h from their respective ground truth (CWL_Int_F1: 5.0 h, CWL_Topo_F1: 6.8 h, CLSM_Int_F2: 3.4 h, CLSM_Topo_F1: 6.1 h). Because only the short-term aging of t_{max} = 24 hours is investigated, a difference of 6.8 hours can already be considered as a comparatively high deviation. Taking furthermore into account that for up to 45.6% of AEFs, t_est cannot be computed at all, the formula-based age estimation approaches investigated in the scope of phase Pk seem to be unreliable. However, in case a valid time t_est can be computed, age estimation within the presented error margin of 3.4 h - 6.8 h is possible.

Formula-based age estimation without prior knowledge of aging characteristics (Npk): In the second experimental phase Npk, a formula-based age estimation approach is investigated, which computes the age of a print solely from a single sequence of AEF values, independently of the general average of the parameters 'a' and 'b'. For such approach, two consecutive age estimation features AEF_n and AEF_{n+1} are used, which are computed in respect to Fr = 1. The estimated age is determined according to formula (80).

During such computation, it can be observed that for a significant amount of samples (between 36% and 68% for the four investigated features CWL_Int_F1, CWL_Topo_F1, CLMS_Int_F2 and CLSM_Topo_F1) the algorithm does no terminate within the specified maximum of t_est_{max} = 1000 h (section 4.4.1). The reason for the formula not producing estimated ages within 1000 hours in so many cases can be seen in its logarithmic nature. If the computed feature values differ to a certain extent from an ideal logarithmic progression, significant differences can occur in the computed results. The median of t_diff for all obtained results are depicted in Figure 82 in respect to the four investigated aging features.

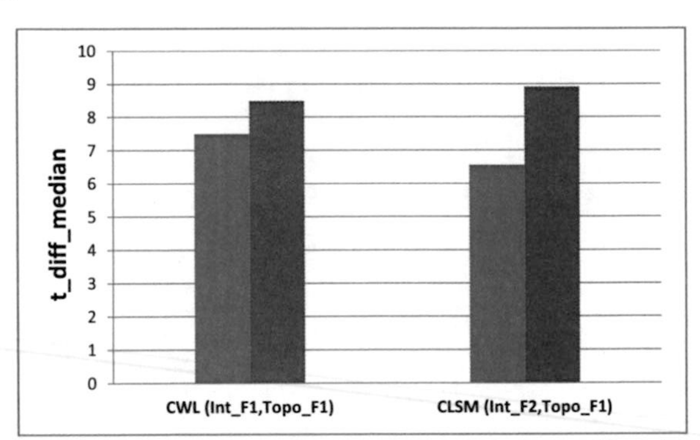

Figure 82: Median experimental value of t_diff computed from TSB with the proposed formula-based age estimation approach of phase Npk in respect to the four different aging features CWL_Int_F1, CWL_Topo_F1, CLMS_Int_F2 and CLSM_Topo_F1.

The results of Figure 82 show median values of t_diff between 6.6 hours and 8.9 hours. Such error is higher than that of experimental phase Pk (3.4 h - 6.8 h), even if the computation is aging curve specific in the scope of Npk. It is therefore assumed that the usage of average values for the parameters 'a' and 'b' provides a certain robustness against inaccurate feature values to the scheme. Similar to the experimental results of Pk for Fr = 2, most of the obtained t_diff values are smaller than or equal to 20 hours (only between 2.7% and 6.7% of t_diff values are greater than 20 hours for the four investigated aging features). Consequently, most of the age estimation features AEF differing from values expected for a logarithmic aging property seem to produce extreme outliers for the estimated ages t_est and are therefore excluded by the abortion of the computation algorithm.

The high amount of aborted computations as well as the comparatively high median values of t_diff demonstrate the limitations of the formula-based age estimation approaches investigated in Pk and Npk. In case a result can be obtained, a statement about the age of a print can be given within the provided error margins. However, only aging periods of t_{max} = 24 hours are investigated so far, leaving the examination of longer periods a subject to future research.

6.4.2 Evaluation of Machine-Learning Based Age Estimation

In this section, the experimental results of the proposed machine-learning based short-term age estimation approach are present. Results are based on the short-term aging test sets TSB_CWL and TSB_CLSM and are evaluated according to the four introduced experimental phases Op1, Op2, Ss and Ap (section 4.4.2). In phases Op1 and Op2, important parameters of the scheme are evaluated and optimized, while phase Ss is designed to experimentally show the statistical significance of the used test sets. Phase Ap presents the final classification results in respect to the investigated features Fk, capturing devices and data types as well as time thresholds Tt. In comparison to the experimental results on this issue published in [79] and [153] (without prior parameter optimization and using an obsolete preprocessing technique), much more reliable results and (in many cases) a significantly improved performance is achieved.

Optimization of parameters Re and Fr (Op1): In phase Op1, the feature representation Fr as well as the approximation type of the regression Re is evaluated and optimized in respect to the best performing features CWL_Int_F1, CWL_Topo_F1, CLSM_Int_F2 and CLSM_Topo_F1 (section 6.3.1). Figure 83 depicts the classification performance kap for the five investigated combinations of the parameters.

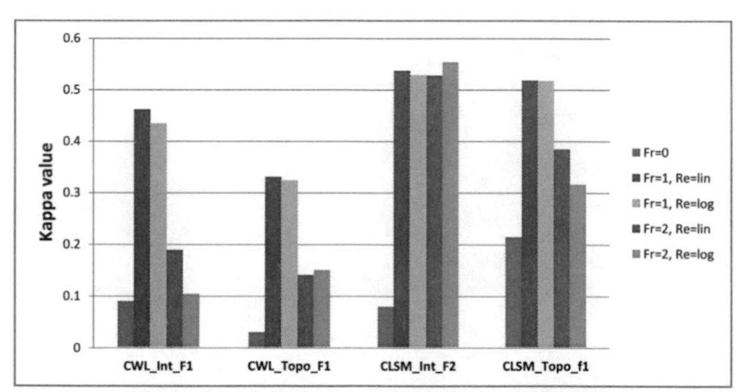

Figure 83: Kappa values of the best performing features CWL_Int_F1, CWL_Topo_F1, CLSM_Int_F2 and CLSM_Topo_F1 in respect to different feature representations Fr and regression types Re.

Analyzing the results of Figure 83, it can be seen that the feature representation of Fr = 0 performs significantly worse than any other representation. Furthermore, the feature representation Fr = 1 performs best considering all four features Fb, while Fr = 2 exhibits a marginally better classification performance in the case of CLSM_Int_F2. However, in all other cases, Fr = 2 fails to deliver a classification performance similar to that of Fr = 1. It is therefore concluded that a feature representation of Fr = 1 performs best in the general case, which is selected here as the optimal parameterization of O_Fr.

When comparing the types of regression Re between the features Fb using O_Fr = 1, no significant difference can be observed between using a linear and a logarithmic approximation. Therefore, either one can be selected for further evaluations. In the scope of this thesis, a linear approximation type Re = lin is chosen as the optimal parameterization of O_Re, because it performs marginally better than Re = log. It is to note that the type of regression Re is different from the overall aging property Pr of a complete aging curve, because Re only approximates a certain number of temporal images Ti as a subset of the overall aging function.

Optimization of parameters Ti and Cl (Op2): The second experimental phase Op2 is designed to optimize the required amount of temporal fingerprint images Ti as well as the used classifier Cl. For an optimal choice of Ti, all specified age classes according to Tt need to be evaluated, because different amounts of temporal images Ti can be optimal for different points in time. Because a total amount of n = 24 temporal samples is present for each complete time series, a total amount of n' = 24 - Ti + 1 feature representations can be approximated using Re = 1 and a sliding window approach. In the experiments, variations of Ti = {2,3,5,7,9,10} and Cl = {NaiveBayes,J48,LMT,RotationForest} are exemplary investigated, leading to an amount of n' = 24 - 10 + 1 = 15 feature representations. Using such 15 representations per time series, a maximum value of Tt = 14 h can be investigated. Table X depicts the classification performance kap for each feature Fb and classifier Cl. Such performance is computed as the mean kappa value of all age classes according to Tt = {1,2,...,14} of the best performing Ti value, which is also depicted.

Table X: Mean classification performance of the best performing Ti value of the investigated variations Ti = {2,3,5,7,9,10}. Results are depicted in respect to the features Fb and the classifiers Cl = {NaiveBayes,J48,LMT,RotationForest}. The upper part of the table presents the classification performance as the mean kappa value of all investigated age classes according to Tt = {1,2,...,14}, while the lower part of the table lists the corresponding best performing Ti values. All kappa values within a 0.015 range of the highest value are marked bold in the upper part. All Ti values in a 0.015 range of the optimal value are listed in parenthesis in the lower part.

Mean kappa value of best performing Ti	NaiveBayes	J48	LMT	RotationForest
CWL_Int_F1	0.327	0.314	**0.366**	**0.360**
CWL_Topo_F1	0.242	0.228	**0.253**	**0.251**
CLSM_Int_F2	**0.486**	**0.487**	**0.486**	**0.487**
CLSM_Topo_F1	**0.448**	**0.437**	**0.441**	**0.438**
Best performing Ti	**NaiveBayes**	**J48**	**LMT**	**RotationForest**
CWL_Int_F1	5	3	5	5 (3)
CWL_Topo_F1	7 (5,9,10)	5 (7)	7 (5,9)	7 (5,9)
CLSM_Int_F2	3 (2)	3 (2)	2 (3)	3 (2)
CLSM_Topo_F1	5 (3)	3 (5)	5 (3)	3 (5)

From the mean kappa values of Table X (upper part), it can be concluded that for the CWL sensor (first and second row of the upper part of Table X), the classifiers LMT and RotationForest perform almost equally well and much better than the NaiveBayes and J48 classifiers. For the CLSM sensor (third and fourth row of the upper part of Table X), all four classifiers perform almost equally well. It can therefore be concluded that out of the four investigated classifiers, the LMT as well as the RotationForest can be used for both capturing devices to deliver optimal results. In the scope of this thesis, O_Cl = LMT is chosen, because it performs marginally better than the RotationForest classifier.

For selecting the best performing value of Ti, Table X (lower part) shows varying results. As mentioned in section 4.4.2, Ti should be chosen as small as possible, to allow for a minimum capturing time in practical applications. It should also be chosen in respect to a certain capturing device, because different devices might exhibit a varying sensitivity to noise and other influences on the captured prints. In Table X (lower part), parenthesized values of Ti indicate possible alternatives, which differ from the kappa performance of the optimal Ti by a maximum kappa of 0.015. Such range is chosen according to the results as being an acceptable loss of accuracy in exchange for selecting a smaller value Ti, decreasing the practical capturing time for age estimation. Comparing the optimal Ti values for the CWL device (first and second row of the lower part of Table X), most classifiers of the intensity as well as the topography data perform best or only marginally worse for Ti = 5. For the LMT, which has been selected as the best performing classifier, this is true for the intensity as well as the topography data. Therefore, and optimal Ti is selected as $O\_Ti_{CWL} = 5$ for the CWL device.

For the CLSM device (third and fourth row of the lower part of Table X), a value of Ti = 3 performs optimal or marginally worse for all classifiers, including the selected LMT. Therefore, an $O\_Ti_{CLSM} = 3$ is chosen as optimized parameterization of Ti for the CLSM device. In case a differentiation is made between the intensity and the topography data of the CLSM device, even Ti = 2 could be chosen for CLSM intensity data. However, to ensure consistent and comparable results for both data types, a general value $O\_Ti_{CLSM} = 3$ is selected for the CLSM device. Figure 84 depicts the classification performance of O_Cl in respect to the required amount of samples Ti as well as the age class pairs according to Tt for the investigated features Fb.

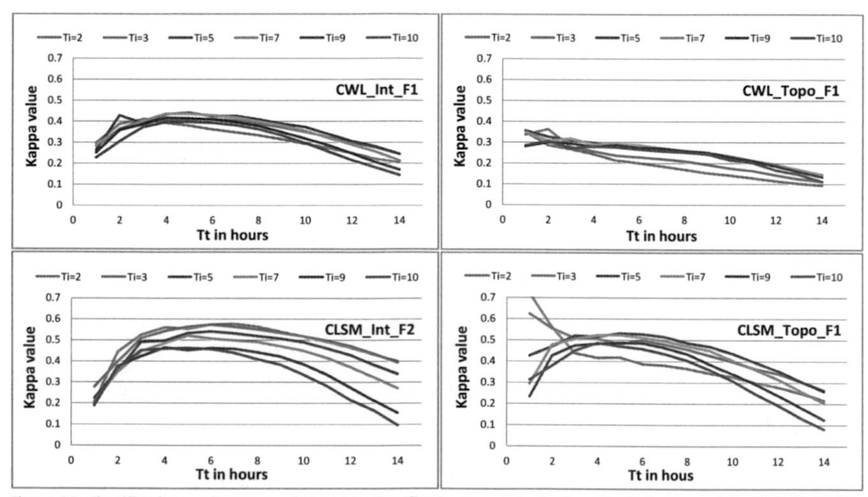

Figure 84: Classification performance of the LMT classifier in respect to the investigated values of Ti = {2,3,5,7,9,10} as well as age class pairs according to Tt = {1,2,...14} for the feature Fb of each data type.

The results depicted in Figure 84 show a comparatively similar trend for all investigated Ti values of a certain feature. Merely in the first two or three hours, significant differences are observed in some cases, which are discussed in the scope of phase Ap. Most investigated configurations perform best for time thresholds Tt between five and seven hours. For higher values of Tt, the classification performance seems to decrease. This might be explained by AEFs becoming more indistinguishable for aged fingerprints, because the drying process slows down and the logarithmic aging curves converge (see discussion in phase Ap). For such increasing time thresholds Tt, differences in the classification performance of the Ti values become more evident. Clear differences between the CWL sensor (Figure 84, upper row) and the CLSM device (Figure 84, lower row) in respect to the optimal value of Ti can be observed. While for the CWL device Ti values of Ti = 3 and Ti = 7 perform best, the CLSM device also performs quite well for Ti = 3. Therefore, the CWL device requires more temporal samples of a captured fingerprint for reliable age estimation than the CLSM device.

Statistical significance determination (Ss): The aim of the third experimental phase Ss is to show that for the investigated pairs of age classes according to Tt the amount of samples in the training set TRS is of sufficient size, to ensure statistical significance of the classification results. For that purpose, TRS is systematically increased, beginning with 1% of its original amount of samples until 100% of its original size is included (section 5.7.2). By using the earlier optimized parameters O_Fr = 1, O_Re = lin, $O\_Ti_{CWL} = 5$, $O\_Ti_{CLSM} = 3$ and O_Cl = LMT, all seven statistical features F1 - F7 are evaluated for each data type and capturing device and in respect to all age class pairs according to Tt = {1,2,...,24-O_Ti}. The classification performance ranges ra of sliding windows including five consecutive kappa values are computed and a configuration is regarded as being of statistical significance, if at a certain split value of TRS (section 5.7.2), ra is smaller than or equal to a maximum fluctuation ra_max. If ra does not fall below the specified value of ra_max until the complete size of TRS is reached, the configuration is regarded as not being of statistical significance. Exemplary learning curves are depicted in Figure 85 for the features Fb. Their corresponding curves of values ra are depicted in Figure 86.

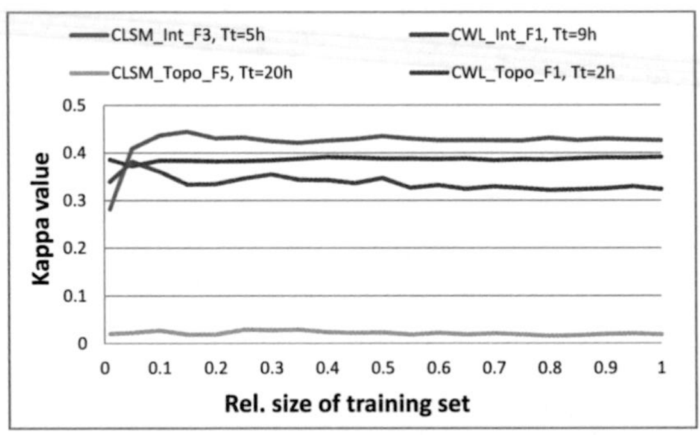

Figure 85: Learning curves of the classifier LMT when systematically increasing the training set size from 1% to 100% of its original size, based on exemplary selected capturing devices, data types, features and time thresholds.

It can be seen from Figure 85 that some learning curves have already converged when only using 1% of the original TRS size (CWL_Int_F1, Th = 9 and CLSM_Topo_F5, Tt = 20 h of Figure 85). Others (CLSM_Int_F3, Tt = 5 h of Figure 85) show a typical logarithmic progression of the learning curve, quickly converging towards its final performance. In some cases (CWL_Topo_F1, Tt = 2 h of Figure 85), the learning curve decreases in the beginning. This phenomenon might be caused by an insufficient amount of test samples, which is later corrected with increasing training set size, leading to a slightly decreased performance. When analyzing the kappa range (ra) curves in Figure 86 of the exemplary chosen learning curves, it can be seen that indeed the range ra (if not done already) quickly decreases to a certain final level, which is then only slightly fluctuating. As introduced in sections 5.7.2 and 5.7.3, maximum ranges of ra_max = {0.01,0.02,...,0.06} are investigated in the scope of phase Ss. Figure 87 provides an overview over the relative amount of investigated configurations fulfilling such quality criterion.

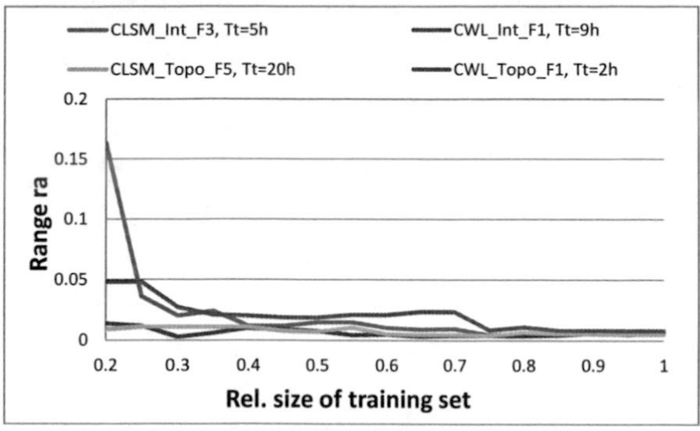

Figure 86: Curves of the kappa sliding window range (ra) corresponding to the learning curves of Figure 85 for systematically increased training set TRS sizes from 1% to 100%.

The results of Figure 87 show that regardless of the used capturing device, most of the curves have been converged for ra ≤ 0.03, whereas almost all curves have been converged for ra ≤ 0.04. Only two configurations using the CLSM sensor have not converged for ra ≤ 0.04, of which one has not converged for ra ≤ 0.05 and all have converged at ra ≤ 0.06. A total of 266 CWL and 294 CLSM configurations have been investigated. A fluctuation of the kappa value of ra ≤ 0.06 is regarded as acceptable in the scope of this thesis. Therefore, the experimental results can be considered as being based on a training set of sufficient (statistically significant) size and therefore Sas = true.

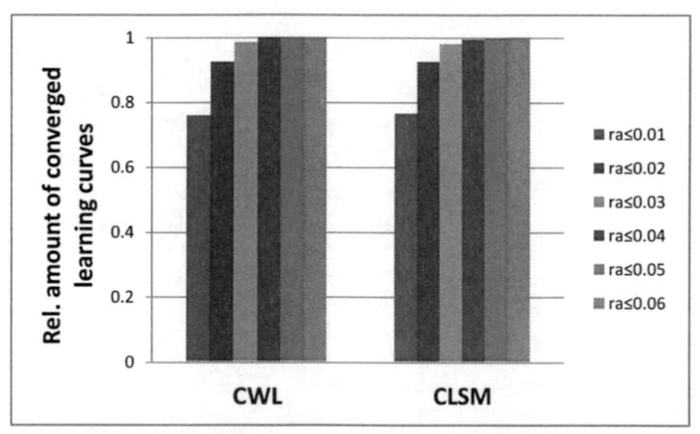

Figure 87: Relative amount of converged learning curves for all investigated configurations of capturing devices, data types, aging features F1 - F7 and age classes for Tt = {1,2,...24-O_Ti} in respect to different maximum range values ra_max = {0.01,0.02,...,0.06}.

Apart from determining if a statistically significant size of the training set TRS is used for the conducted experiments, phase Ss also allows to determine the minimum required amount of training samples in respect to each age class pair of Tt, which is the amount of samples used when ra is for the first time smaller than ra_max. When exemplary defining ra_max = 0.05 as the quality criterion for a training set TRS to be of sufficient size, the minimum amount of required training samples is given in Figure 88 as the average over all seven features of both data types in respect to Tt. The results show that in general, a few more training samples are required when working with the CWL sensor in comparison to the CLSM device.

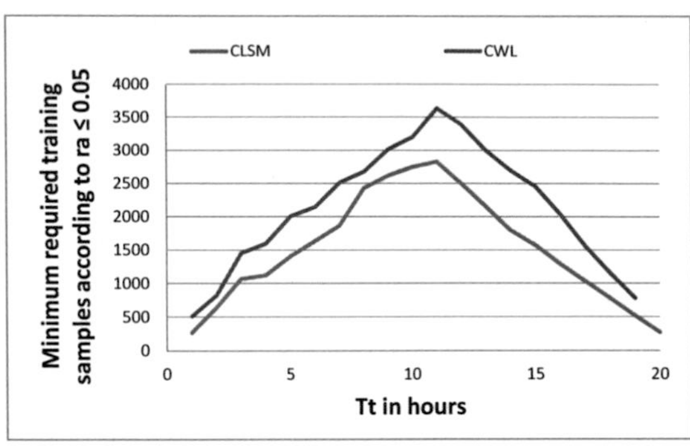

Figure 88: Minimum required amount of training samples in TRS according to ra_max = 0.05 as the mean value of all seven features of both data types in respect to each capturing device and time threshold Tt = {1,2,...,24-O_Ti}.

It can be concluded from Figure 88 that only about 500 or less samples are required for achieving statistical significance in respect to Tt = 1 h. With increasing values of Tt, the required amount of samples increases linear until its maximum of about 2800 (CLSM) or 3600 (CWL) required samples is reached for Tt = 11 h. For further increased values of Tt, the required amount of samples does decrease in a linear fashion, until the maximum investigated value of Tt is reached (Tt = 19 h for the CWL and Tt = 21 h for the CLSM device). This decrease indicates that after Tt = 11 h, the fingerprint ages become less distinguishable (see also discussion of phase Ap).

Age estimation performance evaluation (Ap): Having optimized all necessary parameters in phases Op1 and Op2 as well as having shown the statistical significance of the used test sets in phase Ss, the final age estimation performance is computed in phase Ap. The age estimation performance is evaluated in respect to both capturing devices, data types, all seven features F1 - F7 as well as all variations of age class pairs according to Tt = {1,2,...,24 - O_Ti}. Furthermore, an additional classification is performed using all seven intensity and all seven topography features of a capturing device simultaneously for model training and performance evaluation (referred to as F_ALL). The results are depicted in Figure 89.

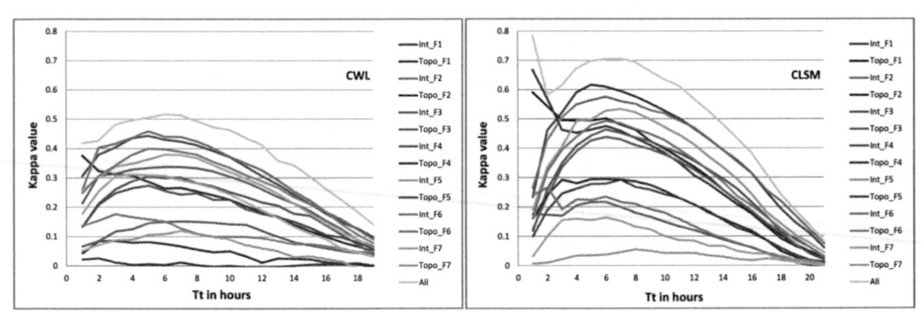

Figure 89: Final classification performance of all seven features F1 - F7 for intensity and topography data of the CWL (left image) and CLSM (right image) device in respect to the investigated age class pairs according to Tt = {1,2,...24-O_Ti}. A combined classification including all seven intensity and all seven topography features of each device is included and referred to as F_ALL.

The results of Figure 89 show a similar general trend of the classification performance in respect to the different time thresholds Tt for both devices. An optimal performance for most features is achieved for time thresholds Tt between 5 hours and 7 hours. For thresholds smaller than 5 hours, the classification performance decreases for decreasing values of Tt. Only for a few features (CWL_Topo_F1, CLSM_All, CLSM_ Topo_F1, CLSM_Topo_F2), a maximum classification performance is achieved for Tt = 1 h. Such phenomenon seems to occur mainly for topography-based images and is in contrast to that of most other features. It is not clear at this point what is the cause of such behavior. A possible source might be the influence of starting a capturing device (section 6.5.2), which is observed to be more dominant for the CLSM sensor (please refer to section 6.5.2 for further reading about this issue). However, also other reasons are possible and need to be investigated in more detail in further studies. For the here evaluated experiments, time thresholds of Tt < 5 h are not considered when determining the maximum classification performance of features.

From the results of Figure 89, it can be concluded that for both devices, the combination of all intensity and topography features performs best (CWL_All: kap(Tt=6h) = 0.52, CLSM_All: kap(Tt =7h) = 0.71). A higher performance can furthermore be observed for the CLSM device in comparison to the CWL sensor. Intensity-based features exhibit a significantly higher classification performance than their topography-based counterparts. When comparing the single features, the intensity-based features Int_F1 and Int_F2 targeting the loss of overall image contrast during aging perform best for both devices (CWL_Int_F1: kap(Tt=5h) = 0.44, CWL_Int_F2: kap(Tt=5h) = 0.46, CLSM_Int_F1: kap(Tt=5h) = 0.62, CLSM_Int_F2: kap(Tt=6h) = 0.58). From the topography-based features, Topo_F1 performs best for both devices, followed by feature Topo_F5 for the CWL sensor and Topo_F2 for the CLSM device (CWL_Topo_F1: kap(Tt=5h) = 0.29, CWL_Topo_F5: kap(Tt=5h) = 0.27, CLSM_Topo_F1: kap(Tt=6h) = 0.50, CLSM_Topo_F2: kap(Tt=6h) = 0.47). Such findings support the earlier results of section 6.3.1, where features F1 and F2 have been observed to exhibit the highest amount of curves showing a characteristic (logarithmic) aging property for intensity as well as topography data of both devices. They therefore seem to be best suited for practical age estimation.

For time thresholds greater than 7 hours, the classification performance is subject to a steady decrease with increasing values of Tt for all features. This decreased performance for increased time thresholds demonstrates that latent prints become less distinguishable with increasing age. To illustrate such finding in more detail, the well performing features CWL_Int_F1, CWL_Topo_F1, CLSM_Int_F2 and CLSM_Topo_F1 are exemplary used. Their corresponding, normally distributed age estimation features AEF computed from test set TSB are depicted in Figure 90 in respect to full hour age classes (Ac1 = [0h,1h], Ac2 = [1h,2h], etc.). Plotting their mean values and standard deviations allows for a comparison of the statistical (normal) distributions of these features within each hour and the identification of their changes over time.

Figure 90 shows a similar trend for all four exemplary investigated aging features. While the mean values of computed age estimation features AEF (which represent the speed of fingerprint degradation) decrease in a logarithmic manner, their standard deviations also become smaller over time. After only a few hours of aging, the difference between consecutive mean values becomes marginal, effectively preventing a further distinction between the different age classes. After such point of time, the mean values as well as the standard deviations remain rather constant. A few fluctuations do occur, which are considered as the natural level of noise and other distortions.

The logarithmic behavior of mean values observed within the first hours seems to represent the drying process of the prints. A significant deviation in the speed of fingerprint degradation can be observed for this first few age classes. Here, some fingerprints exhibit a comparatively fast degradation behavior, and can therefore be determined as being fresh, while others seem to exhibit a significantly slower degradation. Once a few hours have passed, the fingerprints have dried up and the aging speed has significantly slowed down (decreased mean values). The remaining aging speed has also become significantly more similar between the different fingerprints (decreased standard deviation). Therefore, the remaining evaporation and chemical decomposition seems to be much more similar for aged prints.

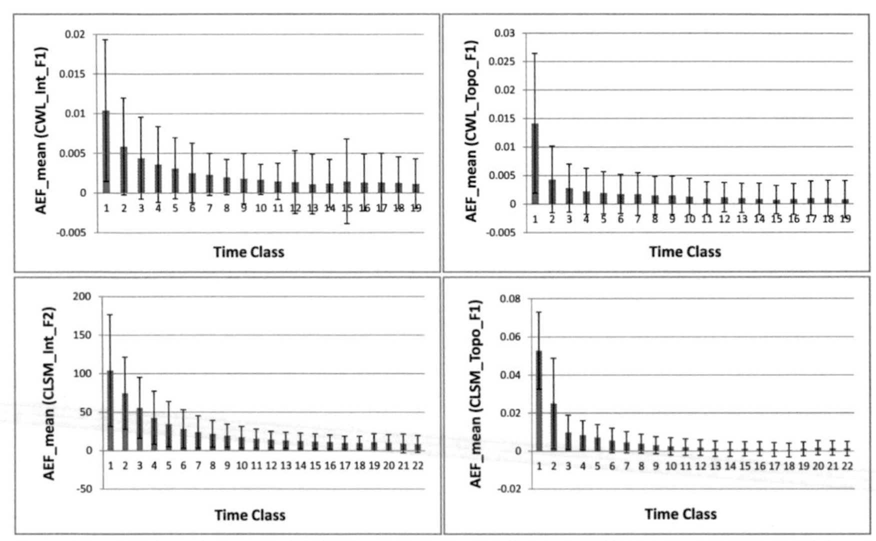

Figure 90: Mean values of the normally distributed age estimation features AEF from CWL_Int_F1 (upper left image), CWL_Topo_F1 (upper right image), CLSM_Int_F2 (lower left image) and CLSM_Topo_F1 (lower right image), computed from test set TSB in respect to full hour age classes (Ac1 = [0h,1h], Ac2 = [1h,2h], etc.). The standard deviations are furthermore visualized using error bars.

Concluding from such findings, a general classification of fingerprints into age classes of only one hour in size does not seem to provide sufficiently accurate results at this point of time. However, distinguishing between fresh and old fingerprints, i.e. those younger as or older than a certain threshold, seems to be possible in the scope of short-term aging with a classification performance of kap = 0.52 to kap = 0.71 (equivalent to an accuracy of 76% - 86% for equally distributed test sets).

6.5 Results for TG5: Evaluation of Selected Short-Term Influences and Model Creation

In this section, the experimental results of different qualitative studies of influences on the fingerprint aging behavior are presented and discussed (TG5). In section 6.5.1, the findings of various CWL-based short-term aging studies published in the scope of this are summarized.

Afterwards, three additional investigations are evaluated in sections 6.5.2 to 6.5.4, based on CWL and CLSM intensity short-term aging studies using the best performing features CWL_Int_F1 and

CLSM_Int_F2 (section 6.3.1). Their evaluated influences seem to belong to the most important ones on the aging of latent fingerprints according to own observations. Experimental findings of investigated influences from the sensor starting process as well as findings concerning the non-invasiveness of the capturing devices are introduced in section 6.5.2. Section 6.5.3 evaluates the experimental results of fingerprint aging under different ambient temperature and humidity conditions. The results of aging selected water-based substances are reported in section 6.5.4. Based on the reported overall findings, a first qualitative influence model on the fingerprint short-term aging is introduced in section 6.5.5.

6.5.1 Summary of Evaluated Influences from own Published CWL Studies

Several experiments are conducted in the scope of this thesis towards the investigation of influences on the short-term aging based on the CWL sensor. The experimental results of these studies are published in [95], [153], [154], [155], [156], [157], [158], [159] and are briefly summarized here. They are based on obsolete preprocessing techniques. However, because the CWL device is observed to achieve comparatively accurate results even when using such obsolete preprocessing methods (which is not the case for the CLSM device), the obtained results are considered of sufficient accuracy for extracting first qualitative findings on aging influences from them. These findings are furthermore included into the influence model proposed in section 6.5.5. The reader may refer to the given references for detailed information about the studies. Experimental results are categorized into the earlier introduced groups of influences, which are scan influences, environmental influences, sweat influences, application influences as well as surface influences.

6.5.1.1 Scan Influences Published in the Scope of this Thesis

To demonstrate that the characteristic fingerprint aging properties can be observed independent of the used capturing device, the aging behavior of the well performing Binary Pixels feature (CWL_Int_F1) is reproduced for a common flat bed scanner in [95]. It is shown that the property can be produced even with such basic capturing device, however only when using fingerprints of high quality. From 250 captured fingerprint time series (25 different test subjects), 24.0% are subjectively rated by two scientist as being of high quality, whereas 20.0% are rated as being of medium quality and 56.0% as being of low quality. In comparison, 74.4%, 5.6% and 20.0% of 250 CWL time series are manually assigned to these respective classes. In regard to the image quality, the CWL sensor therefore significantly outperforms the flat bed scanner and can furthermore produce characteristic aging properties for time series of medium and low quality. In such respect, the investigated flat bed scanner performs poorly.

Different resolutions and measured area sizes are investigated in [153], [156], [157], [159] and are presented and discussed in the scope of TG1 (section 6.1). Apart from the scan influence of the capturing process on the aging behavior of fingerprints, also the capturing devices themselves seem to be sensitive to certain influences, as exemplary shown in [154] for fluctuating ambient temperatures. Such influences might be propagated and potentially lead to distortions of the captured fingerprint time series. Additional experiments on the influence of starting a capturing device as well as the non-invasiveness of the CWL and CLSM devices are investigated in section 6.5.2.

6.5.1.2 Environmental Influences Published in the Scope of this Thesis

First qualitative short-term aging studies on the influence of different ambient climatic conditions are conducted in [153] and [154]. The test set is limited, using only one test subject and three different fingerprints. The major findings of these experiments are summarized in Figure 91, based on the intensity time series. The topography-based results are inconclusive and not of sufficient quality for an accurate evaluation.

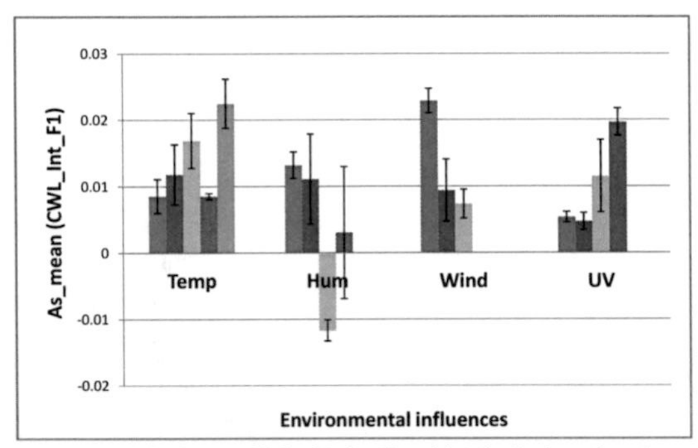

Figure 91: Summarized experimental findings published in [153] and [154] of selected environmental influences on the aging of latent fingerprints using the CWL_Int_F1 feature. From left to right: Temperature (30°C, 35°C, 40°C, 45°C, 50°C), Relative humidity (50%, 60%, 75%, 80%), Wind (2.93 m/s, 3.45 m/s, 3.95 m/s), UV radiation (0.6 mW/cm², 1.3 mW/cm², 1.7 mW/cm², 2.8 mW/cm²). The mean aging speeds As are depicted, standard deviations are given in the form of error bars.

The summarized results of Figure 91 show (although sometimes distorted) first tendencies of changes in the aging speed in respect to changed environmental conditions: an increased aging speed with increased temperature or UV radiation as well as a decreased aging speed with increasing relative humidity or wind. Furthermore, a high ambient humidity seems to lead to a negative aging speed and therefore a refreshment of the print. This issue is further discussed in section 6.5.3, where additional studies are performed on the influence of ambient temperature and humidity on the aging speed of latent prints using a bigger test set and both capturing devices.

An increased aging speed for increased ambient temperature or UV radiation seems to be a consequence of the higher amount of energy introduced to the system and therefore the acceleration of evaporation as well as chemical degradation processes. Furthermore, higher amounts of humidity decrease the ability of the ambient air to absorb water, leading to a decreased aging speed. However, the decreased aging speed during increased air circulation (wind) cannot be explained at this point of time. An increased amount of wind is expected to accelerate the evaporation of water from the print residue and therefore to lead to an increased aging speed. Further investigations on this issue are required in future work, to adequately explain such phenomenon. Potential explanations include a cooling effect (and therefore a decrease of the aging speed) in the direct vicinity of a print due to the increased air circulation.

6.5.1.3 Sweat Influences Published in the Scope of this Thesis

The qualitative and quantitative composition of sweat seems to have an important influence on the aging behavior of fingerprints. In [153] and [154], different types of sweat as well as sweat from different activities and ambient conditions are investigated based on three different prints of one test subject. In particular, prints of eccrine sweat (where nothing is touched for 30 minutes after cleaning the finger), sebaceous sweat (touching the forehead before print application) as well as sweat from daily work (six hours of normal office activity) are studied. Furthermore, fingerprints after normal (thermal) sweating under high and low ambient temperatures, after sweating from sports (high and low ambient temperatures) as well as after the consumption of alcohol are investigated. The major findings of this experiment are summarized in Figure 92, based on the intensity time series. The topography-based results are inconclusive and not of sufficient quality for an accurate evaluation.

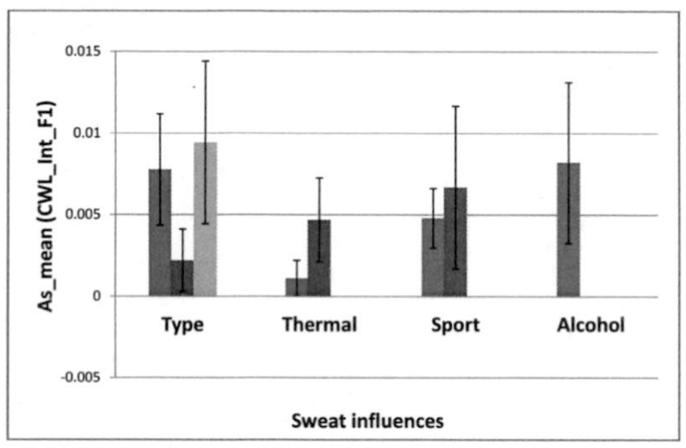

Figure 92: Summarized experimental findings published in [153] and [154] of selected sweat influences on the aging of latent fingerprints using the CWL_Int_F1 feature. From left to right: Sweat type (eccrine, sebaceous, daily work), Thermal sweating (15°C, 35°C), Sweating from sport (15°C, 20°C), Sweating after alcohol consumption (approximately 1.53 per mil). The mean aging speed As is depicted, standard deviations are given in the form of error bars.

Significant differences in the aging speeds As are observed from Figure 92 for the different types of sweat, indicating that the sweat influence has an important impact on the aging of latent prints. However, it is particularly hard to derive conclusions from them. While sebaceous sweat is observed to age significantly slower than eccrine sweat and sweat from daily work, sweating under different ambient conditions seems to lead to a lower aging speed for low ambient temperatures (for thermal sweating as well as sweating from sports). No significant difference in the aging speed is observed for sweating after the consumption of alcohol in respect to eccrine and daily work sweat. It might be presumed that the amount of water present within a fingerprint contributes to its aging speed, where in the cases of sebaceous sweat as well as sweat under low ambient temperatures, a decreased amount of water present within the prints would lead to a decreased aging speed. However, such presumption needs to be studied in more detail in future work, including the evaluation of substance-specific aging characteristics.

In [153] and [155], a significant difference in aging speed is shown between fingerprints contaminated with skin lotion and those contaminated with cooking oil in respect to the CWL_Int_F1 feature (five fingerprints of a single test subject are investigated). While topography-based time series are inconclusive, intensity-based series exhibit an approximately nine times higher aging speed for the water-containing skin lotion in respect to the cooking oil. Also, the aging behavior of a single droplet of water, skin-lotion and oil is evaluated in [153] and [155]. Results exhibit a comparatively high aging speed for water, a medium aging speed of skin lotion and a low aging speed for cooking oil in respect to intensity as well as topography time series. Summarizing the three conducted experiments, a strong influence of the amount of water contained within a latent print on its observed aging behavior is indicated by the results.

When comparing the aging speed of CWL time series between 50 males and 27 females in [153], only minor differences are observed in respect to the corresponding standard deviation. The results are depicted for the features CWL_Int_F1 and CWL_Topo_F1 in Figure 93.

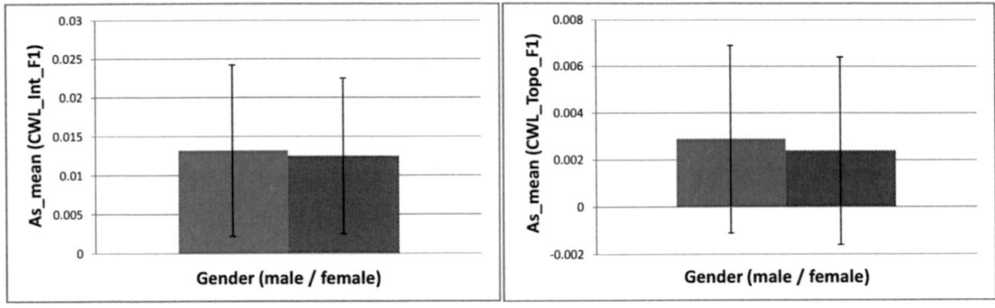

Figure 93: Summarized experimental findings published in [153] of the gender influence on the aging of latent fingerprints using the CWL_Int_F1 (left image) and the CWL_Topo_F1 (right image) feature. The mean aging speed As is depicted, standard deviations are given in the form of error bars.

Figure 93 indicates that gender has no significant influence on the aging speed of a latent fingerprint in respect to the high standard deviation of observed aging speeds. While fingerprints of females seem to age marginally slower than those of males, this influence is observed to be negligible in respect to the other influences contributing to the overall standard deviation of aging speeds.

Based on an extension of this test set, differences between test subjects (inter-person variation), between the same finger of one person applied at different points in time (intra-person variation), between different fingers of a person applied at the same point of time (inter-finger variation) as well as between different regions of a single fingerprint (intra-finger variation) are observed. The results are depicted in Figure 94 for the CWL_Int_F1 as well as the CWL_Topo_F1 feature. The mean standard deviation is provided in respect to each of the four variations. For more details about the experimental evaluations please refer to [153].

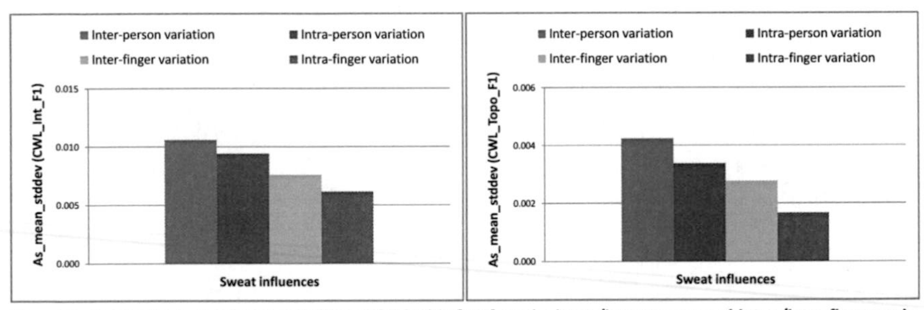

Figure 94: Summarized experimental findings published in [153] on the inter-/intra person and inter-/intra-finger variation of the latent fingerprint aging speed As using the CWL_Int_F1 (left image) and the CWL_Topo_F1 (right image) feature. The mean standard deviation of the observed aging speeds As is depicted.

In Figure 94, the inter-person variation is reported to be the highest for intensity- as well as topography-based time series, followed by the intra-person variation. The inter-finger variation is observed to be second smallest and the intra-finger variation the smallest. These findings confirm expectations towards the variation of aging speeds. While the highest variation is assumed between different people (inter-person variation), the intra-person variation is expected to be second highest, because the sweat composition of a specific finger at different points in time is depending on diet, type of sweating and others, which might change considerably over time. Differences in the sweat composition (and therefore the aging speed) between different fingers of a person (inter-finger variation) and those from different regions within a finger (intra-finger variation) are expected to be comparatively small. Reasons for this can be seen in the fact that most influences on the sweat composition of a human are similar for all fingers as well as different regions within a finger. It can be concluded from this experiment that the sweat composition represents a major influence on the aging speed of latent fingerprints. Its qualitative and quantitative composition therefore needs to be further explored in future work using spectroscopic techniques. In section 6.5.4, additional studies on the aging behavior of certain water-based substances are conducted, such as distilled water, tap water and artificial sweat.

6.5.1.4 *Surface Influences Published in the Scope of this Thesis*

The aging behavior of latent prints is investigated for ten different surfaces based on CWL intensity and topography images in [153] and [158]. Examined surfaces are veneer, car door, glass, furniture, socket cover, brushed scissor blade, hard disk platter, 5 Euro-Cent coin, mobile phone display and CD case (three fingerprints of different test subjects per surface). It can be concluded from the study that the amount of overlap of fingerprint and background (surface) pixels in the histogram seems to be the decisive factor for how well the fingerprint aging property can be extracted. In case fingerprint and background pixels are to a certain extent represented by different gray values in the histogram (as is the case for well reflecting surfaces, such as glass, car door, hard disk platter, mobile phone display and furniture), the aging property can be well extracted. In case fingerprint and surface pixels do significantly overlap in the histogram (as is the case for less well reflecting surfaces, such as veneer, socket cap, brushed scissor blades, 5 Euro-Cent coin and CD case), severe distortions might be observed. Exemplary histograms of a glass as well as a veneer surface are depicted in Figure 95.

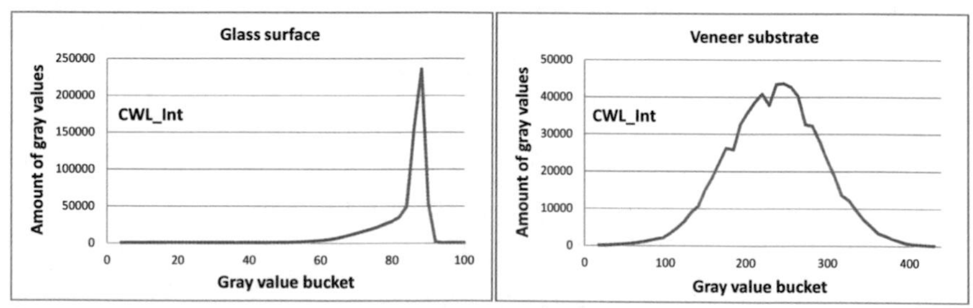

Figure 95: Exemplary CWL intensity image histograms of a glass (left image) and a veneer (right image) substrate.

As can be seen from Figure 95, the histograms are different between the glass substrate (left image) and the veneer surface (right image). For the glass substrate, most fingerprint pixels are presented by gray value buckets in the interval [0,84]. For buckets higher than 84, the amount of gray values sharply increases, indicating that the majority of these pixel values belongs to the background. When binarizing such image with a threshold of thresh = 84, a good separation of fingerprint from background pixels is achieved. For the veneer surface, fingerprint and background pixels are overlaid to one combined normal distribution. A clear separation of fingerprint from background pixels is not possible. Therefore, an optimal threshold would assume values in the interval $150 \leq thresh \leq 350$, achieving a medium or bad separation performance.

When analyzing the aging behavior of these exemplary prints using the CWL_Int_F1 feature (Binary Pixels), the observed characteristic aging property is caused by fingerprint pixels from the left part of the histogram crossing the threshold to the right part of the histogram, therefore becoming background pixels. In case of the glass surface, most background pixels are located at the right side of the threshold. Slight changes in their values (e.g. due to environmental fluctuations) do barely influence the observed aging behavior, because most pixels do not cross the threshold. For the veneer surface, fingerprint and background pixels are widely overlapped in the histogram, assuming positions at both sides of the optimal threshold. Because the amount of background pixels in an image is usually much higher than that of fingerprint pixels, slight fluctuations of background pixel values would lead to a high amount of pixels crossing the threshold in arbitrary directions, significantly distorting the observed fingerprint aging behavior. In such case, age estimation is still possible, if influences on the background pixels are kept constant, e.g. by controlling the environmental conditions during the capturing process.

Concluding from such findings, the expected aging curve quality of a certain surface can be determined by the inspection of the degree of overlap of fingerprint and background pixels in the histogram. Reliable results can be expected if only few pixels overlap. In any other case, reliable results might still be obtained, if influences on the surface as well as the capturing process are significantly understood and reduced. In future work, additional studies should be conducted using quantitative test sets of different surfaces, to further specify the quality of different surfaces for a practical age estimation.

For selected substrates, surface specific properties can be used to develop substrate specific aging features. For example, corrosion artifacts of fingerprints applied to copper surfaces might be used. Such effect is investigated in [157], where sigmoid and logarithmic aging properties are observed for the amount and size of corrosion artifacts. Such artifacts occur mainly at fingerprint pores and could be observed on 5 Cent-Euro coins fresh from the mint as well as copper substrates. The study is conducted over a time period of 15 weeks and therefore regarded as being out of the scope of the here investigated short-term aging. However, additional substrate-specific features might be derived in future work from the chemical interaction of fingerprints with certain substrates for short- as well as long-term aging periods.

6.5.1.5 Application Influences Published in the Scope of this Thesis

Application influences on the speed of fingerprint degradation are studied in [153] and [155], based on five fingerprints of one test subject. Only eccrine sweat is used, because fingerprints have been washed before the experiments and nothing was touched for 30 minutes. For CWL intensity and topography images of features CWL_Int_F1 and CWL_Topo_F1, no significant influence of low contact pressure with short contact time (i.e. only tipping the surface), high contact pressure with long contact time (e.g. pressing the finger to the surface for several seconds) as well as smearing of the prints could be observed. Therefore, the application influence seems to be rather negligible in the scope of fingerprint short-term aging. Here, the evaporation and degradation of the residue material seems to be more important than its specific shape. Due to such assumed negligibility of the application influence, it is not further investigated in the scope of this thesis. However, only eccrine sweat is used in the experiments, leading to comparatively low aging speeds. The studies should therefore be repeated in future work using also prints containing sebaceous components.

6.5.2 Evaluation of Selected Scan Influences

In this section, the influence of starting a capturing device as well as its non-invasiveness in respect to radiation energy are investigated in the scope of fingerprint short-term aging. As described in section 5.8.2, seven (CWL) and ten (CLSM) fingerprint samples are used (test sets TSD_CWL and TSD_CLSM), from each of which six different time series Ts1 - Ts6 are captured from three different fingerprint areas Ar1 - Ar3 during three different time periods Tp1 - Tp3 (Table VI). The experimental aging curves are exemplary depicted in Figure 96 for one sample of the CWL and CLSM sensor. As discussed in section 5.8.2, the valid pixels mask VP of CWL images is not dilated in this particular test set, leading to all CWL aging curves starting at an y-offset of zero.

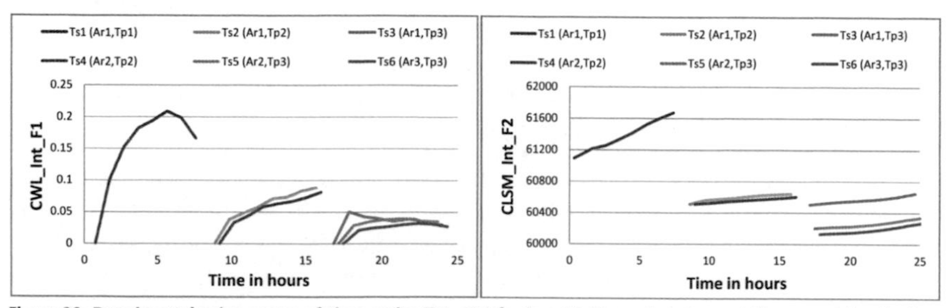

Figure 96: Experimental aging curves of time series Ts1 - Ts6 for investigating selected scan influences, using the best performing intensity-based features CWL_Int_F1 and CLSM_Int_F2. Curves are exemplary depicted for one fingerprint sample of the CWL sensor (left image) and the CLSM device (right image).

When studying the exemplary aging curves presented in Figure 96, it can be observed that during period Tp1 (fresh prints), the experimental curves of both devices exhibit the highest slope and the corresponding features therefore compute the highest aging speed. The curve of the CWL device (left image) exhibits a buckle at the end. Such distortion might be caused by an additional influence factor occurring near the end of the capturing period, such as a change in the ambient climatic conditions. Comparing the curves between time periods Tp2 and Tp3, a decrease in the slopes and therefore the aging speeds is observed for the CWL sensor (left image), while for the CLSM device, the slopes remain constant or even increase slightly between Tp2 and Tp3. Such observation indicates that the aging speed is continuously decreasing and eventually declines below a certain noise level. When such point is reached, the aging speed seems to be no longer of use for age estimation. To investigate the differences of aging speeds between periods Tp1 - Tp3 in more detail, their mean values from all captured time series Ts1, Ts2 and Ts3 of both devices are compared in Figure 97.

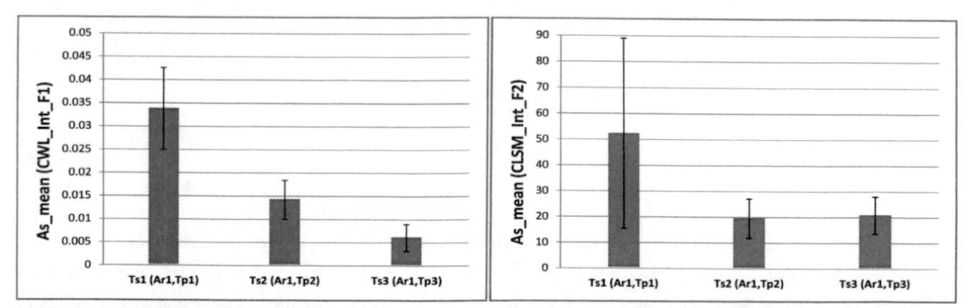

Figure 97: Aging speeds As of the experimental curves for selected scan influences, compared between time series Ts1, Ts2 and Ts3. Mean results are depicted, based on features CWL_Int_F1 and CLSM_Int_F2 for the CWL sensor (TSD_CWL, left image) and the CLSM device (TSD_CLSM, right image). Standard deviations are given as error bars.

Even though the capturing devices are started separately for each time period Tp1 - Tp3, it is expected that the aging speeds of its corresponding time series Ts1 - Ts3 decrease with increasing age, confirming the earlier investigated logarithmic property for the short-term aging. In case a significant influence of starting a capturing device on the observed aging curves does exist (e.g. a certain increase in temperature when starting the device), the expected behavior might be distorted. Such distortion can manifest itself in the form of a certain minimum aging speed, which is always observed when starting a device, even if the captured fingerprint has been aged for a significant amount of time. When evaluating the experimental results of Figure 97, it can be seen that for the CWL sensor (left image), the experimental aging speeds of all three time periods Tp1, Tp2 and Tp3 exhibit the expected decreasing tendency very well. While a high aging speed is observed for Tp1 (fresh prints), such speed is significantly decreased for period Tp2 and further decreased for Tp3. Therefore, no significant influence of starting the capturing device can be observed for the investigated time period. Furthermore, it is to note that the standard deviation of the aging speeds is biggest for period Tp1 and decreases with increasing time, confirming the findings of section 6.4.2.

In regard to the CLSM device, similar results than for the CWL sensor are expected, with an even higher accuracy, due to the significantly higher resolution of the device. The results of Figure 97 (right image) show indeed a comparatively high aging speed for period Tp1 (fresh prints) in comparison to Tp2 and Tp3. However, when comparing Tp2 and Tp3 between each other, it is observed that the mean aging speeds are similar and even slightly higher for Tp3 than for Tp2. This observation suggests that at some point during period Tp2, the fingerprint aging speed cannot be distinguished any more from the overall noise level induced by different influence factors. Two reasons might be given for explaining such phenomenon: either the precision of the capturing device and features is smaller than for the CWL device or additional influences when starting the device are present, being more dominant for the CLSM than for the CWL sensor. Because the CLSM device has a significantly higher resolution than the CWL sensor and is therefore expected to capture images with a much higher precision, the observed phenomenon seems to rather be caused by the latter reason. It seems to be the case that a certain portion of the observed aging speed is caused by specific influences present during and after the start of a measurement run. Such influence could be an increase in temperature after starting the device (e.g. heating up of the device or its light source). Its influence might be significantly higher in comparison to the CWL sensor, which uses a white light LED continuously being switched on.

The standard deviations of the CLSM mean aging speeds exhibit a similar tendency than for the CWL device, where a much higher variation of speeds is present for fresh fingerprints (Tp1), which significantly decreases after the prints have dried up (Tp2 and Tp3). Summarizing the findings of Figure 97, the CWL sensor seems to be of a higher precision than the CLSM device when capturing time series, even though using a significantly lower resolution. The CLSM device in contrast seems to be subject to certain systematic influences when starting the device, which might lead to an overlay of the observed aging property with systematic, sensor specific distortions, especially in the first hours after starting the device.

To investigate the non-invasiveness of the capturing devices in respect to the emitted radiation energy, the different fingerprint time series captured during similar time periods are compared. In case the capturing devices influence the fingerprint aging (e.g. by increasing the aging speed by introducing additional energy from the light source), the different curves of a single time period would vary in their aging speeds, because they have been exposed to the light source radiation for different amounts of time. However, if their aging speeds are similar for each time period, the devices can be considered as being non-invasive. In particular, Ts2 and Ts4 can be compared for period Tp2, while Ts3, Ts5 and Ts6 can be compared for period Tp3. Figure 98 depicts these comparisons for both capturing devices.

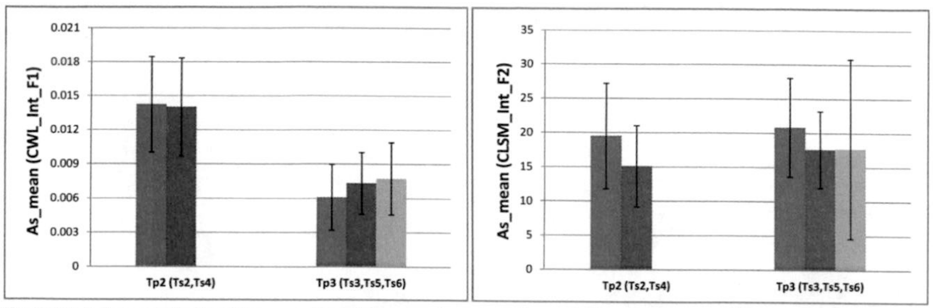

Figure 98: Aging speeds As of the experimental curves for selected scan influences, compared between time series Ts2 and Ts4 (period Tp2) as well as Ts3, Ts5 and Ts6 (period Tp3). Mean results are depicted, based on features CWL_Int_F1 and CLSM_Int_F2 for all samples captured with the CWL sensor (TSD_CWL, left image) and the CLSM device (TSD_CLSM, right image). Standard deviations are given as error bars.

When evaluating the non-invasiveness of both capturing devices based on Figure 98, it can be observed that Ts2 and Ts4 of the CWL device (left image) exhibit almost equal aging speeds. Ts3, Ts5 and Ts6 exhibit a certain difference. However, such difference is much smaller than the corresponding standard deviations of the three time series and therefore seems to be rather insignificant. Similar results are observed for the CLSM device (right image of Figure 98). A certain difference does exist between Ts2 and Ts4 for period Tp2 as well as between Ts3, Ts5 and Ts6 for period Tp3. However, also this difference is much smaller than the standard deviations of the time series. It can be concluded that no significant difference exists between the aging speeds of different time series of a certain time period and therefore both capturing devices are non-invasive. The minor fluctuations which do exist between the speeds of corresponding time series might be attributed to the different regions Ar1 - Ar3, from which the series were recorded, because the aging speed within a fingerprint is not completely similar for different regions (section 6.5.1.3). The exemplary aging curves of Figure 96 confirm such finding, exhibiting a general similarity of the aging curves between Ts2 and Ts4 as well as Ts3, Ts5 and Ts6 for both capturing devices.

The mean aging speed of time series Ts6 of the CLSM device (rightmost row of the right image of Figure 98) seems to exhibit a comparatively higher standard deviation that its counterparts of the same time period. It is unclear at this point of time which is the cause of such higher variation. Because fingers are not pre-treated prior to their application to the surface, a higher variation of the qualitative or quantitative sweat composition of the fingers for this particular series could be a possible source (e.g. a particularly inhomogeneous mixture of dry and wet fingerprints).

The experimental results presented in this section are of great importance for the reliability of the earlier reported preprocessing, feature extraction and age estimation methods in test goals TG2 - TG4. Because CLSM intensity images seem to be overlaid by an additional, most likely sensor-related influence, CWL-based findings are considered as the more reliable results. Furthermore, the assumption of the non-invasiveness of both capturing devices is confirmed, which is an important foundation for all results reported in this thesis.

6.5.3 Evaluation of Selected Environmental Influences

The potential influence of different ambient climatic conditions on the short-term aging of latent prints is investigated in this section. Based on the test sets TSE_CWL and TSE_CLSM (section 5.8.3), three different variations of ambient temperature as well as relative humidity are investigated for each capturing device. A total amount of 20 (CWL) and 30 (CLSM) time series is evaluated for each climatic condition over a total time period of t_{max} = 24 hours. The experimental results are depicted in Figure 99.

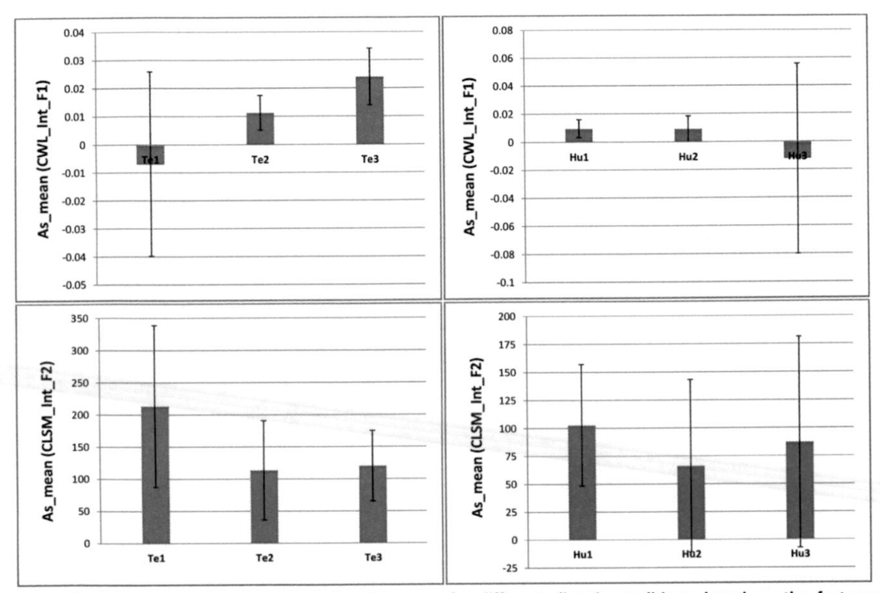

Figure 99: Aging speeds As of the experimental curves for different climatic conditions, based on the features CWL_Int_F1 and CLSM_Int_F2. Three temperature variations Te (left images) as well as three variations of the relative humidity Hu (right images) are investigated. Mean results are depicted for all samples captured with the CWL sensor (TSE_CWL, upper row) and the CLSM device (TSE_CLSM, lower row). Standard deviations are given as error bars.

When evaluating the experimental results of Figure 99, the climatic variations listed in Table VII need to be taken into account, because they are not always equidistant, due to the limited (but more realistic) means of controlling temperature and humidity in the used indoor environments. The results of ambient temperature variation for the CWL sensor (upper left image) shows a direct correlation between the temperature and the aging speed. The higher the ambient temperature, the higher the measured speed. Therefore, a direct dependency of the aging speed on the ambient temperature is assumed (confirming the findings reported in section 6.5.1.2). Warmer conditions seem to lead to a significantly faster degradation of the latent prints, whereas lower temperatures significantly slow down such process. Therefore, latent prints are assumed to be sustained much longer in a cold environment and might even be completely preserved, if temperatures are sufficiently low.

For the investigated temperature variation Te1 (mean temperature of 16.7°C), the mean aging speed is even slightly negative, however accompanied by a high variation. Such high variation renders the results for Te1 unreliable, because it does not seem realistic according to prior observations that the fingerprint degradation has already come to a halt at 16.7°C. Here, additional influences need to be taken into account, such as the impact of the ambient temperature on the capturing device as well as local fluctuations due to air circulation (e.g. when opening windows). A temperature influence on the speed of fingerprint degradation seems to exist, however, more reliable experimental setups (e.g. using climatic chambers), need to confirm such findings in future studies.

The results of a varying ambient humidity observed for the CWL sensor (upper right image) exhibit a comparatively similar aging speed for the variations Hu1 (mean relative humidity of 33.0%) and Hu2 (mean relative humidity of 37.9%). Therefore, it is assumed that small changes in the relative humidity (in this case a difference of 4.9%) do not seem to significantly influence the speed of short-term fingerprint degradation. Such finding is vital for the earlier presented indoor scenario results, because it renders small humidity fluctuations rather negligible and therefore might allow for a reliable practical age estimation in indoor scenarios, if only minor changes in ambient humidity do occur.

The highest ambient humidity is present in Hu3. Here, a mean relative humidity of 51.2% is measured. However, as has been reported in section 5.8.3, an air humidifier is used for this variation, which distributes humidity unequally over the measurement table and leads to different local humidity values of up to 100%. The climatic logging device is positioned at a distance of approximately 20 cm from the prints. It might therefore not be able to appropriately capture local fluctuations of the humidity. Consequently, the resulting variation of the fingerprint aging speed is comparatively high, not allowing for reliable conclusions using Hu3. However, an important finding can be reported here. The observed negative aging speed is caused by an absorption or even condensation of water particles at the latent prints, which is visualized in Figure 100 for extreme cases. Such finding limits the scope of a reliable fingerprint age estimation to an ambient humidity lower than a certain (so far unknown) threshold. When such threshold is reached, latent prints start absorbing ambient water, leading to a refreshment of the prints.

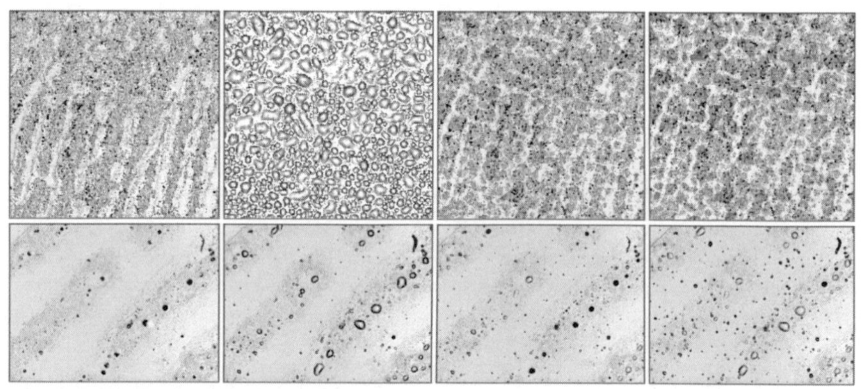

Figure 100: Visualization of exemplary fingerprint time series under high ambient humidity (Hu3), based on the CWL sensor (upper row, t_1 = 1 h, t_2 = 6 h, t_3 = 15 h, t_4 = 24 h) as well as the CLSM device (lower row, t_1 = 1 h, t_2 = 9 h, t_3 = 16 h, t_4 = 24 h).

The absorption or even condensation of water at a latent print under a high ambient humidity can be very well observed in Figure 100. For the CWL sensor (upper row), condensation of ambient water at the hard disk platter can be seen for time t_2, where the complete fingerprint has been covered by condensed water. This is a very extreme case. For points t_3 and t_4, the ambient humidity has been decreased, leading to the evaporation of the condensed water. It can be observed from Figure 100 that the fingerprint has not been completely destroyed during the condensation and subsequent evaporation of ambient water, which might be an interesting finding for the research of fingerprints submerged under water (e.g. as conducted in [142] and [146]). However, the form of the ridges has been heavily altered during the condensation process.

A less extreme case is observed in Figure 100 for the CLSM device (lower row). Here, the surface is not completely covered by condensed water. However, the absorption of ambient water by the print can be clearly seen for times t_2 and t_4, where certain fingerprint droplets exhibit a significant increase in size. The reversibility of such process can be seen for time t_3, where most of the ambient water absorbed in t_2 has evaporated, to be again absorbed for time t_4. Therefore, a high ambient humidity can significantly distort the observed aging speed and needs to be taken into account when estimating a fingerprints age.

When evaluating the experimental results of Figure 99 for the CLSM device (lower row), inconclusive results are observed for the variation of ambient temperature as well as humidity, which are in contrast to the earlier findings of the CWL sensor. For example, the mean aging speed of the curves is significantly higher for low temperatures (Te1) in comparison to higher temperatures (Te2, Te3). Furthermore, no clear trend can be observed for the investigation of humidity changes. In section 6.5.2, the CLSM device has been found to produce less reliable results than the CWL sensor, potentially being distorted by certain influences from the starting process of the capturing device. It is possible that such influences significantly alter the observed aging curves, especially their slopes and consequently lead to the here observed inconsistency of the expected aging speeds. Therefore, the CLSM device seems to be less reliable than the CWL sensor and its higher performance in the earlier feature evaluation (section 6.3) and age estimation (section 6.4) experiments might be partially attributed to systematical influences of the capturing device.

Summarizing the overall experimental findings of varied ambient temperature and humidity conditions for both capturing devices, it is concluded that the influence of starting the CLSM device (section 6.5.2) seems to introduce a significant amount of distortions to the resulting aging curves. A second source of influences can be seen in the inability of the used experimental setup to produce stable climatic conditions, especially for low temperature values (Te1) and high humidity values (Hu3), which lead to a comparatively high variation. Here, climatic chambers seem to be a feasible improvement for future research. Apart from these inaccuracies, three qualitative findings can be derived from the experiments. First, the (assumingly more reliable, see section 6.5.2) CWL device exhibits a certain correlation between an increased ambient temperature and the speed of fingerprint degradation. Second, according to the CWL results, the ambient humidity seems to be of minor influence on the short-term aging when only slightly varied (up to 4.9% of relative humidity in the scope of this experiment). Third, the relative humidity needs to remain below a certain maximum threshold, to avoid the absorption of significant amounts of water from the ambient air by the print. It is to note that these findings are of limited, qualitative nature only and need to be confirmed in future experiments using quantitative and more accurate test setups.

6.5.4 Evaluation of Selected Sweat Influences

This section presents the experimental results of investigating the short-term aging behavior of different water-based substances. Based on test sets TSF_CWL and TSF_CLSM (section 5.8.4), the aging behavior of distilled water DW, tap water TW and artificial sweat AS is recorded and compared over a time period of t_{max} = 24 hours. A total amount of 20 (CWL) and 30 (CLSM) time series is investigated for each substance. The experimental results are depicted in Figure 101. For an adequate evaluation, 20 real fingerprint time series are randomly chosen for each capturing device from the earlier investigated test sets TSB_CWL and TSB_CLSM and are depicted for comparison.

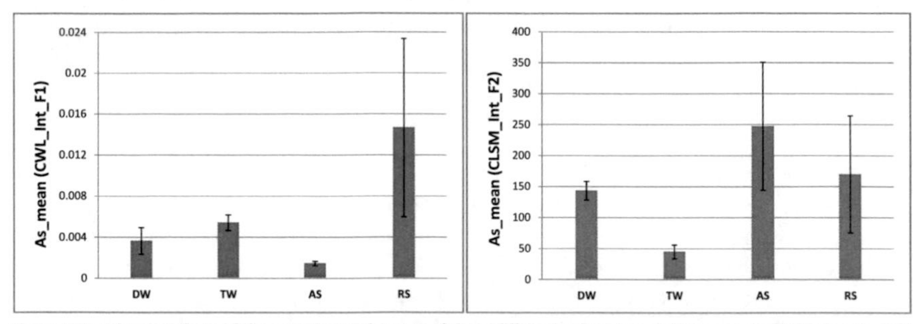

Figure 101: Aging speeds As of the experimental curves of three different substances. Substances are distilled water DW, tap water TW and artificial sweat AS based on the features CWL_Int_F1 and CLSM_Int_F2. Mean results are depicted for all samples captured with the CWL sensor (TSF_CWL, left image) and the CLSM device (TSF_CLSM, right image). The results of 20 randomly selected time series from TSB_CWL and TSB_CLSM are also depicted for comparison, representing real fingerprint sweat RS. Standard deviations are given as error bars.

Comparing the results of Figure 101, it is observed that widely varying aging speeds do occur. Furthermore, different tendencies can be observed between the CWL and the CLSM device, e.g. for the artificial sweat AS, whose mean aging speed is the lowest of all substances for the CWL sensor and highest for the CLSM device. Because no clear tendencies can be identified, the results are considered inconclusive and no reliable information about the aging speed of the different substances can be derived from them.

For the CWL device (left image) all three substances DW, TW and AS exhibit a comparatively low aging speed, while real fingerprint sweat RS seems to age significantly faster. When taking the substance images of Figure 45 into consideration, the comparatively low resolution of the CWL device might be a possible source of such behavior. It can be seen from Figure 45 that the three investigated substances are represented as small black dots in the CWL intensity images. When comparing these images to those of real fingerprints captured with the CWL device (e.g. from Figure 22 or Figure 34), it can be seen that much broader areas of the image are covered by fingerprint pixels for real prints. It is therefore assumed that the significantly smaller density of dots resulting from the substance printing process are not adequately captured with the resolution used for the CWL sensor. In such case, the dots can still be seen in the captured images, however, the small changes occurring within them might not be recognizable any more.

From the investigations conducted with the CLSM device (right image of Figure 101), two of three substances exhibit a mean aging speed similar to (DW), or higher as (AS), real fingerprint sweat RS. As can be seen from Figure 45, CLSM substances are captured with a much higher resolution and form a ring-like shape of particles around darker cores. Such rings seem to consist of particles remaining after the evaporation of the outer area of droplets observed in section 5.8.4 during the first 60 seconds directly after the printing process. These rings are most visible for DW and AS and are only barely visible for TW. Their presence therefore seems to correlate to the computed aging speeds and the observed substance degradation can be regarded as a result of the disappearance of such outer rings over time. This would also explain why no significant aging behavior of the three substances is observed for the CWL device, because the outer rings could not be adequately captured with the used CWL resolution.

Concluding from such findings, the observed degradation behavior of latent prints within this experimental setup seems to be significantly influenced by the shape and distribution of the different particles, which might be more dominant than the influence of the specific substance compositions. However, the substance composition could also have a significant impact on the formation and visibility of such rings around droplets (e.g. caused by the different contained chemical compounds).

The conducted experiments expose a major limitation of the used capturing devices. Because the acquisition of latent prints is based on the optical shape and structure of a substance rather than its chemical composition, important information like the qualitative and quantitative sweat composition cannot be sufficiently included into the evaluation. In future work, such limitation should be overcome by the application of FTIR spectroscopes, which can non-invasively acquire information about the qualitative and quantitative chemical composition of substances.

6.5.5 Creation of a First Qualitative Influence Model

Based on all experiments conducted in the scope of this thesis, a first qualitative influence model on the short-term aging of latent fingerprints is proposed. Results from the published studies [95], [153], [154], [155], [156], [157], [158] and [159] summarized in section 6.5.1 as well as the additional experimental findings of sections 6.5.2 to 6.5.4 are included into such model. The model is depicted in Figure 102. It is to note that it is based merely on first qualitative studies of selected capturing devices and therefore needs to be confirmed by quantitative test sets and using additional capturing devices in future work.

The model proposed in Figure 102 is of preliminary nature. However, it provides a first qualitative overview over the experimental findings on potential influences conducted in the scope of this thesis. The application influence (application time, application pressure and smearing) is found to be negligible, because the specific form of the residue does not seem to be of importance for the observation of print material degradation processes (section 6.5.1.5). It is therefore visualized using dotted lines in Figure 102. However, theses finding need to be confirmed in future studies, using fingerprints containing sebaceous components (section 6.5.1.5).

The surface influence is of importance. However, it can be determined from the histogram of a captured fingerprint image if the fingerprint pixels do significantly overlap the background pixels in the histogram and therefore if the age estimation results are reliable or potentially subject to severe distortions (section 6.5.1.4). It can therefore be controlled by discarding age estimation results of the latter case and is marked in Figure 102 using dashed lines. However, if fingerprint and background pixels do significantly overlap, an age estimation still seems possible, if additional influences are investigated and understood in future work, such as the influences of ambient conditions on the surface and the capturing device, which seem to overlay the aging property in many cases. Furthermore, the absorption of fingerprint material by porous surfaces needs to be considered. Additionally, physical or chemical reactions with the substrate should be considered as well, such as corrosion processes on metals (section 6.5.1).

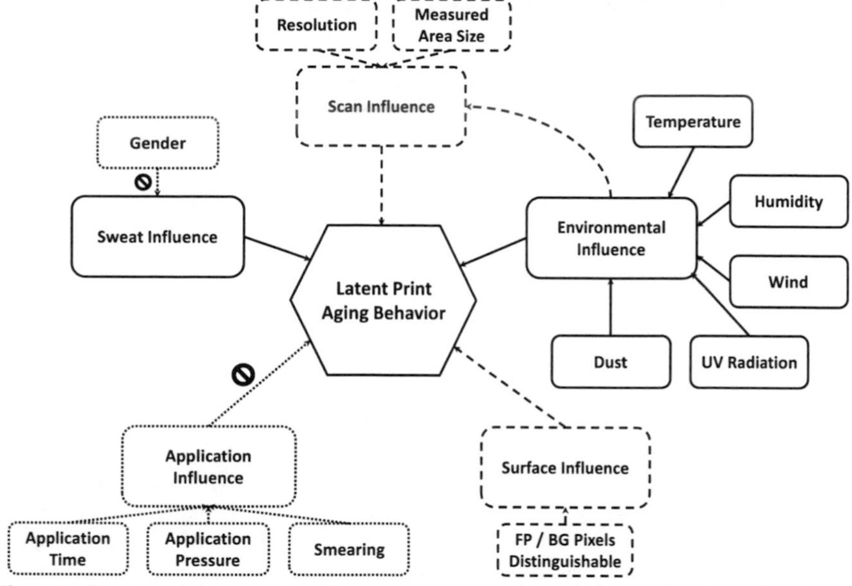

Figure 102: The proposed short-term aging influence model based on first qualitative experiments of sections 6.5.1 to 6.5.4. Dotted lines represent influences which have been found to be negligible. Dashed lines represent influences which can be controlled or reproduced and therefore be considered during an evaluation. Continuous lines represent influences which can hardly be controlled or even recorded.

Environmental influences impact the aging behavior of fingerprints, because they are able to increase or decrease the speed of print degradation (sections 6.5.1.2 and 6.5.3). Especially temperature, humidity, wind and UV radiation are identified as impacting the aging behavior. Furthermore, the exclusion of dust seems to be an important issue (section 4.2.5.6). When the application is limited to indoor scenarios (as is the case in the scope of this thesis), UV radiation and wind seem to be of minor importance, because they are comparatively small in closed rooms. It is to note that environmental influences cannot be controlled at practical application scenarios, such as crime scenes. Furthermore, it is hard to even reconstruct these conditions. The environmental influence is therefore marked in Figure 102 using continuous lines. Here, additional environment-independent features need to be found in future work or the environmental conditions need to be reconstructed and taken into account, which is a challenging task.

Another important influence on the aging of fingerprints is the used capturing device (scan influence). Apart from the capturing resolution and measured area size, which can be controlled by fixation to a certain configuration, environmental influences seem to impact the devices used in the scope of this thesis. Especially the ambient temperature and vibrations (section 4.1.2, PC4 and section 6.5.1) seem to have a significant influence on the captured images. Furthermore, the devices themselves introduce severe distortions, such as changes of the overall image brightness over time (section 4.1.2, PC2 and PC5) as well as influences when starting a device (section 6.5.2). However, the different scan influences might be controlled by the design of more accurate devices (e.g. emitting a constant amount of light from the light source at all times) or by operating the devices only in climatized laboratories (excluding environmental fluctuations). The scan influence is therefore visualized in Figure 102 using dashed lines and considered to be controllable in future age estimation schemes.

The qualitative and quantitative sweat composition (sweat influence) is furthermore considered an important influence on the fingerprint aging process. Gender is identified as having a rather negligible impact (section 6.5.1, visualized in Figure 102 using dotted lines). However, the general sweat composition seems to be subject to significant variations (sections 6.5.1 and 6.5.4). Such variations cannot be controlled at practical crime scenes and the composition of print residue at the time of fingerprint application can barely be reconstructed. Therefore, the sweat influence is visualized in Figure 102 using continuous lines. The capturing devices used in the scope of this thesis cannot determine the qualitative and quantitative composition of recorded latent prints, which remains an important issue for future research, e.g. using spectroscopic devices, such as FTIR.

Summarizing the proposed qualitative influence model on the short-term aging of latent fingerprints, several conclusions might be drawn, which need to be confirmed in future quantitative tests using also additional capturing devices. While the application influences seem to be negligible, the surface and scan influences exhibit the potential to be controlled in future work, if adequately understood and if additional, more accurate capturing devices and conditions can be provided. Therefore, the potential of a future age estimation does not seem to be limited by these factors. However, the impact of environmental and sweat influences presents a challenge to future age estimation schemes, because they can barely be recorded or reconstructed from practical crime scenes. Here, either influence-independent features need to be designed or adequate methods for the digital removal or reduction of influences need to be established. Also, this challenge could be addressed by limiting the application scenario, e.g. using indoor crime scenes, where environmental conditions are much more constant than in outdoor scenarios. Including chemical analysis methods, such as FTIR spectroscopy, can potentially contribute towards this goal. The here proposed model might be regarded as the basis for more detailed and quantitative studies in future work.

7 Summary, Discussion and Future Work

In this chapter, the scientific results are discussed in respect to the earlier identified research gaps and aims of this thesis (section 7.1). Furthermore, prospects and limitations of non-invasive capturing devices and digital processing methods for the age estimation are summarized (section 7.2). Implications to practical age estimation applications (section 7.3) as well as to the general forensic domain (section 7.4) are derived. Future work is presented in section 7.5.

7.1 Summary and Discussion of Results in Respect to the Addressed Research Gaps and Aims of this Thesis

In the scope of this thesis, the potential of non-invasive capturing devices and digital automated processing techniques for the challenge of latent fingerprint age estimation has been investigated. The findings are here discussed in respect to the five main research gaps and aims of this thesis:

(1) The first research gap is focused on the lack of using non-invasive, high-resolution capturing devices to create time series of fingerprints without prior development or other alterations. A CWL and a CLSM device have been successfully applied to this challenge, capturing intensity as well as topography data. Adequate basic sensor settings were defined, including a minimal measured area size (CWL: 4 x 4 mm, CLSM: 1.3 x 1 mm) and a minimal lateral dot distance (CWL: 20 µm, CLSM: 1.3 µm). Five main preprocessing challenges have been identified and were addressed by designed methods in the preprocessing step (spatial alignment, changes of the overall image brightness over time, image segmentation, ambient environmental influences and sensor-specific distortions). As a result, latent fingerprint degradation behavior could be adequately studied from print time series. While the first research gap is therefore considered to have been adequately addressed by meeting the aims of O1 and TG1, also limitations of this concept have been identified. Especially distortions of the computed aging curves from the capturing devices and other influences lead to a decreased reliability of some of the achieved results. The distortions manifest themselves in the form of changes of the overall image brightness over time, temporal fluctuations caused by environmental conditions, scan artifacts at droplets, line-wise fluctuations in image gray values due to the measurement table propulsion (CWL), barrel distortions (CLSM) and a partial bias of the computed aging speeds caused by starting a device (CLSM). Although these limitations have been addressed in the scope of time series preprocessing, their complete exclusion cannot be guaranteed. They should therefore be addressed in future work, to further increase the reliability of the approach.

(2) The lack of using digital preprocessing methods has been identified in the second research gap as a major limitation to successful age estimation. In this thesis, a digital processing pipeline has therefore been designed and applied to fingerprint age estimation for the first time. It allows for an automated processing of fingerprint time series with increased accuracy. Different preprocessing methods have been designed or adapted, to achieve adequate spatial alignment, tem-

poral normalization (reduction of changes of the overall image brightness over time), segmentation as well as reduction of distortions (such as dust and different sensor artifacts). For an adequate binarization, 17 different intensity-based techniques (mainly from the literature) as well as three different topography-based techniques (designed in respect to each device) are compared. The resulting preprocessing step realizes the functionally aimed at in the scope of O2 and T2. Adequately preprocessed and segmented time series are provided for a later feature extraction. Although the second research gap is therefore regarded as being adequately addressed in the scope of this thesis, there is room for future improvements. Especially for the topography images, which exhibit a significantly worse performance in the scope of binarization (CWL_Int: b_err = 0.11, CWL_Topo: b_err = 0.29, CLSM_Int: b_err = 0.07, CLSM_Topo: b_err = 0.15), additional preprocessing techniques might be investigated in future work.

(3) The third research gap highlights the lack of feasible features for age estimation. Seven different statistical measures have been proposed and applied to the short-term aging of 770 different fingerprint time series from each device for the first time. Best features include the Binary Pixels feature as well as the mean image gray value, which exhibit a strong correlation to a logarithmic function (Pearson correlation coefficient $r \geq 0.8$) in 88% (CWL, Binary Pixels) and 92% (CLSM, mean image gray value) of cases. Such performance is regarded as a promising result in respect to the high variability of sweat compositions from different donors. The research gap is therefore considered to have been adequately addressed in the scope of O3 and T3. The long-term aging investigations have furthermore shown that statistical features perform equally well or better than their morphological counterparts, clearly highlighting the advantages of statistical approaches. However, significant systematic influences of the capturing devices are observed for the long-term aging investigation, indicating the necessity of further studies on this issue.

(4) According to the fourth research gap, no digital fingerprint age estimation scheme exists at this point of time. With the successful design and evaluation of four digital formula-based age estimation approaches (median deviation from original age for best approach: 5 hours for CWL and 3.4 hours for CLSM) and one machine-learning based approach (best performance for CWL: kap = 0.52, equivalent classification accuracy: 76%; best performance for CLSM: kap = 0.71, equivalent classification accuracy: 86%), this research gap is addressed for the first time within this thesis, achieving promising results. Several parameters have been optimized and the statistical significance of the sample set size is experimentally shown. As a result of optimization, five (CWL) and three (CLSM) consecutive scans of a print are required for age estimation, equaling total capturing times of five (CWL) and three (CLSM) hours. Limitations of the approach can be seen in the age estimation performance, which is far from being reliable. Still, a separation between fresh and aged prints seems possible in many cases, which is a significant advancement in this area. A second limitation can be seen in the finding of the age estimation performance becoming worse with increasing age, where prints become less distinguishable over time.

(5) In the scope of the fifth research gap, the insufficient knowledge about influences on the aging process as well as the initial state of aging features is identified as a major issue. Consequently, the influence of different factors from the capturing process, environmental conditions, sweat composition, surface type as well as fingerprint application process is investigated in the scope of O5 and T5. From the evaluated influences, a qualitative and preliminary influence model is created. While contact time, contact pressure, smearing as well as the gender of a donor are found to be rather negligible, influences of the capturing device as well as the used substrate need to be considered for a reliable age estimation. The sweat composition and environmental influences cannot be controlled at all. The fifth research gap therefore is far from being adequately ad-

dressed, even with the various experiments conducted in the scope of TG5. A significant amount of quantitative studies seems necessary in future work for the extraction of reliable information about the impact of different influences. Non-invasive devices capturing substance-specific properties should also be included, such as FTIR spectroscopes. Because the here used capturing devices have been identified as a significant source of error, also improvements of these devices are of importance in respect to precision and constant climatic capturing conditions.

Concluding from the above discussion, the proposed non-invasive capturing devices and the designed processing pipeline seem to adequately address the identified research gaps (1) - (4) in the scope of this first study. However, a significant performance improvement for each step of the processing pipeline might be achieved, if the applied methods are further enhanced. The here achieved advancements are therefore regarded as a first step only. Concerning the different influences on the aging behavior (fifth research gap), a significant amount of additional research seems necessary for an adequate understanding. However, such increased understanding would also significantly benefit the improvement of the other four research gaps.

7.2 Prospects and Limitations of Non-Invasive Capturing Devices and Digital Processing Methods for Age Estimation of Latent Fingerprints

The approach proposed in this thesis for the non-invasive capture and digital automated processing of latent print time series is novel and promising. The numerous prospects of the approach can be summarized as follows:

- The general potential of non-invasive capturing devices and digital processing methods for deriving information about the age of a latent print has been confirmed by the characteristic aging properties observed for different features. Non-invasive devices are therefore shown to be able to detect characteristic aging properties of latent prints.
- The proposed digital processing pipeline is able to capture, preprocess, extract features and classify latent prints in an automated, objective way based on high-resolution data. such automated and objective age estimation is regarded as a significant improvement to computer-assisted forensics.
- Required capturing times for short-term age estimation are found to be five hours (CWL) and three hours (CLSM), where 20 different prints can be analyzed in parallel using one capturing device. Such performance is regarded as comparatively fast in regard to common fingerprint processing times of a few days in criminal laboratories.
- Age estimation is found to be a privacy protective technique, because the captured areas are too small for identification [21], [22]. Consequently, the here proposed fingerprint age estimation approach captures data, which is not person-related and therefore has the potential to be applied in many different practical applications without being subject to data protection regulations. Potential applications include crime scene investigations, daily police work as well as preventive applications.
- The provided objective quality measures offer the potential for a systematic comparison and improvement of age estimation strategies in future research. So far, no benchmarking criteria have been presented to rate age estimation schemes, rendering this a novel approach.

- A significant amount of changes happens within the first 24 hours of aging. The here proposed age estimation approach allows for the investigation of short-term aging periods of a few hours, apart from the long-term periods of several days, months or years. It is furthermore especially useful for the differentiation between fresh and aged fingerprints.

Apart from the identified prospects of the new non-invasive, high-resolution, digital and automated age estimation approach, the numerous conducted experiments have also identified significant limitations of the scheme. These limitations are in particular:

- Several influences of the capturing devices on the observed aging curves have been identified, which in some cases significantly distort the experimental results. The overall image brightness needs to be constant throughout a complete time series and also throughout all regions of an image. Here, severe limitations are observed especially for the CLSM device. The measurement principle based on the reflection of light leads to a refraction at droplets and invalid values at sharp edges in some cases. The propulsion of the CWL measurement table introduces further distortions. For the age estimation as a process subject to many different influences, the capturing devices are required to work as accurate as possible, not introducing additional distortions. At this point of time, the distortions introduced by the capturing devices seem to significantly limit the reliability of the proposed age estimation strategies.
- The topographic domain of both devices is promising, because specific height values of fingerprints can be derived. However, the precision of the topographic scans does not seem to be of sufficient quality at this point of time for reliably extracting the tiny changes happening at fingerprint particles during aging. Topography therefore seems to be only of use if the capturing quality can be significantly improved.
- Both investigated devices are not able to capture substance-specific properties of a print. This represents a severe limitation, however might be overcome in future work by using non-invasive spectroscopic devices, such as FTIR spectroscopes.
- Influences of the environment, sweat composition and substrate on the aging of latent prints leads to a significant variation of the observed aging speeds. They need to be understood in more detail and either controlled or removed from the observed aging properties for increased age estimation accuracy. Alternatively, features independent of such influences might also be identified in future research.
- At this point of time, the achieved age estimation accuracy is insufficient for a practical usage in court. A significant performance improvement is required for achieving such accuracy.
- Because the most characteristic aging processes seem to follow a logarithmic decrease of their aging speed, the age estimation performance becomes worse with increased age of the prints, eventually leading to their indistinguishability. Therefore, much longer capturing intervals are required for long-term age estimation approaches.
- In total, a few hundred test subjects donated their fingerprints for the investigations. However, the representativeness of the fingerprint aging speeds acquired from them cannot be shown to be valid for the complete world. Although some international students were included into the study, people on other continents or even in other countries might exhibit differences in the aging speed of their fingerprints. However, also craftsmen could exhibit different aging speeds than students, or smokers others than non-smokers. The variety of test subjects therefore needs to be increased in future work. However, this is regarded as a matter of effort and therefore only as a temporary limitation.

From the listed prospects and limitations of the proposed age estimation scheme, the general potential of the approach is confirmed. However, new challenges have been identified from the specific properties of the capturing devices, which need to be adequately addressed before acceptable performances can be achieved. Furthermore, the high variability of the speed of print degradation caused by different sweat compositions, environmental influences and surface properties remains an issue, which requires a substantial amount of additional research in future work.

7.3 Implications to Practical Age Estimation Application Scenarios

In this section, the numerous findings throughout the conducted experiments are discussed in respect to their implications for practical age estimation applications in the future. Especially their impact on crime scene forensics (section 7.3.1), data privacy protection in daily police work (section 7.3.2), preventive application scenarios (section 7.3.3) as well as the separation and sequencing of overlapped fingerprints (section 7.3.4) are discussed.

7.3.1 Age Estimation for Crime Scene Forensics

Linking a latent fingerprint to the time of a crime is the most important implication of a practical age estimation scheme. It is an important part of providing evidence that a suspect has committed a certain crime, by not only showing that the suspect has been at the scene of the crime (identification), but also during the crime (age estimation). Furthermore, additional information might be extracted from the age of a latent print, such as the sequence in which objects where touched and therefore the sequence of certain events.

When investigating a crime scene, latent prints are often searched for on objects potentially handled by the criminal, such as doorknobs or used weapons. Fingerprints applied to areas not likely to have been touched or those on uncooperative surfaces are often not acquired at all. However, when using contactless, non-invasive capturing techniques, such prints could also be detected in the scope of a coarse scan, as introduced by Hildebrandt et al. in [20] and [175]. Therefore, a substantial increase of the amount of detected latent prints at crime scenes is expected when using such devices. A detailed evaluation, including identification of each of these traces, might be very time consuming, especially for public crime scenes, where many people unrelated to the crime have been present.

Therefore, the age estimation of latent fingerprints could be used in such case for a preselection of prints based on their age. If the age of a latent print can be estimated in an automated way using the proposed digital processing pipeline, a distinction between potentially relevant and not relevant traces can be performed. In such scenario, prints applied during the time of a crime or in a certain time period before or after the crime are considered as relevant, while other prints are regarded as of lower priority (and can still be investigated at a later point of time if necessary). Consequently, age estimation seems to offer the possibility of pre-selecting high-priority fingerprints at crime scenes in respect to relevant time periods and to process these fingerprints first. Therefore, a significant amount of manpower and time might be saved. However, such application is based on the assumption of a highly reliable fingerprint age estimation scheme, which remains a challenge at this point of time.

When analyzing the additional time effort introduced by fingerprint age estimation, a total capturing time of five hours (CWL) and three hours (CLSM) is identified in this thesis, with the potential of further optimizations by decreasing the time difference between consecutive scans. When comparing this effort to the time span of several days usually required for classical fingerprint development and lifting, age estimation performs comparatively fast. Furthermore, the age of at least 20 prints can be investigated in parallel with a single capturing device, if they can be fitted onto the measure table.

The mobility of a potential age estimation device is furthermore of importance. The here investigated capturing devices seem to be sensitive to certain environmental conditions (such as temperature) and therefore require a climatic chamber. Although devices could be transported to a potential crime scene, they are bulky and require the scanned object to be fixed onto a measurement table, leading to its removal from the original context (which is not always possible, e.g. when a fingerprint is found at a wall). However, the mobility aspect is dependent on the capturing device and the usage of mobile devices seems to be possible in the future.

When including the age estimation of latent fingerprints into crime scene investigations, the general line of actions would have to be adapted. At this point of time, the discovery of a latent print is often dependent on its chemical or physical development. Age estimation needs to be conducted prior to a potential development on the original trace. Therefore, additional detection methods are required, such as the earlier introduced coarse scan. A general change of the line of action is expected in crime scene forensics in the next decades, if novel contactless capturing methods are transferred from research into practical application at crime scenes. Such new lines of action could include elements such as coarse scan, fine scan and even separation of overlapped fingerprints [20]. Age estimation would fit well into such automated, digital evidence acquisition process. It might be conducted after the detection of latent prints (coarse scan) and before the selection of specific prints captured in the scope of fine scans, assisting in the selection of relevant traces. In classical lines of action, age estimation might still be performed for latent prints, if detected without a prior chemical or physical development or by using small areas of a print, which have not been subject to prior development.

7.3.2 Data Privacy Protection in Daily Police Work

Data privacy protection is an important aspect in daily police work. The loss of reputation or psychological damage possibly occurring in case a subject is falsely accused of a crime as well as the potential misuse of captured fingerprints when leaked from a police database justifies the strict regulations of many countries in this regard. A discussion of this topic has been published in [21] in the scope of this thesis. Because the here investigated age estimation techniques capture only very small fingerprint regions, which are considered as not being linkable to a specific person [21], [22], age estimation can be regarded as a privacy conform technique, not capturing person-related data. As such, it might be applied to daily police work for the enhancement of privacy preserving routines.

For example, a preselection of latent prints in respect to time periods of interest could be performed. In such scenario, the age of all prints found in a certain location is determined without capturing person-related data. Afterwards, only fingerprints dating from the time period of interest are selected, captured and identified, while all other prints are discarded. Such procedure would significantly enhance the privacy protection in daily police work, because prints of uninvolved people would not be identified at all. However, an accurate age estimation performance is the basic prerequisite for such approach, which requires significant additional research on the issue. Furthermore, adaptations are required to certain parts of the current line of action of daily forensic fingerprint work.

7.3.3 Enabling of Fingerprint Scans for Preventive Police Applications

So far, the preventive capture of latent fingerprints is not allowed in most European Countries, because person-related data cannot be recorded and stored without an adequate suspicion. Please refer to [21] for more detailed information about this issue. However, it might be of interest in certain application scenarios to capture fingerprints for preventive purposes. For example, if fingerprints are found in highly restricted areas or in unusual locations, applied during suspicious time periods (e.g. when access to the area was forbidden), it could be of interest to further investigate these prints to prevent a potentially planned crime.

Age estimation can be performed on found latent prints without violating data protection laws, because a person cannot be identified from the captured partial prints at this point of time. If the print is found to have been applied during a suspicious time period (e.g. during the night when nobody was granted access to the area), a well-founded suspicion is given, under which a capture and identification of the complete print seems justifiable. Here, an adequate age estimation performance as well as an additional understanding of potential preventive application scenarios has to be achieved.

7.3.4 Separation and Sequencing of Overlapped Fingerprints

The separation of overlapped fingerprints is a well known challenge for forensic investigators. However, this challenge is accompanied by additional questions, such as which of the two overlapped fingerprints has been applied first and which was placed afterwards. If age estimation of adequate accuracy is possible, such question could be answered using the relative age. If the prints can be adequately separated, the proposed age estimation scheme can be applied to each of them and the aging speed of both prints can be compared to each other. The print exhibiting the higher aging speed can then be considered as the younger one. Such method might also be applicable if more than two prints are overlapped, if they can be separated successfully. First results of sequencing two separated, formerly overlapped latent prints have been published in the scope of this thesis [26].

For the challenge of separating overlapped fingerprints, age information could also be of interest. If the overlapping ridge lines of different prints can be adequately segmented into belonging to one of the prints or to the crossing of ridges from both prints, a mapping of each region to its corresponding print by the usage of age information might be possible in future work.

7.4 Implications to the Forensic Domain

Apart from their relevance for practical age estimation schemes, the findings of this thesis have significant implications to the work of forensic experts investigating latent prints as well as other trace types. Similar to latent fingerprints, other forensic traces are subject to degradation processes over time. Depending on its state of degradation, a trace can be of high quality, barely visible or not detectible at all. When capturing and processing such traces, their ages should be considered, to select adequate physical as well digital enhancement techniques. Furthermore, various influences might significantly accelerate or decelerate the aging speed, which need to be considered when working with forensic evidence. Especially organic substances, but also those which are moved by wind or contact (e.g. fibers) or react with a substrate (e.g. on corroding substrates) are often subject to significant changes over time, which cannot be neglected.

The studied CWL and CLSM capturing devices provide a great potential for the non-invasive acquisition of various forensic traces beyond fingerprints [145]. Their capability of producing high-resolution images of different surfaces creates a wide range of potential applications. Even firearms and tool marks can be studied [176], [177], [178]. However, the identified limitations need to be addressed before a reliable application of the devices. The proposed digital processing pipeline might be transferred and adapted to suit the specific needs of different trace types, where the here designed pre-processing methods, statistical and morphological features as well as age estimation approaches are regarded as a starting point for such studies. Especially the concept of creating time series of a trace and the approximation of corresponding aging functions using regression seems to be promising for other forensic trace types.

In respect to the fields of digitized and computational forensics, the proposed pipeline could be adapted for an automated processing and age estimation of various digitized forensic traces, which can significantly support the work of forensic experts, by providing digital tools to assist the so far mainly manual and subjective trace assessment with automated and objective measures.

7.5 Future Work

Because the application of non-invasive capturing devices and digital processing methods to the challenge of age estimation is a novel approach, a wide variety of future work does exist. Subjects of future research are summarized here in respect to the general concept, the four different steps of the designed processing pipeline, different influences on the aging process, potential fusion approaches as well as future applications.

Concept: The general concept of using non-invasive capturing devices and the designed digital processing techniques seems to be feasible for a practical age estimation in the future. However, several adaptations might be required. In particular, the basic assumptions (section 3.2.1) need to be reviewed concerning their validity and adapted if necessary. Additional measures need to be taken if certain assumptions are not met. For example, if the required constant capturing properties cannot be realized by certain devices, e.g. if systematic influences of the capturing devices impact the observed aging properties and cannot adequately be removed by digital techniques, improvements of the devices need to be performed in order to meet the specific assumption.

Furthermore, the assumptions themselves need to be refined. For example, if the relative humidity has to remain below a certain threshold to prevent the absorption of water from the ambient air by the print (as found in section 6.5.3), the assumed indoor application scenario should be specified more precisely by defining a maximum allowed ambient humidity. However, also an extension of the assumed indoor scenario is possible, e.g. to certain outdoor scenarios of similar ambient conditions. Assumptions might even be dropped or added. In general, a re-evaluation of assumptions should be conducted especially in respect to all identified influences on the aging behavior and is an iterative process which should be repeated if new findings are obtained. This is furthermore an important measure to assure the representativeness of the underlying sample set, if for newly discovered influences an adequate amount of respective fingerprints is added to the set. To comprehensively evaluate if assumptions are well-specified and met by the used age estimation application is a basic requirement for a reliable future age estimation.

The specific realization of the general concept can be improved in a variety of ways. While the main steps of data acquisition, preprocessing, feature extraction and age estimation are found to be feasible, the various sub-steps and digital methods for their realization might be significantly improved by a systematic evaluation of additional techniques and the replacement with better performing methods. Potential future work for each step of the processing pipeline is summarized below.

Data acquisition: Several limitations of the used capturing devices have been identified in the scope of this thesis, some of which do significantly challenge the assumption of constant capturing properties (section 3.2.1). Future work should be focused on the improvement of the capturing devices and the capturing process towards the reduction of distortions and the achievement of constant capturing properties. Measures for improving the capturing process include adaptations to the hard- and software of the capturing devices as well as the investigation of digital techniques for removing such distortions. In particular, improvements need to be conducted in regard to changes of the overall image brightness of captured time series, sensor noise, reflection and refraction artifacts at droplets due to the measurement principle as well as the sensitivity of the capturing devices to ambient temperature fluctuations, vibrations and measurement table propulsion distortions. Furthermore, so far unidentified influences need to be investigated in more detail and excluded, such as the influence of starting the measurement device (especially for the CLSM sensor) and potential, proprietary (not publicly known) internal preprocessing techniques of the hardware devices. An important starting point for such kind of investigations seems to be the operation of the devices in climatized environments, such as climatic chambers. Only if the captured changes over time are exclusively caused by the fingerprint aging process and not by any sensor-specific distortion, a reliable age estimation can be performed.

Apart from improving the here investigated CWL and CLSM sensor, additional capturing devices and data types need to be examined in future work towards their feasibility for the age estimation. Constant capturing properties as well as preciseness seem to be main requirements for suitable future devices and data types. The discussion of section 2.2 provides an adequate overview of potential devices for such future evaluation. Investigations of capturing devices should especially include different acquisition techniques, targeting different fingerprint properties. For example, spectroscopy might be feasible for non-invasively capturing substance-specific properties of a print. Different capturing devices can also be combined in the scope of a sensor fusion approach.

If capturing devices with adequately constant capturing properties are identified, their preciseness should also be improved in future work. For example, the topographic height information captured with the CWL as well as the CLSM device seems to be promising, yet limited in precision, with topography images being subject to severe distortions, preventing their effective applicability at this point of time. If the preciseness of topographic height information or even the acquisition of 3D volume data can be achieved in future work, additional improvements to the age estimation performance can be expected. Also, higher capturing resolutions should be investigated towards their possibility to achieve an increased preciseness, if limitations of capturing time and data storage requirements can be overcome. The specific surface types from which devices can reliably capture fingerprint images are another important factor when evaluating the preciseness of a device.

Future devices also have to be optimized towards their usability in terms of capturing time, mobility and cost. For the practical work at crime scenes or in other forensic application scenarios, these factors are important and therefore need to be adequately addressed in the scope of future studies.

Preprocessing: In addition to optimizing the used capturing devices, the designed preprocessing techniques can significantly be improved in future work. The performance of the here exemplary investigated methods for spatial alignment, temporal normalization, segmentation as well as the reduction of influences from the capturing devices and other distortions (e.g. dust) might significantly be enhanced by additional investigations of their performance, which has only been evaluated for selected methods in the scope of this thesis. In particular, the propagation of the error introduced by image binarization and its potential increase or decrease throughout all preprocessing sub-steps should be systematically examined. Furthermore, optimal values for the dilation operations used for masking image regions, the performance of segmenting fingerprint structures, the correctness of temporal normalization as well as the performed distortion reduction are of interest.

Apart from improving the designed preprocessing methods, alternative methods could be evaluated in future work, replacing the here designed ones in case of better performance. Such methods can be adapted from the variety of algorithms existing in other image processing areas or the design of new methods specifically addressing the characteristics of captured fingerprint time series. For example, phase congruency used in the area of palm print recognition for the detection of lines [148] might be investigated towards its applicability to latent prints enhancement using high-resolution images. Methods should be optimized in respect to specific capturing devices, data types, aging periods and features.

Feature Extraction: Several statistical and morphological features are proposed and investigated in the scope of this thesis. The evaluation of these features for additional combinations of capturing devices, data types and aging periods is required for a comprehensive analysis of their performance in future work. For the long-term aging, the sample set should furthermore be increased to obtain statistically reliable results. Also, the observed aging properties have to be understood in more detail in future studies, including the examination of the specific causes of variations in the experimental aging speeds and correlations to mathematical functions (even different monotonies in some cases) as well as differences observed between short- and long-term aging periods. More detailed investigations of the specific causes of certain aging properties need to be conducted and compared between the different features, including the analysis of the substance-specific properties of the prints. Additional approximation types should also be investigated in addition to linear, logarithmic and exponential aging functions, such as the usage of polynomial, moving average and power approximations. A more detailed understanding of the sources of certain aging properties and influences on the progression of experimental aging curves might allow for the improvement of these features, including an increased age estimation accuracy and the time period specific application of features.

Additional aging features could be designed in future work, based on characteristic changes during fingerprint degradation. While feature spaces of digital applications include a wide variety of potential metrics, features independent of certain influences from the sweat composition and the environmental conditions are of particular interest. Especially from the area of biometrics, features commonly used for matching exemplary prints (or even digit prints or palm prints) can be evaluated in regard to their behavior during the degradation of a print. Also, the methods of principal component analysis, most discriminant features as well as regularized-direct linear discriminant analysis might be applied, as described by Pavesic and Ribaric et al. in [92]. Texture-based, second order features might furthermore be investigated, e.g. based on co-occurence matrices [179]. The identification of characteristic features is regarded as a promising source of age estimation performance improvements in future work.

Age estimation: Exemplary age estimation strategies are investigated in this thesis for the short-term aging based on formulae as well as machine-learning. These investigations should be extended in future work, by investigating longer time periods as well as additional age classes, based on an increased amount of long-term aging series. Furthermore, the specific causes of age estimation performance fluctuations between different features for age classes smaller than five hours (section 6.4.2) need to be examined. The various age estimation parameters might be further extended, e.g. by investigating additional classifiers and optimizing the time offset between consecutive scans, which could significantly reduce the capturing time required for practical age estimation schemes in the future.

Additional aging features can be included, to further improve the age estimation performance. However, also a feature reduction might be performed, by identifying and selecting most characteristic metrics and excluding others. Apart from improving the proposed age estimation strategies, additional approaches could be designed and evaluated. However, improvements of the performance of an age estimation approach depends largely on the quality of features provided as input.

Influences: A wide variety of potential influences on the aging behavior of fingerprints exists. The here proposed preliminary influence model is regarded as a starting point for more detailed and quantitative studies of this topic. However, the investigation of single influences can barely scratch the surface of the issue, because influences (e.g. from the environment) are often highly interdependent. Furthermore, information about the initial state of a print and the ambient aging conditions are often fuzzy (they might be roughly estimated depending on the characteristics of a crime scene). To properly model and address such high complexity and fuzziness of aging conditions and to derive conclusions about the speed of fingerprint degradation, additional digital techniques need to be investigated in future work. For example, methods from the field of artificial intelligence and fuzzy knowledge representation might be evaluated towards their potential for addressing this challenge, e.g. as proposed by Ribaric and Pavesic in [180].

Particular research questions of future work include the dependence of the specific fingerprint composition and degradation behavior on the different investigated influences as well as the distribution of certain chemical components within the residue of a latent print. Furthermore, influences need to be analyzed in respect to different aging periods, because their impact can vary between short- and long-term periods. For example, if the fingerprint has dried up, influences on the evaporation might be less relevant in comparison to influences on the chemical degradation of compounds. An interdisciplinary study seems to be required, because only a comprehensive understanding of influences seems to allow for a precise age estimation.

Evidence pooling (fusion): In general, fingerprint age estimation could be fused during all four steps of the proposed processing pipeline, as proposed in the scope of this thesis in [79]. Different capturing devices can be investigated and combined during the acquisition step, especially those from different capturing domains, e.g. devices targeting physical as well as those targeting substance-specific properties of a trace (section 2.2). Fingerprint images from different capturing devices (potentially capturing different characteristics of a print or having different resolutions) might also be combined during the preprocessing step, similar to the fusion of infrared and visible images of facades proposed by Ribaric et al. in [181].

Furthermore, fusion could be performed at the feature extraction step, combining the information of different capturing devices or data types into a single feature. Age estimation results of different features might be combined to a final decision, e.g. by averaging or majority voting. In the field of biometrics, features are often fused at the matching-score level, which is considered here as a part of the age estimation step. For example, such approach has been used by Ribaric and Pavesic et al. to fuse live palm and hand geometry features in [182] as well as palm and face features in [183] and might be adapted to the field of age estimation.

In case several trace types of an individual are found at a crime scene, the age information of both of them seems usable for a combined age estimate. For example, if imprints of several parts of a hand are found at an object (e.g. fingerprints, digit prints or palm prints), age information of these evidence types might be fused in a multimodal approach, similar to the one presented by Pavesic and Ribaric et al. in [184] for biometric hand-based verification. Evidence pooling is considered as a promising tool to achieve increased age estimation performance in future work.

Applications: Extending the proposed age estimation approach to scenarios outside the here investigated indoor locations is an important research issue for future work. Outdoor environments [9], [10], [11] as well as fingerprints submerged under water [142], [146] represent potential scenarios. Furthermore, the concept of digital fingerprint age estimation as well as selected processing methods might be used within different practical applications, which are discussed in sections 7.3 and 7.4. These applications would require numerous adaptations of the proposed algorithms in the future. However, when applying age estimation to the general fields of crime scene investigation, data privacy improvement in daily police work, preventive print acquisition as well as potential separation and sequencing of overlapped fingerprints, changes to the application scenarios themselves could also be required. For example, the current line of action when investigating crime scenes might need to be adapted to the usage of non-invasive capture and digital processing of fingerprint data, where prints first need to be located in a coarse scan, followed by a more precise acquisition in a fine scan, subsequently allowing for the digital age estimation and potential separation of overlapped prints, as described by Hildebrandt et al in [20] and [175]. A detailed understanding and adaptation of these specific applications is therefore required beforehand. Most important, an increased reliability of the proposed age estimation approach is a necessary prerequisite for a potential practical application in the future. The development of a fully functioning, accurate age estimation scheme that achieves acceptance in court, e.g. by fulfilling Daubert-compliance [33], remains subject to a substantial amount of future research.

Concluding from the possibilities of future studies on the digital acquisition and age estimation of latent fingerprints, the here proposed approach opens up a novel research field of significant size. Such field is not limited to computer science, but also includes sensor design, chemical substance analysis, daily police work and even legal aspects. Due to the novelty of addressing the age estimation challenge in the digital domain, significant enhancements on the issue can be expected in the next decades. Solving the challenge of fingerprint age estimation in the future would lead to a significant improvement in the evidentiary value of latent prints as well as the capabilities of forensic investigations.

8 Appendix

In this appendix, the capturing parameters of all test sets TSA - TSF used within the experimental evaluations of this thesis are given. They are listed in Table XI and Table XII in respect to the CWL and the CLSM device and allow for a more detailed insight into the used capture settings. Furthermore, selected capturing parameters of the exemplary fingerprint images displayed throughout the thesis are given in Table XIII.

Table XI: Detailed capturing parameters of the test sets TSA_CWL - TSF_CWL.

Parameter	Value					
	TSA_CWL	TSB_CWL	TSC_CWL	TSD_CWL	TSE_CWL	TSF_CWL
Captured time series	100	770	20	42	120	60
Approximate total capturing time	1 day - 2.3 years	1 day	2.3 years	8 hours	1 day	1 day
Approximate repetition offset	1 hour - 1 week	1 hour	1 week	1 hour	1 hour	1 hour
Sensor head	CWL600	CWL600	CWL600	CWL600	CWL600	CWL600
Numerical aperture	0.5	0.5	0.5	0.5	0.5	0.5
Z-resolution	20 nm	20 nm	20 nm	20 nm	20 nm	20 nm
Z-distance	660 µm	660 µm	660 µm	660 µm	660 µm	660 µm
Double-scan-function	Off	Off	Off	Off	Off	Off
Brightness	100	100	100	100	100	100
Minimal intensity	50	50	50	50	50	50
Measurement frequency	2000 Hz	2000 Hz	2000 Hz	2000 Hz	2000 Hz	2000 Hz
Number of layers	1	1	1	1	1	1
Measured area size	15 x 20 mm - 2 x 2 mm	4 x 4 mm	8 x 8 mm	4 x 4 mm	4 x 4 mm	4 x 4 mm
Pixels	1500 x 2000 - 200 x 200	200 x 200	800 x 800	200 x 200	200 x 200	200 x 200
Resolution	1270 dpi - 12700 dpi	1270 dpi	2540 dpi	1270 dpi	1270 dpi	1270 dpi
Dot distance	20 µm - 2 µm	20 µm	10 µm	20 µm	20 µm	20 µm

Table XII: Detailed capturing parameters of the test sets TSA_CLSM - TSF_CLSM.

Parameter	Value					
	TSA_CLSM	TSB_CLSM	TSC_CLSM	TSD_CLSM	TSE_CLSM	TSF_CLSM
Captured time series	100	770	20	60	180	90
Approximate total capturing time	1 day - 30 weeks	1 day	30 weeks	8 hours	1 day	1 day
Approximate repetition offset	1 hour - 1 week	1 hour	1 week	1 hour	1 hour	1 hour
Sensor head	VK-X110	VK-X110	VK-X110	VK-X110	VK-X110	VK-X110
Objective lens	10x - 100x	10x	10x	10x	10x	10x
Numerical aperture	0.3	0.3	0.3	0.3	0.3	0.3
Mode	Surface Profile	Surface Profile	Surface Profile	Surface Profile	Surface Profile	Surface Profile
Real peak detection	Off	Off	Off	Off	Off	Off
Quality	High Precision	High Precision	High Precision	High Precision	High Precision	High Precision
Z-pitch	0.20 µm	0.20 µm	0.20 µm	0.20 µm	0.20 µm	0.20 µm
Z-distance	40 µm - 160 µm	150 µm - 160 µm	150 µm	157 µm	40 µm	150µm - 160 µm
Double-scan-function	Off	Off	Off	Off	Off	Off
Brightness	7600 - 8000	7300 - 8000	8000	7200	8000	7000 - 8000
ND-filter	3%	3%	3%	30%	3%	3%
Optical zoom	1.0x	1.0x	1.0x	1.0x	1.0x	1.0x
Measurements for mean	1	1	1	1	1	1
Filter	Off	Off	Off	Off	Off	Off
Fine mode	On	On	On	On	On	On
Measured area size	1.3 x 1 mm - 0.13 x 0.1 mm	1.3 x 1 mm	1.3 x 1 mm	1.3 x 1 mm	1.3 x 1 mm	1.3 x 1 mm
Pixels	1024x768	1024x768	1024x768	1024x768	1024x768	1024x768
Resolution	20007 dpi - 195385 dpi	20007 dpi	20007 dpi	20007 dpi	20007 dpi	20007 dpi
Dot distance	1.3 µm - 0.13 µm	1.3 µm	1.3 µm	1.3 µm	1.3 µm	1.3 µm

Table XIII: Selected capturing parameters of displayed exemplary fingerprint images.

Figure	CWL		CLSM	
	Measured area size	Dot distance	Measured area size	Dot distance
4	4 x 4 mm, 8 x 8 mm	20 µm, 10 µm	1.3 x 1 mm	1.3 µm
5	10 x 10 mm	10 µm	1.3 x 1 mm	1.3 µm
7	-	-	1.3 x 1 mm	1.3 µm
9	4 x 4 mm	20 µm	-	-
10	-	-	1.3 x 1 mm	1.3 µm
11	4 x 4 mm	20 µm	-	-
12	-	-	1.3 x 1 mm	1.3 µm
13	4 x 4 mm	20 µm	-	-
14	-	-	1.3 x 1 mm	1.3 µm
15	2.5 x 2.5 mm	3 µm	-	-
16	2.5 x 2.5 mm	3 µm	-	-
17	2.5 x 2.5 mm	5 µm	-	-
18	-	-	1.3 x 1 mm	1.3 µm
19	-	-	1.3 x 1 mm	1.3 µm
20	-	-	1.3 x 1 mm	1.3 µm
21	-	-	1.3 x 1 mm	1.3 µm
22	4 x 4 mm	20 µm	1.3 x 1 mm	1.3 µm
23	4 x 4 mm	20 µm	1.3 x 1 mm	1.3 µm
25	8 x 8 mm	10 µm	-	-
26	8 x 8 mm	10 µm	-	-
27	-	-	1.3 x 1 mm	1.3 µm
28	-	-	1.3 x 1 mm	1.3 µm
29	-	-	1.3 x 1 mm	1.3 µm
30	-	-	1.3 x 1 mm	1.3 µm
31	8 x 8 mm	10 µm	-	-
32	8 x 8 mm	10 µm	1.3 x 1 mm	1.3 µm
34	8 x 8 mm	10 µm	1.3 x 1 mm	1.3 µm
35	4 x 4 mm	20 µm	1.3 x 1 mm	1.3 µm
36	8 x 8 mm	10 µm	-	-
37	8 x 8 mm	10 µm	-	-
38	8 x 8 mm	10 µm	-	-
39	-	-	1.3 x 1 mm	1.3 µm
40	-	-	1.3 x 1 mm	1.3 µm
41	-	-	1.3 x 1 mm	1.3 µm
42	-	-	1.3 x 1 mm	1.3 µm
45	4 x 4 mm	20 µm	1.3 x 1 mm	1.3 µm
71	8 x 8 mm	10 µm	-	-
75	8 x 8 mm	10 µm	-	-
100	4 x 4 mm	20 µm	1.3 x 1 mm	1.3 µm

Bibliography

[1] M. Fairhurst and M. Erbilek, "Analysis of physical ageing effects in iris biometrics," *IET Computer Vision,* vol. 5, no. 6, pp. 358-366, 2011.

[2] P. Gnanasivam and S. Muttan, "Estimation of Age Through Fingerprints Using Wavelet Transform and Singular Value Decomposition," *International Journal of Biometrics and Bioinformatics (IJBB),* vol. 6, no. 2, pp. 58-67, 2012.

[3] H. Howorka, "Questions relating to the determination of the age of objects assuming relevance in criminal investigations," *Fingerprint Whorld,* no. 15, pp. 23-28, 1989.

[4] S. Rankin, "Forensic Science Central: Trace Evidence," [Online]. Available: http://forensicsciencecentral.co.uk/traceevidence.shtml. [Accessed 20 1 2014].

[5] J. Dittmann, C. Vielhauer, M. Ulrich, C. Kraetzer and T. Hoppe, "Digi-Dak Digitale Fingerspuren," Project Proposal, Magdeburg, Germany, 2008.

[6] R. Heindl, System und Praxis der Daktyloskopie und der sonstigen technischen Methoden der Kriminalpolizei, 3 ed., Walter de Gruyter & Co, 1927.

[7] J. De Alcaraz-Fossoul, C. M. Patris, A. B. Muntaner, C. B. Feixat and M. G. Badia, "Determination of latent fingerprint degradation patterns — a real fieldwork study," *International Journal of Legal Medicine,* vol. 127, no. 4, pp. 857-870, 2013.

[8] W. W. Clements, "Latent Fingerprints - One Year Later," *Reprinted in "THE PRINT",* vol. 10, no. 7, p. 6, 1994.

[9] W. J. O'Brien, "My fingerprints? Of course, I lived there," *Australian Police Journal,* pp. 142-147, 1984.

[10] D. Greenlees, "Age Determination - Case Report," *Reprinted in "THE PRINT",* vol. 10, no. 7, pp. 4-5, 1994.

[11] J. F. Schwabenland, "Determining the Evaporation Rate of Latent Impressions on the Exterior Surfaces of Aluminum Beverage Cans," *Journal of Forensic Identification,* vol. 42, no. 2, pp. 84-90, 1992.

[12] W. C. Sampson and G. C. Moffett, "Lifetime of a Latent Print on Glazed Ceramic Tile," *Journal of Forensic Identification,* vol. 44, no. 4, pp. 379-386, 1994.

[13] S. R. Balloch, "The life of a latent," *Identification News,* vol. 6, p. 10, 1977.

[14] G. Thielemann, "Altersbestimmung einer daktyloskopischen Spur," *Kriminalistik,* vol. 12, no. 82, p. 630, 1982.

[15] H. U. Koch, "Altersbestimmung daktyloskopischer Spuren," *der kriminalist,* vol. 2, no. 96, pp. 81-83, 1994.

[16] P. L. Johnson, "Life of Latents," *Identification News,* vol. 23, no. 4, pp. 10-13, 1973.

[17] A. L. McRoberts and K. E. Kuhn, "A Review of the Article - 'Determining the Evaporation Rate of Latent Impressions on the Exterior Surfaces of Aluminum Beverage Cans'," *Journal of Forensic Identification,* vol. 42, no. 3, pp. 213-218, 1992.

[18] C. R. Midkiff, "Lifetime of a Latent Print: How long? Can you tell?," *Journal of Forensic Identification,* vol. 43, no. 4, p. 386–392, 1993.

[19] S. N. Srihari, "Computing the scene of a crime," *IEEE Spectrum,* vol. 47, no. 12, pp. 38-43, 2010.

[20] M. Hildebrandt, S. Kiltz, I. Grossmann and C. Vielhauer, "Convergence of digital and traditional forensic disciplines: a first exemplary study for digital dactyloscopy," in *Proceedings of the 13th ACM Workshop on Multimedia and Security (MM&Sec)*, New York, USA, 2011.

[21] R. Merkel, M. Pocs, J. Dittmann and C. Vielhauer, "Proposal of Non-invasive Fingerprint Age Determination to Improve Data Privacy Management in Police Work from a Legal Perspective Using the Example of Germany," in *Lecture Notes in Computer Science 7731: Data Privacy Management and Autonomous Spontaneous Security*, Pisa, Italy, 2013.

[22] A. Roßnagel, S. Jandt, M. Desoi and B. Stach, "Kurzgutachten: Fingerabdrücke in wissenschaftlichen Datenbanken," Projektgruppe verfassungsverträgliche Technikgestaltung (provet), Kassel, Germany, 2013.

[23] Federal Republic of Germany, "German Federal Data Protection Act (Bundesdatenschutzgesetz)," as amended in the Bundesgesetzblatt I 2814, 2009.

[24] EU Commission, "Proposal for a Directive of the European Parliament and of the Council on Data Protection in the Police Sector," COM(2012) 10 final, Brussels, Belgium, 2012.

[25] EU Commission, "Proposal for a Regulation of the European Parliament and of the Council on the Protection of Individuals with Regard to the Processing of Personal Data and on the Free Movement of Such Data (General Data Protection Regulation)," COM(2012) 11 final, Brussels, Belgium, 2012.

[26] M. Schott, R. Merkel and J. Dittmann, "Sequence detection of overlapping latent fingerprints using a short-term aging feature," in *Proceedings of the IEEE International Workshop on Information Forensics and Security (WIFS)*, Tenerife, Spain, 2012.

[27] Constitutional Court of Germany, "Decision of 04.04.2006, no. 1 BvR 518/02 ('Rasterfahndung II')," Collection of Decisions of the Constitutional Court of Germany, BVerfGE 115, 2006.

[28] M. Leich, S. Kiltz, J. Dittmann and C. Vielhauer, "Non-destructive forensic latent fingerprint acquisition with chromatic white light sensors," in *Proceedings of SPIE 7880, Media Watermarking, Security, and Forensics III, 78800S*, San Francisco, USA, 2011.

[29] M. Hildebrandt, R. Merkel, M. Leich, S. Kiltz, J. Dittmann and C. Vielhauer, "Benchmarking contact-less surface measurement devices for fingerprint acquisition in forensic investigations: results for a differential scan approach with a chromatic white light sensors," in *17th IEEE International Conference on Digital Signal Processing (DSP)*, Corfu, Greece, 2011.

[30] K. M. Antoine, S. Mortazavi, A. D. Miller and L. M. Miller, "Chemical differences are observed in children's versus adults' latent fingerprints as a function of time," *Journal of Forensic Science,* vol. 55, no. 2, pp. 513-518, 2010.

[31] D. K. Williams, C. J. Brown and J. Bruker, "Characterization of children's latent fingerprint residues by infrared microspectroscopy: Forensic implications," *Forensic Science International,* vol. 206, no. 1-3, pp. 161-165, 2011.

[32] P. Watson, R. J. Prance, S. T. Beardsmore-Rust and H. Prance, "Imaging Electrostatic Fingerprints with Implications for a Forensic Timeline," *Forensic Science International,* vol. 209, no. 1, pp. 41-45, 2011.

[33] E. H. Holder, L. O. Robinson and J. H. Laub, "The Fingerprint Sourcebook," US Department of Justice, Office of Justice Programs, National Institute of Justice, 2011. [Online]. Available: https://www.ncjrs.gov/pdffiles1/nij/225320.pdf. [Accessed 17 2 2014].

[34] C. Champod, C. Lennard, P. Margot and M. Stoilovic, Fingerprints and Other Ridge Skin Impressions, CRC Press, 2004.

[35] J. Aehnlich, "Altersbestimmung von datkyloskopischen Spuren mit Hilfe der Laser-Fluoreszenzspektroskopie," Gottfried Wilhelm Leibniz University, Hannover, Germany, 2001.

[36] C. Kohlepp, "Möglichkeiten und Grenzen der Altersbestimmung daktyloskopischer Spuren und ihre kriminalistische Bedeutung," Wiesbaden, Germany, 1994.

[37] V. Stamm, "Methoden zur Altersbestimmung daktyloskopischer Spuren," Wiesbaden, Germany, 1997.

[38] B. Holyst, "Kriminalistische Abschätzung des Spurenalters bei Fingerpapillarlinien," *Archiv für Kriminologie,* pp. 94-103, 1983.

[39] K. Baniuk, "Determination of age of fingerprints," *Forensic Science International,* vol. 46, no. 1, pp. 133-137, 1990.

[40] M. Marcinowski, "Evaluation of fingerprint age determination methods," *Problemy Kryminalistyki,* no. 228, pp. 62-68, 2000.

[41] K. Baniuk, "Importance of laboratory experiment in fingerprint age determination," *Problemy Kryminalistyki,* no. 228, pp. 69-71, 2000.

[42] K. Baniuk, "Fingermark ageing process. Diagnostic examinations and age indicators," *Problemy Kryminalistyki,* no. 243, pp. 35-39, 2004.

[43] M. Y. Omar and L. Ellsworth, "Possibility of using fingerprint powders for development of old fingerprints," *Sains Malaysiana,* vol. 41, no. 4, pp. 499-504, 2012.

[44] M. Azoury, E. Rozen, Y. Uziel and Y. Peleg-Shironi, "Old latent prints developed with powder: A rare phenomenon?," *Journal of Forensic Identification,* vol. 54, no. 5, pp. 534-541, 2004.

[45] T. Knabe, "Einflußfaktoren auf die Lebensdauer von Papillarleistenspuren und die Verwendbarkeit einiger ausgewählter mechanischer Sicherungsmittel zur Sichtbarmachung," Berlin, Germany, 1992.

[46] G. A. Granowski, "Über die Dauer der Erhaltung von Schweißspuren und die Möglichkeit der Feststellung ihres Alters," Halle, Germany, 1960.

[47] U. Ramminger, "Mechanistische Untersuchungen zum Verlauf der Ninhydrinreaktion bei der Entwicklung latenter Fingerspuren," Erlangen, Germany, 2000.

[48] A. A. Moenssens, Fingerprint Techniques, Chilton Book Company, 1971.

[49] E. Horschler, "Wie lange sind Fingerspuren haltbar?," München, Germany, 1943.

[50] W. R. Scott and R. D. Olsen Sr, Scott's Fingerprint Mechanics, Thomas, 1978.

[51] L. A. Lewis, R. W. Smithwick 3rd, G. L. Devault, B. Bolinger and S. A. Lewis Sr, "Processes involved in the development of latent fingerprints using the cyanoacrylate fuming method," *Journal of Forensic Science,* vol. 46, no. 2, pp. 241-246, 2001.

[52] K. D. Edelmann, "Untersuchungen zur Altersbestimmung daktyloskopischer Spuren," Wiesbaden, Germany, 1982.

[53] C. R. Midkiff, "Fingerprints - Determination of Time of Placement. An Age Old Problem," *Fingerprint Whorld,* vol. 18, no. 70, pp. 125-128, 1992.

[54] P. Schottenheim, "Im Rahmen der Erörterung der Tatrelevanz stellt sich vor Gericht immer wieder die Frage nach dem Alter einer daktyloskopischen Spur. Nehmen Sie zu den Möglichkeiten des daktyloskopischen Sachverständigen Stellung, dahingehend Aussagen zu machen," Wiesbaden, Germany, 1996.

[55] R. Herrmann, "Altersbestimmung von Fingerabdrücken - Sachstandsbericht und Lösungsansatz," Hannover, Germany, 2000.

[56] K. Oppermann, Der daktyloskopische Identitätsnachweis, Schmidt-Römhild, 2000.

[57] K. Wertheim, "Fingerprint Age Determination: Is There Any Hope?," *Journal of Forensic Identification,* vol. 53, no. 1, pp. 42 - 49, 2003.

[58] S. Barcikowski, J. Bunte, A. Ostendorf, J. Aehnlich and R. Herrmann, "Contribution to the age determination of fingerprint constituents using laser fluorescence spectroscopy and confocal laser scanning microscopy," in *Proceedings of SPIE 5621, Optical Materials in Defence Systems Technology*, London, UK, 2004.

[59] A. Girod, R. Ramotowski and C. Weyermann, "Composition of fingermark residue: A qualitative and quantitative review," *Forensic Science International,* vol. 223, no. 1, pp. 10-24, 2012.

[60] E. Angst, "Untersuchungen zur Bestimmung des Alters von daktyloskopischen Spuren auf Papier," *Kriminalistik,* pp. 1-12, 1961.

[61] R. S. Croxton, M. G. Baron, D. Butler, T. Kent and V. G. Sears, "Variation in amino acid and lipid composition of latent fingerprints," *Forensic Science International,* vol. 199, no. 1, pp. 93-102, 2010.

[62] C. Weyermann, C. Roux and C. Champod, "Initial results on the composition of fingerprints and its evolution as a function of time by GC/MS analysis," *Journal of Forensic Science,* vol. 56, no. 1, pp. 102-108, 2011.

[63] Y. S. Dikshitulu, L. Prasad, J. N. Pal and C. V. Rao, "Aging studies on fingerprint residues using thin-layer and high performance liquid chromatography," *Forensic Science International,* vol. 31, no. 4, pp. 261-266, 1986.

[64] G. M. Mong, C. E. Petersen and T. R. W. Clauss, "Advanced Fingerprint Analysis Project Final Report – Fingerprint Constituents," Pacific Northwest National Laboratory, Richland, USA, 1999.

[65] K. G. Asano, C. K. Bayne, K. M. Horsman and M. V. Buchanan, "Chemical composition of fingerprints for gender determination," *Journal of Forensic Science,* vol. 47, no. 4, pp. 805-807, 2002.

[66] N. E. Archer, Y. Charles, J. A. Elliott and S. Jickells, "Changes in the lipid composition of latent fingerprint residue with time after deposition on a surface," *Forensic Science International,* vol. 154, no. 2-3, pp. 224-239, 2005.

[67] G. De Paoli, S. A. Lewis Sr, E. L. Schuette, L. A. Lewis, R. M. Connaster and T. Farkas, "Photo- and thermal-degradation studies of select eccrine fingerprint constituents," *Journal of Forensic Sciences,* vol. 55, no. 4, pp. 962-969, 2010.

[68] R. Wolstenholme, R. Bradshaw, M. R. Clench and S. Francese, "Study of latent fingermarks by matrix-assisted laser desorption/ionisation mass spectrometry imaging of endogenous lipids," *Rapid communications in mass spectrometry,* vol. 23, no. 19, pp. 3031-3039, 2009.

[69] G. Mong, S. Walter, T. Cantu and R. Ramotowski, "The Chemistry of Latent Prints from Children and Adults," *Fingerprint Whorld,* vol. 27, no. 104, pp. 66-69, 2001.

[70] J. M. Duff and E. R. Menzel, "Laser-assisted thin-layer chromatography and luminescence of fingerprints: an approach to fingerprint age determination," *Journal of Forensic Sciences,* vol. 23, no. 1, pp. 129-134, 1978.

[71] E. R. Menzel and J. M. Duff, "Laser detection of latent fingerprints-Treatment with fluorescers," *Journal of Forensic Sciences,* vol. 24, no. 1, pp. 96-100, 1979.

[72] E. R. Menzel, Fingerprint Detection with Lasers, Marcel Dekker Inc, 1999.

[73] E. R. Menzel, "Letter to the Editor: Fingerprint Age Determination by Fluorescence," *Journal of Forensic Sciences,* vol. 37, no. 5, pp. 1212-1213, 1992.

[74] U. R. Bernier, D. L. Kline, D. R. Barnard, C. E. Schreck and R. A. Yost, "Analysis of human skin emanations by gas chromatography/mass spectrometry. 2. Identification of volatile compounds that are candidate attractants for the yellow fever mosquito (Aedes aegypti)," vol. 72, no. 4, pp. 747-756, 2000.

[75] G. Popa, R. Potorac and N. Preda, "Method for fingerprints age determination," *Romanian Journal of Legal Medicine,* vol. 18, no. 2, pp. 149-154, 2010.

[76] N. C. Crane, E. G. Bartick, R. Schwartz Perlman and S. Huffman, "Infrared Spectroscopic Imaging for Noninvasive Detection of Latent Fingerprints," *Journal of Forensic Science,* vol. 52, no. 1, pp. 48-53, 2007.

[77] R. Bhargava, R. Schwartz Perlman, D. C. Fernandez, I. W. Levin and E. G. Bartick, "Non-invasive detection of superimposed latent fingerprints and inter-ridge trace evidence by infrared spectroscopic imaging," *Analytical and Bioanalytical Chemistry,* vol. 394, no. 8, pp. 2069-2075, 2009.

[78] A. Grant, T. J. Wilkinson, D. R. Holman and M. C. Martin, "Identification of Recently Handled Materials by Analysis of Latent Human Fingerprints Using Infrared Spectromicroscopy," *Applied Spectroscopy,* vol. 59, no. 9, pp. 1182-1187, 2005.

[79] R. Merkel, S. Gruhn, J. Dittmann, C. Vielhauer and A. Braeutigam, "General fusion approaches for the age determination of latent fingerprint traces: results for 2D and 3D binary pixel feature fusion," in *Proceedings of SPIE 8290, Three-Dimensional Image Processing (3DIP) and Applications II, 82900Y*, Burlingame, USA, 2012.

[80] FRT GmbH, "Fries Research & Technology (FRT)," [Online]. Available: www.frt-gmbh.com/en/, www.frtofamerica.com/us, www.frt-gmbh.com/en/microprof-200.aspx. [Accessed 21 1 2014].

[81] S. K. Dubey, D. S. Mehta, A. Anand and C. Shakher, "Simultaneous topography and tomography of latent fingerprints using fullfield swept-source optical coherence tomography," *Journal of Optics A: Pure and Applied Optics,* vol. 10, no. 1, pp. 015307-0153015, 2008.

[82] K. Kuivalainen, A. Oksman and K. Peiponen, "Definition and measurement of statistical gloss parameters from curved objects," *Applied Optics,* vol. 49, no. 27, pp. 5081-5086, 2010.

[83] N. S. Claxton, T. J. Fellers and M. W. Davidson, "Laser Scanning Confocal Microscopy," Olympus, Tallahassee, USA, 2006.

[84] Keyence Corporation, "Keyence Color 3D Laser Scanning Microscope," [Online]. Available: www.keyence.com/topics/vision/vk/guide.php?lf=KD, www.keyence.com/products/ microscope/laser-microscope/vk-x100_x200/models/vk-x110/index.jsp. [Accessed 21 1 2014].

[85] N. Saitoh and N. Akiba, "Ultraviolet Fluorescence Spectra of Fingerprints," *The Scientific World Journal,* no. 5, pp. 355-366, 2005.

[86] G. Williams and N. McMurray, "Latent fingermark visualisation using a scanning Kelvin probe," *Forensic Science International,* vol. 167, no. 2, pp. 102-109, 2007.

[87] G. S. Watson and J. A. Watson, "Potential Applications of Scanning Probe Microscopy in Forensic Science," *Journal of Physics: Conference Series,* vol. 61, no. 1, p. 1251, 2007.

[88] A. J. Goddard, A. R. Hillman and J. W. Bond, "High Resolution Imaging of Latent Fingerprints by Localized Corrosion on Brass Surfaces," *Journal of Forensic Sciences,* vol. 55, no. 1, pp. 58-65, 2010.

[89] G. Williams, H. N. McMurray and D. A. Worsley, "Latent fingerprint detection using a scanning Kelvin microprobe," *Journal of Forensic Science,* vol. 46, no. 5, pp. 1085-1092, 2001.

[90] C. D. Frisbie, L. F. Rozsnyai, A. Noy, M. S. Wrighton and C. M. Lieber, "Functional Group Imaging by Chemical Force Microscopy," *Science,* vol. 265, no. 5181, pp. 2071-2074, 1994.

[91] GFMesstechnik GmbH. [Online]. Available: http://www.gfm3d.com/, http://spectronet.de/portals/visqua/story_docs/intern_spectronet/vortraege/100826_08_collab_vortraege/100827_30_terboven_gfm.pdf. [Accessed 10 31 2013].

[92] N. Pavesic, S. Ribaric and B. Grad, "Finger-Based Personal Authentication: a Comparison of Feature-Extraction Methods Based on Principal Component Analysis, Most Discriminant Features and Regularised-Direct Linear Discriminant Analysis," *IET Signal Processing,* vol. 3, no. 4, pp. 269-281, 2009.

[93] S. Ribaric and I. Fratric, "A Biometric Identification System Based on Eigenpalm and Eigenfinger Features," *IEEE Transactions on Pattern Analysis and Machine Intelligence,* vol. 27, no. 11, pp. 1698-1709, 2005.

[94] S. S. Lin, K. M. Yemelyanov, E. N. Pugh Jr and N. Engheta, "Polarization- and specular-reflection-based, non-contact latent fingerprint imaging and lifting," *Journal of the Optical Society of America A,* vol. 23, no. 9, pp. 2137-2153, 2006.

[95] R. Merkel and C. Vielhauer, "On using flat bed scanners for the age determination of latent fingerprints: first results for the binary pixel feature," in *Proceedings of the 14th ACM Workshop on Multimedia and security (MM&Sec),* Coventry, UK, 2012.

[96] Seiko Epson Corporation, "Epson Perfection 1660 Photo," [Online]. Available: http://www.epson.de/download.php?file=/files/brochures/b11b/b11b153-engperfection-1660-photo-brochure-en.pdf&name=Epson-Perfection-1660-Photo-Brochures-2.pdf. [Accessed 1 4 2012].

[97] M. Hildebrandt, A. Makrushin, K. Qian and J. Dittmann, "Visibility Assessment of Latent Fingerprints on Challenging Substrates in Spectroscopic Scans," in *IFIP CMS 2013, LNCS 8099,* Magdeburg, Germany, 2013.

[98] S. E. Mudiyanselage, M. Hamburger, P. Elsner and J. J. Thiele, "Ultraviolet A Induces Generation of Squalene Monohydroperoxide Isomers in Human Sebum and Skin Surface Lipids In Vitro and In Vivo," *Journal of Investigative Dermatology,* vol. 120, no. 6, pp. 915-922, 2003.

[99] A. Richards and O. P. Partner, "Reflected Ultraviolet Imaging for Forensics Applications," 2010. [Online]. Available: www.uvcorder.com/pdf/Reflected_UV_Imaging_for_Forensics_V2.pdf. [Accessed 24 2 2014].

[100] A. C. Lin, H. M. Hsieh, L. C. Tsai, A. Linacre and J. C. I. Lee, "Forensic Applications of Infrared Imaging for the Detection and Recording of Latent Evidence," *Journal of Forensic Science,* vol. 52, no. 5, pp. 1148-1150, 2007.

[101] FLIR Systems Inc, "FLIR SC305," [Online]. Available: www.flir.com/DE/. [Accessed 31 10 2013].

[102] R. M. Connaster, S. M. Prokes, O. J. Glembocki, R. L. Schuler, C. W. Gardner, S. A. Lewis and L. A. Lewis, "Toward Surface-Enhanced Raman Imaging of Latent Fingerprints," *Journal of Forensic Science,* vol. 55, no. 6, pp. 1462-1470, 2010.

[103] R. C. Gonzalez and R. E. Woods, Digital Image Processing, 3 ed., Prentice Hall, 2007.

[104] itseez, "OpenCV (Open Source Computer Vision Library)," [Online]. Available: http://opencv.org/. [Accessed 6 11 2013].

[105] A. Bjoerck, Numerical methods for least squares problems, Siam, 1996.

[106] J. Schindelin, I. Arganda-Carreras, E. Frise, V. Kaynig, M. Longair, T. Pietzsch, S. Preibisch, C. Rueden, S. Saalfeld, B. Schmid, J. Y. Tinevez, D. J. White, V. Hartenstein, K. Eliceiri, P. Tomancak and A. Cardona, "Fiji: an open-source platform for biological-image analysis," *Nature Methods,* vol. 9, no. 7, pp. 676-682, 2012.

[107] Fiji Project, "Fiji Is Just ImageJ," [Online]. Available: http://fiji.sc/Fiji. [Accessed 31 10 2013].

[108] W. S. Rasband, "ImageJ," [Online]. Available: http://imagej.nih.gov/ij/. [Accessed 24 2 2014].

[109] M. D. Abramoff, P. J. Magalhaes and S. J. Ram, "Image Processing with ImageJ," *Biophotonics International,* vol. 11, no. 7, pp. 36-42, 2004.

[110] C. A. Schneider, W. S. Rasband and K. W. Eliceiri, "NIH Image to ImageJ: 25 years of image analysis," *Nature Methods,* vol. 9, no. 7, pp. 671-675, 2012.

[111] Fiji Project, "Fiji Auto Threshold Function," [Online]. Available: http://fiji.sc/wiki/index.php/Auto_Threshold#Default. [Accessed 31 10 2013].

[112] L. K. Huang and M. J. J. Wang, "Image thresholding by minimizing the measure of fuzziness," *Pattern Recognition,* vol. 28, no. 1, pp. 41-51, 1995.

[113] J. M. S. Prewitt and M. L. Mendelsohn, "The analysis of cell images," *Annals of the New York Academy of Sciences,* vol. 128, no. 3, pp. 1035-1053, 1966.

[114] T. W. Ridler and S. Calvard, "Picture thresholding using an iterative selection method," *IEEE Transactions on Systems, Man and Cybernetics,* vol. 8, no. 8, pp. 630-632, 1978.

[115] C. H. Li and P. K. S. Tam, "An Iterative Algorithm for Minimum Cross Entropy Thresholding," *Pattern Recognition Letters,* vol. 18, no. 8, pp. 771-776, 1998.

[116] J. N. Kapur, P. K. Sahoo and A. C. K. Wong, "A New Method for Gray-Level Picture Thresholding Using the Entropy of the Histogram," *Graphical Models and Image Processing,* vol. 29, no. 3, pp. 273-285, 1985.

[117] C. A. Glasbey, "An analysis of histogram-based thresholding algorithms," *CVGIP: Graphical Models and Image Processing,* vol. 55, no. 6, pp. 532-537, 1993.

[118] J. Kittler and J. Illingworth, "Minimum error thresholding," *Pattern Recognition,* vol. 19, no. 1, pp. 41-47, 1986.

[119] W. Tsai, "Moment-preserving thresholding: a new approach," *Computer Vision, Graphics, and Image Processing,* vol. 29, no. 3, pp. 377-393, 1985.

[120] N. Otsu, "A threshold selection method from gray-level histograms," *IEEE Transactions on Systems, Man and Cybernetics,* vol. 9, no. 1, pp. 62-66, 1979.

[121] W. Doyle, "Operation useful for similarity-invariant pattern recognition," *Journal of the Association for Computing Machinery,* vol. 9, no. 2, pp. 259-267, 1962.

[122] A. G. Shanbhag, "Utilization of information measure as a means of image thresholding," *Graphical Models Image Processeing,* vol. 56, no. 5, pp. 414-419, 1994.

[123] G. W. Zack, W. E. Rogers and S. A. Latt, "Automatic measurement of sister chromatid exchange frequency," *Journal of Histochemistry & Cytochemistry,* vol. 25, no. 7, pp. 741-753, 1977.

[124] J. C. Yen, F. J. Chang and S. Chang, "A New Criterion for Automatic Multilevel Thresholding," *IEEE Transactions on Image Processing,* vol. 4, no. 3, pp. 370-378, 1995.

[125] L. Hong, Y. Wan and A. K. Jain, "Fingerprint image enhancement: Algorithm and performance evaluation," *IEEE Transactions on Pattern Analysis and Machine Intelligence,* vol. 29, no. 8, pp. 777-790, 1998.

[126] P. Kovesi and R. Thai, "Fingerprint enhancement (matlab-code)," [Online]. Available: http://www.csse.uwa.edu.au/~pk/research/matlabfns/index.html#fingerprints. [Accessed 30 1 2013].

[127] A. K. Jain, Y. Chen and M. Demirkus, "Pores and ridges: High-resolution fingerprint matching using level 3 features," *IEEE Transactions on Pattern Analysis and Machine Intelligence,* vol. 29, no. 1, pp. 15-28, 2007.

[128] International Organization for Standardization (ISO), "ISO 4287:1997 Geometrical Product Specifications (GPS) – Surface texture: profile method – terms, definitions and surface texture parameters," 1997. [Online]. Available: http://www.iso.org/iso/iso_catalogue/catalogue_tc/catalogue_detail.htm?csnumber=10132. [Accessed 24 2 2014].

[129] D. Maltoni, D. Maio, A. K. Jain and S. Prabhakar, Handbook of Fingerprint Recognition, 2 ed., Springer, 2009.

[130] K. D. Toennies, Grundlagen der Bildverarbeitung, Pearson, 2005.

[131] J. Fox, Applied regression analysis and generalized linear models, 2 ed., Sage, 2008.

[132] R. O. Duda, P. E. Hart and D. G. Stork, Pattern Classification, 2 ed., John Wiley & Sons Inc, 2001.

[133] Machine Learning Group at the University of Waikato, "Waikato Environment for Knowledge Analysis (WEKA)," [Online]. Available: http://www.cs.waikato.ac.nz/ml/index.html. [Accessed 6 11 2013].

[134] E. D. Hamm, "The Problem with Latents," *The Detective (US Army),* vol. 7, no. 3, pp. 32-34, 1979.

[135] S. Bleay, "Fingerprint Development and Imaging – Fundamental Research to Operational Implementation," Home Office Scientific Development Branch, 2009. [Online]. Available: http://www.heacademy.ac.uk/assets/ps/documents/FORREST/2009/presentations/k2_bleay.pdf. [Accessed 5 4 2013].

[136] C. A. Barnum and D. R. Klasey, "Factors Affecting the Recovery of Latent Prints on Firearms," *Reprinted in "THE PRINT",* vol. 13, no. 3, pp. 6-9, 1997.

[137] G. Popa, R. C. Ionitescu and R. Potorac, "Fingerprints Evaluation in Systemic Sclerosis," *International Journal of Criminal Investigation,* vol. 1, no. 1, pp. 31-36, 2011.

[138] K. Habermann and H. Otte, "Die Konsistenz von Papillarleistenspuren auf ausgewählten Spurenträgern unter Verwendung verschiedener mechanischer Spurensicherungsmittel und einige theoretische Aspekte zum Problem der Beständigkeit von Fingerabdrücken," Berlin, Germany, 1982.

[139] Unknown author (provided by B. Corson), "Fingerprints. Are they there and for how long? Vital Considerations in Developing Latents," *Printed in "THE PRINT",* vol. 14, no. 4, p. 7, 1998.

[140] Y. Cohen, M. Azoury and M. L. Elad, "Survivability of Latent Fingerprints Part II: The Effect of Cleaning Agents on the Survivability of Latent Fingerprints," *Journal of Forensic Identification,* vol. 62, no. 1, pp. 54-61, 2012.

[141] P. D. Barnett and R. A. Berger, "The effects of temperature and humidity on the permanency of latent fingerprints," *Journal of the Forensic Science Society,* vol. 16, no. 3, pp. 249-254, 1976.

[142] B. E. Devlin, "Recovery of Latent Fingerprints after Immersion in Various Aquatic Conditions," Fairfax, USA, 2011.

[143] Y. Cohen, E. Rozen, M. Azoury, D. Attias, B. Gavrielli and M. L. Elad, "Survivability of Latent Fingerprints, Part 1: Adhesion of Latent Fingerprints to Smooth Surfaces," *Journal of Forensic Identification,* vol. 62, no. 1, pp. 47-53, 2012.

[144] T. Ito, N. Tamiya, H. Takahashi, K. Yamazaki, H. Yamamoto, S. Sakano, M. Kashiwagi and S. Miyaishi, "Factors that prolong the 'postmortem interval until finding' (PMI-f) among community-dwelling elderly individuals in Japan: analysis of registration data," *BMJ Open,* vol. 2, no. 5, pp. 1-8, 2012.

[145] J. Dittmann, C. Vielhauer, G. Saake and S. Kiltz, "DigiDak+ Sicherheits-Forschungskolleg," Project Proposal, Magdeburg, Germany, 2011.

[146] M. A. Wood and I. T. James, "Latent Fingerprint Persistence and Development Techniques on Wet Surfaces," *Fingerprint Whorld,* vol. 35, no. 135, pp. 90-100, 2009.

[147] C. Vielhauer, Biometric User Authentication for IT Security: From Fundamentals to Handwriting, Springer, 2006.

[148] V. Struc and N. Pavesic, "Phase congruency features for palm-print verification," *IET Signal Processing,* vol. 3, no. 4, pp. 258-268, 2009.

[149] S. Boslaugh, Statistics in a Nutshell, 2 ed., O'Reilly Media Inc, 2012.

[150] J. Carletta, "Assessing Agreement on Classification Tasks: The Kappa Statistic," *Computational Linguistics,* vol. 22, no. 2, pp. 249-254, 1996.

[151] R. Merkel, K. Otte, R. Clausing, J. Dittmann, C. Vielhauer and A. Braeutigam, "First Investigation of Latent Fingerprints Long-term Aging using Chromatic White Light Sensors," in *Proceedings of the 1st ACM Workshop on Information Hiding and Multimedia Security (IH&MMSec)*, Montpellier, France, 2013.

[152] R. Merkel, J. Dittmann and C. Vielhauer, "Novel Fingerprint Aging Features Using Binary Pixel Sub-Tendencies: A Comparison of Contactless CLSM and CWL Sensors," in *Proceedings of the IEEE International Workshop on Information Forensics and Security (WIFS)*, Tenerife, Spain, 2012.

[153] R. Merkel, S. Gruhn, J. Dittmann, C. Vielhauer and A. Braeutigam, "On non-invasive 2D and 3D Chromatic White Light image sensors for age determination of latent fingerprints," *Forensic Science International*, vol. 222, no. 1, pp. 52-70, 2012.

[154] R. Merkel, A. Breuhan, M. Hildebrandt, C. Vielhauer and A. Braeutigam, "Environmental impact to multimedia systems on the example of fingerprint aging behavior at crime scenes," in *Proceedings of SPIE 8436, Optics, Photonics, and Digital Technologies for Multimedia Applications II, 84360Y*, Brussels, Belgium, 2012.

[155] R. Merkel, J. Dittmann and C. Vielhauer, "How Contact Pressure, Contact Time, Smearing and Oil/Skin Lotion Influence the Aging of Latent Fingerprint Traces: First Results for the Binary Pixel Feature using a CWL Sensor," in *Proceedings of the IEEE International Workshop on Information Forensics and Security (WIFS)*, Foz do Iguacu, Brazil, 2011.

[156] R. Merkel, J. Dittmann and C. Vielhauer, "Approximation of a Mathematical Aging Function for Latent Fingerprint Traces Based on First Experiments Using a Chromatic White Light (CWL) Sensor and the Binary Pixel Aging Feature," in *IFIP CMS 2011, LNCS 7025*, Ghent, Belgium, 2011.

[157] R. Merkel, A. Krapyvskyy, M. Leich, J. Dittmann and C. Vielhauer, "A first framework for the development of age determination schemes for latent biometric fingerprint traces using a chromatic white light (CWL) sensor," in *Proceedings of SPIE 8189, Optics and Photonics for Counterterrorism and Crime Fighting VII, Optical Materials in Defence Systems Technology VIII, and Quantum-Physics-based Information Security, 81890T*, Prague, Czech Republic, 2011.

[158] R. Merkel, A. Braeutigam, C. Kraetzer, J. Dittmann and C. Vielhauer, "Evaluation of Binary Pixel Aging Curves of Latent Fingerprint Traces for Different Surfaces Using a Chromatic White Light (CWL) Sensor," in *Proceedings of the 13th ACM Workshop on Multimedia and Security (MM&Sec)*, New York, USA, 2011.

[159] R. Merkel and J. Dittmann, "Resolution and Size of Measured Area Influences on the Short- and Long-Term Aging of Latent Fingerprint Traces Using the Binary Pixel Feature and a High-Resolution Non-Invasive Chromatic White Light (CWL) Sensor," in *Proceedings of the 7th International Symposium on Image and Signal Processing and Analysis (ISPA)*, Dubrovnik, Croatia, 2011.

[160] A. Krapyvskyy, "Long-term Aging Analysis of Latent and Printed Fingerprints Using Contactless Confocal Laser Scanning Microscopy (CLSM)," Otto-von-Guericke-University, Magdeburg, Germany, 2014.

[161] L. Schwarz, "An amino acid model for latent fingerprints on porous surfaces," *Journal of Forensic Sciences,* vol. 54, no. 6, pp. 1323-1326, 2009.

[162] J. Dittmann and M. Hildebrandt, "Context Analysis of Artificial Sweat Printed Fingerprint Forgeries: Assessment of Properties for Forgery Detection," in *Proceedings of the 2nd International Workshop on Biometrics and Forensics (IWBF)*, Valletta, Malta, 2014.

[163] FRT GmbH, "FRT Acquire Automation XT," [Online]. Available: http://www.frt-gmbh.com/frt-acquire-automation-xt.aspx. [Accessed 6 11 2013].

[164] M. Hildebrandt, "DD+Acquire," Workgroup Multimedia & Security, Faculty of Computer Science, Otto-von-Guericke-University, Magdeburg, Germany, 2012.

[165] M. Hildebrandt, "VKscript," Workgroup Multimedia & Security, Faculty of Computer Science, Otto-von-Guericke-University, Magdeburg, Germany, 2012.

[166] Microsoft Corporation, "Microsoft Windows Script," [Online]. Available: http://www.microsoft.com/de-de/download/details.aspx?id=1406. [Accessed 6 11 2013].

[167] International Organization for Standardization (ISO), "ISO/IEC 23270:2006 Information technology - Programming languages - C#," [Online]. Available: http://www.iso.org/iso/home/store/catalogue_ics/catalogue_detail_ics.htm?csnumber=42926. [Accessed 6 11 2013].

[168] M. Leich, "QTFPX," Workgroup Multimedia & Security, Faculty of Computer Science, Otto-von-Guericke-University, Magdeburg, Germany, 2010.

[169] Digia, "Qt Project," [Online]. Available: http://qt-project.org/. [Accessed 6 11 2013].

[170] International Organization for Standardization (ISO), "ISO International Standard ISO/IEC 14882:2011(E) – Programming Language C++," [Online]. Available: http://isocpp.org/std/the-standard. [Accessed 6 11 2013].

[171] U. Rathmann and J. Wilgen, "Qwt User's Guide," [Online]. Available: http://qwt.sourceforge.net/. [Accessed 6 11 2013].

[172] Oracle Corporation, "Java," [Online]. Available: http://www.java.com. [Accessed 6 11 2013].

[173] Beurer GmbH, "beurer Air Humidifier LB50," [Online]. Available: http://www.beurer.com/web/en/products/air/air_humidifier.php?pid=3520. [Accessed 9 12 2013].

[174] Canon GmbH, "Canon PIXMA iP4500," [Online]. Available: http://www.canon.de/Support/Consumer_Products/products/printers/InkJet/PIXMA_iP_series/PIXMA_iP4500.aspx?type=faq&page=1. [Accessed 9 12 2013].

[175] M. Hildebrandt, J. Dittmann, M. Pocs, M. Ulrich, R. Merkel and T. Fries, "Privacy preserving challenges: New Design Aspects for Latent Fingerprint Detection Systems with contact-less Sensors for Preventive Applications in Airport Luggage Handling," in *LNCS 6583: Biometrics and ID Management*, Brandenburg an der Havel, Germany, 2011.

[176] R. Fischer and C. Vielhauer, "Forensic ballistic analysis using a 3D sensor device," in *Proceedings of the 14th Workshop on Multimedia and Security (MM&Sec)*, Conventry, UK, 2012.

[177] A. Makrushin, M. Hildebrandt, J. Dittmann, E. Clausing, R. Fischer and C. Vielhauer, "3D imaging for ballistics analysis using chromatic white light sensor," in *Proceedings of SPIE 8290, Three-Dimensional Image Processing (3DIP) and Applications II, 829016*, Burlingame, USA, 2012.

[178] E. Clausing, C. Kraetzer, J. Dittmann and C. Vielhauer, "A first approach for the contactless acquisition and automated detection of toolmarks on pins of locking cylinders using 3D confocal microscopy," in *Proceedings of the 14th Workshop on Multimedia and Security (MM&Sec)*, Coventry, UK, 2012.

[179] R. M. Haralick, K. Shanmugam and I. H. Dinstein, "Textural features for image classification," *IEEE Transactions on Systems, Man and Cybernetics,* no. 6, pp. 610-621, 1973.

[180] S. Ribaric and N. Pavesic, "Inference Procedures for Fuzzy Knowledge Representation Scheme," *Applied Artificial Intelligence,* vol. 23, no. 1, pp. 16-43, 2009.

[181] S. Ribaric, D. Marcetic and D. S. Vedrina, "A knowledge-based system for the non-destructive diagnostics of facade isolation using the information fusion of visual and IR images," *Expert Systems with Applications,* vol. 36, no. 2, pp. 3812-3823, 2009.

[182] S. Ribaric and N. Pavesic, "A Biometric Identification System Based on the Fusion of Hand and Palm Features," in *Proceedings of Cost 275 Workshop : The Advent of Biometrics on the Internet*, Rome, Italy, 2002.

[183] S. Ribaric, I. Fratric and K. Kis, "A Biometric Verification System Based on the Fusion of Palmprint and Face Features," in *Proceedings of the 4th International Symposium on Image and Signal Processing and Analysis (ISPA)*, Zagreb, Croatia, 2005.

[184] N. Pavesic, T. Savic, S. Ribaric and I. Fratric, "A multimodal hand-based verification system with an aliveness-detection module," *Annales des Telecommunications,* vol. 62, no. 1-2, pp. 130-155, 2007.